229
147.
125,7
89
77
69

ROOT DISEASES AND SOIL-BORNE PATHOGENS

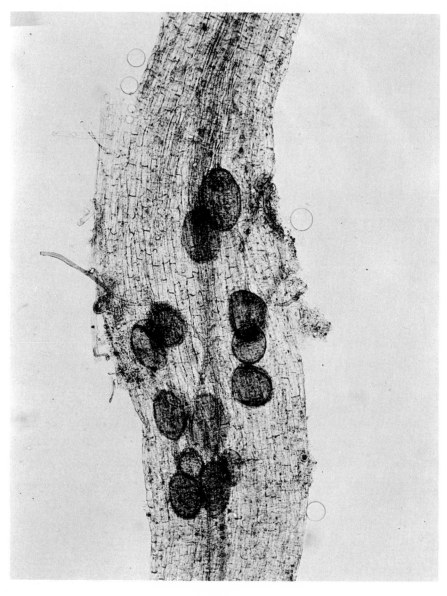

A cleared strawberry rootlet showing mycelium and vesicles of the mycorrhizal fungus *Rhizophagus*, in the cortex. (Courtesy S. Wilhelm.)

ROOT DISEASES
AND SOIL-BORNE PATHOGENS

◀

Second International Symposium on
Factors Determining the Behavior of Plant Pathogens in Soil
Held at Imperial College, London, July 14–28, 1968
in Conjunction with the First International Congress
of Plant Pathology

Edited by
T. A. TOUSSOUN, ROBERT V. BEGA, *and*
PAUL E. NELSON

◀

UNIVERSITY OF CALIFORNIA PRESS
Berkeley, Los Angeles, London: 1970

University of California Press
Berkeley and Los Angeles, California

University of California Press, Ltd.
London, England

ISBN: 0520–01582–7
Library of Congress Catalog Card Number: 73–84531

Printed in the United States of America

Foreword

S. D. GARRETT—*Botany School, University of Cambridge, England*

I am greatly privileged to have been invited by the editors to write a foreword to this symposium volume, which I am confident will receive quite as warm a welcome from plant pathologists, soil microbiologists, and all others interested in root diseases of crop plants as did its predecessor, *Ecology of Soil-borne Plant Pathogens*, edited by Kenneth F. Baker and William C. Snyder and published by the University of California Press in 1965.

This first symposium volume was the product of the first truly international and representative meeting of workers in our subject, which was held at the University of California at Berkeley over the period 7–13 April 1963. This was the most successful, stimulating, and enjoyable meeting I had ever attended; in the pleasant surroundings of the University campus, I was able to meet many old friends, and to get to know others whom before I had known only through their writings. No wonder that it was with great though private reluctance that I agreed to organize a second international meeting of root disease workers at Cambridge in 1968; I despaired of being able to maintain so high a standard. But fortune was kind to me, and I was spared so heavy a responsibility, as I shall now briefly recount.

On a Saturday morning of August 1964, during the Tenth International Botanical Congress, I was briefly basking in the sunshine in one of the ancient courts of Edinburgh University, comfortably reflecting that my not very arduous labours as Chairman of the Plant Pathology Sectional Committee were nearly over, when I noticed three people walking towards me. They were W. C. Snyder, J. G. Horsfall, and R. K. S. Wood, and they had come to tell me that there was a growing opinion amongst the plant pathologists attending this Congress that we should organize a First International Congress of Plant Pathology; we therefore called a meeting of all plant pathologists attending the Edinburgh Congress to discuss this proposition. The issue was never in doubt; in any meeting of this kind, where only a small minority feel themselves at risk of heavy administrative involvement in the undertaking, a substantial majority can always be relied upon to pass the proposal with acclamation, and so they did. In due course I was elected Chairman of the Executive Committee for the First International Congress of Plant Pathology, and I was thus in a strong position to suggest, though perhaps I should more truthfully say to declare, that the successor to the Berkeley meeting should be incorporated with the First International Congress as one of its sections.

In order to prepare a programme for our section of Root Diseases and Soil-borne Pathogens, I first caught up with my reading of the literature, which is always in arrears, and then sought the advice of various friends who had helped to organize the Berkeley meeting. This advice was generously given, and from it I eventually selected titles for seven symposia, for which I chose seven introducers. Each introducer undertook to assume sole responsibility for his own symposium, with such help in advice as I could give, and I am deeply indebted to all of them for their excellent programmes; they were W. C. Snyder, R. Baker, N. T. Flentje, D. M. Griffin, J. Rishbeth, R. A. Fox, and S. Wilhelm. The chairmen of the various sessions are listed separately, and I should like to say that the outstanding success of the disscussions in all the sessions (and I did not miss a single one of them) owed not a little to their skilful conduct of the meetings. In making these acknowledgements, however, I must not omit a more general one; the success of the Congress as a whole was due in large measure to the prolonged and heroic labours of R. K. S. Wood, as Secretary of the Executive Committee. In this he was ably assisted by B. E. J. Wheeler, as Assistant Secretary and Treasurer. An account of the Congress prepared by R. K. S. Wood will be published by the Commonwealth Mycological Institute in the Review of Applied Mycology.

For anyone of my generation, it is reassuring to be able to reflect that our branch of plant pathology, for so long held back by the formidable complexity of its problems, is now rapidly advancing. Although many workers and laboratories all over the world have taken part in this leap forward, no one can deny that much of the credit for these recent rapid advances, particularly over the last decade, must go to the Department of Plant Pathology of the University of California at Berkeley, under its Chairman, W. C. Snyder. It is thus particularly appropriate that this second international symposium volume on soil-borne plant pathogens is being published, like its predecessor, by the University of California Press. It has my good wishes, though it will have no need of them, and I am sure it will enjoy an equally resounding success.

Preface

This volume is the product of the Second International Symposium on Factors Determining the Behavior of Plant Pathogens in Soil, held in London, England, July 14–28, 1968, in conjunction with the First International Congress of Plant Pathology. It is the successor to the volume *Ecology of Soil-Borne Plant Pathogens: Prelude to Biological Control*, edited by Kenneth F. Baker and William C. Snyder, and presents new knowledge that has accumulated in the interim as well as subject matter not treated in the first volume.

We wish especially to thank Dr. S. D. Garrett, Chairman of the Executive Committee of the Congress and section organizer of the Root Disease and Soil-Borne Pathogen Symposium. His untiring efforts in organizing the symposium and assisting us in preparation of this volume were invaluable.

We wish to acknowledge the encouragement and help of Dr. K. F. Baker, who lightened our task considerably. We also thank Mr. Ernest Callenbach and his staff at the University of California Press in seeing the project through. We are likewise indebted to the Institute for Fungus Research, San Francisco, for financial support of one of us (T.A.T.), the defrayment of expenses for the preparation and mailing of notices, announcements, et cetera, to the authors, and for the indexing of this volume.

In the last analysis, it is the authors who are responsible for the high quality of this volume, and we thank them for their cooperation and their thoroughness in the preparation of manuscripts.

T. A. Toussoun, Institute for Fungus Research, San Francisco, California

R. V. Bega, Pacific Southwest Forest and Range Experiment Station, Forest Service, U. S. Department of Agriculture, Berkeley, California

P. E. Nelson, The Pennsylvania State University, University Park

June 16, 1969

▶

Chairmen of Symposium Sessions

▶

KENNETH F. BAKER, Department of Plant Pathology, University of California, Berkeley

P. W. BRIAN, Department of Botany, University of Glasgow, Scotland

J. L. HARLEY, Department of Botany, University of Sheffield, England

GEORGE H. HEPTING, Southeastern Forest Experiment Station, U.S. Department of Agriculture, Asheville, North Carolina

JAMES TAMMEN, Department of Plant Pathology, The Pennsylvania State University, University Park

WILLIAM C. SNYDER, Department of Plant Pathology, University of California, Berkeley

GEORGE A. ZENTMYER, Department of Plant Pathology, University of California, Riverside

Contents

PART I
◀
INTRODUCTION

Recent Advances in the Study of the Ecology of Soil-borne Plant Pathogens

WILLIAM C. SNYDER—*Department of Plant Pathology, University of California, Berkeley.*

▶

It was timely in 1963 that an International Symposium be arranged on soil-borne pathogens, with emphasis on their biological control. A revolution in thinking was taking place, particularly in relation to the behavior of fungi in soil and to the factors influencing their existence, survival, and activity. At the first Berkeley Symposium the attempt was made to set a secure foundation for this new thinking by focusing the variously related disciplines on the topic of biological control of soil-borne pathogens. A broad look was taken at the biological, chemical, and physical environments of these pathogens as they are influenced by the beneficial and nonbeneficial constituents of the soil flora and fauna.

This 1968 Symposium will point up the dramatic progress which has taken place in recent years. Some of the conjecture of 1963 has become established fact and now represents a base in our thinking today. The historical aspects of this topic have been covered by Garrett and by other participants in 1963 and in various reviews. Only a few more recent highlights will be referred to here.

THE DIRECTION OF RECENT RESEARCH EFFORTS.—Progress since 1963 has largely been in defining more specifically some of the important factors and relationships in the ecology of soil-borne plant pathogens. We are moving from the study of general concepts and observations toward the study of specific interactions and mechanisms. Inevitably this demands a recognition of the microhabitat approach in soil investigations. Already this approach has stimulated work with many new techniques in old and new areas of study, which we may best classify as concerning (a) the form in which the fungus exists in soil, (b) factors affecting formation and survival of these resting structures, (c) factors affecting resumption of activity and suscept-detection by the pathogen, and (d) factors influencing the suscept-pathogen disease interaction.

Not all progress in these areas has come in the last 5 or even 10 years. But much supporting data and confirmations of earlier work have followed the stimulus provided by the first symposium.

Mode of pathogen existence.—Only in recent years has it been generally recognized that pathogenic fungi are for the most part inactive in soil and are not vigorously moving about in a continual search for roots of a suscept or for other food sources. Yet Park (1959) made this observation some 10 years ago, and Nash, Christou, and Snyder (1961) have provided overwhelming data in support of it. To be sure, there are fungi like *Armillaria mellea* which do move in the form of rhizomorphs and which may grow for several feet through certain soils from a well-established food base provided by a partially invaded root when conditions of moisture and temperature are favorable. But other pathogens like *Fusarium, Thielaviopsis, Olpidium* and *Plasmodiophora,* and probably *Rhizoctonia, Aphanomyces,* and perhaps *Pythium* and *Phytophthora* spend much of their existence in soil in a state of rest, either as spores, sporangia, sclerotia, or as fragments of mycelium. Such resting structures may be reactivated by the appearance in the environment of a proper food source, and may flare into temporary vegetative activity or, in the presence of the host, enter a sustained period of pathogenic development. A return to the resting state takes place with the depletion of nutrients or moisture, or in response to other unfavorable factors.

Our knowledge of a pathogen's particular survival structure in soil, recently summarized by Warcup (1967), has enabled us to make more valid observations on pathogen population dynamics (Baker, 1968; and in this volume) to infest soil in a biologically acceptable manner (Nash, Christou, and Snyder, 1961; Lacy and Horner, 1965; Linderman, 1967; Stanghellini, unpublished data), and to study relevant factors affecting the survival of the pathogen in soil.

Ecology of formation and survival of pathogen resting structures.—Certain fungal pathogens may survive well in one soil type but not in another adjacent soil subject to the same climatic conditions. The destructive Panama disease of banana may eliminate the culture of susceptible varieties in one soil type but may be unimportant in another on the same farm, and this is true for many other vascular Fusarium wilts. A seedling disease of cotton caused by *Thielaviopsis* is far more abundant and serious on some soils in a growing area than on other soils subjected to similar farming practices.

3

In addition, cultural practices may greatly influence the survival and the population increases of *Macrophomina phaseoli* (Watanabe, 1967). It has been well demonstrated that barley straw accumulating in soil during crop sequences decreases the disease development in beans caused by *F. solani* f. sp. *phaseoli* (Snyder, Schroth, and Christou, 1959). Yet a barley crop preceding potatoes may increase the amount of Streptomyces scab (Weinhold et al., 1964). Alfalfa meal amendments have been effective in reducing disease caused by *Phytophthora* spp. (Zentmyer, 1963), and other examples may be added to illustrate the fact that a crop sequence may be useful in reducing the incidence of some diseases but not necessarily others.

These are but a few of many observations where soil types, vegetational cover, crop sequence, or substrate amendments have been shown to affect inoculum potentials of plant pathogens. Work is now directed to determining the mechanisms responsible for these differences, operating at the microhabitat level.

Alexander et al. (1966), Bourret (1966), and Ford (1969) have elucidated the delicate interaction between nutrients and inhibitors in the soil involved in the induction of chlamydospore formation by *Fusarium solani*. It has been suggested that *Fusarium* spp. dependent on specific requirements for chlamydospore formation will not survive in a soil which does not supply them. Menzinger, Toussoun, and Smith (1966) found extracts from forest litter caused *Fusarium oxysporum* chlamydospores to germinate and lyse without formation of new chlamydospores. They suggested that this mechanism may be responsible for the very low populations of *F. oxysporum* found in forest soils.

The survival of a resting structure once formed is in part dependent on its failure to germinate spontaneously when no substrate is available. Again it seems that the mechanism determining whether or not a resting structure germinates must be a balance in the microhabitat between nutrient and inhibition factors. Lockwood (1964) and Park (1967) have reviewed recent work on fungistasis, and Lingappa and Lockwood (1964), Ko and Lockwood (1967), and Vaartaja and Agnihotri (1967) have begun to elucidate the mechanisms involved. The mystery of fungistasis may be solved in time for the next symposium.

Other advances have been made in the study of factors influencing successive germination and reformation of resting structures and the intervention of lysis in these cycles (Cook and Snyder, 1965; Stanghellini, unpublished data). The mechanisms bringing about lysis are as yet little known.

The effect of nutrients on the saprophytic survival of fungal mycelium in wheat straw, brilliantly elucidated by Garrett (1963; 1966), demonstrates the relationship between rate of substrate utilization and saprophytic survival of mycelium.

Resumption of pathogen activity and suscept detection.—Recent research has centered on the nature of the stimuli which initiate germination of resting propagules and their detection of the suscept. Exudates from roots and seeds have long been implicated in the release of fungistasis and subsequent germination of resting structures of fungi and nematodes (Barton, 1957; Jackson, 1957; Wallace, 1958; Schroth and Snyder, 1961; and Buxton, 1962). In their review article on root exudates, Schroth and Hildebrand (1964) suggest that generally the action of exudates on fungal germination is not specific; that is, both hosts and nonhosts may supply the stimulus.

However, Coley-Smith (1960) demonstrated the host specificity, under natural conditions, of the stimulus for germination of sclerotia of *Sclerotium cepivorum*. It now seems clear that other propagules such as resting spores of *Ceratocystis* (Muñoz, 1967), and microsclerotia of *Verticillium* and of *Macrophomina* are similar, germination being restricted by their apparent requirements for rather specific substances. Such fungi are in contrast to *Fusarium*, *Rhizoctonia*, *Pythium*, and most *Phytophthoras*, which seem to be activated by a fairly broad range of nutrients.

The role of exudates in stimulating tropic and tactic responses in soil fungi has been the subject of recent studies. Particular attention has been given to zoospore taxis toward roots (Zentmyer, 1961; Cunningham and Hagedorn, 1962; Royle and Hickman, 1964; Carlile, 1966; Hickman and Ho, 1966).

A degree of host specificity was reported by Zentmyer (1961) for *Phytophthora cinnamomi*, but no specificity has been observed in other *Phytophthora* spp., *Pythium* nor *Aphamomyces*. The mechanism whereby zoospores are attracted to plant roots represents a fascinating unsolved problem of the root microhabitat; many clues are available, but the solution is not known. Indeed, the biological significance of zoospore attraction has not been conclusively proven.

Factors influencing disease development.—The events which lead directly to the establishment of a pathogen-suscept disease relationship occur in the soil microhabitat immediately surrounding the susceptible tissue bathed in the exudate from that tissue. The function of carbohydrate and nitrogen nutrients in the establishment of *F. solani* f. sp. *phaseoli* as a pathogen was pointed out by the brilliant work of Toussoun, Nash, and Snyder (1960). Since then, Kamal and Weinhold (1967) and Weinhold, Bowman, and Dodman (1969) have demonstrated a similar dependence on external nutrition in the physiology of *Rhizoctonia* pathogenesis; and Garraway and Weinhold (1968) have investigated the nutritional dependence of rhizomorph morphogenesis in *Armillaria mellea*. The location of pathogen infection structures also points to the importance of external nutrition in pathogenesis, which may be more important than the nutrition of the host (Flentje, Dodman, and Kerr, 1963).

Interactions which take place in this microhabitat responsible for failure of the pathogen to establish a disease relationship with the suscept have recently been elucidated. They include lysis of *Fusarium* germ tubes (Cook and Snyder, 1965; Cook and Flentje, 1967), hyperparasitism of *Ophiobolus graminis* in the rhizosphere (Siegle, 1961), accumulation of antibiotic materials (Wright, 1956) and protection by mycorrhizal associations (Zak, 1964; Marx, 1969). While discussing mycorrhizal associations we might mention the recent work summarized by Gerdemann (1968) on vesicular-arbuscular mycorrhiza. Gerdemann suggests that the vast majority of roots in the field are infected by these fungi, which fact further defines the suscept root microhabitat.

Further evidence to support the importance of preinfection nutrition on pathogenesis by soil pathogens can be found in the studies of Schroth, Weinhold, and Hayman (1966), Hayman (1967), and Linderman and Toussoun (1968). In these studies, increasing pathogenesis could be correlated with factors which brought about an increase in host exudation.

DEVELOPMENT OF SUITABLE TECHNIQUES.—The complexity of the soil and the desire to measure conditions at the microhabitat level demands new or modified techniques for the study of pathogen ecology in soil.

Biological measurements.—Methods for observing the behavior of pathogen propagules in field soil have been devised for a number of fungi. Selective media have been developed to estimate populations of *Fusarium* (Nash and Snyder, 1962), *Phytophthora* (Haas, 1964; Ocaña and Tsao, 1966), *Thielaviopsis* (Tsao, (1964), *Verticillium* (Nadakavukaren and Horner, 1959; Evans, Wilhelm, and Snyder, 1967), and *Pythium* (Singh and Mitchell, 1961; Vaartaja, 1968).

Certainly in the studies of *Fusarium* ecology the techniques of soil smears in weak agar solution followed by acid fuchsin stain (Nash, Christou, and Snyder, 1961) and of plating out soil dilutions on selective isolation media have provided qualitative and quantitative measures of a soil population which have been basic and essential in advancing our knowledge of the biology of *Fusarium* in soil.

The stage was then set for the discovery by Toussoun, Nash, and Snyder (1960) that these fungi have a requirement for exogenous nutrients for germination and host infection and the correlation of these results with the findings of Rovira (1958) regarding exudation of nutrients from underground plant parts.

Physical measurements.—In recent years our appreciation of the physical parameters of the soil has greatly expanded, together with our ability to measure some of these parameters and to record them in meaningful units. Burges (1967) has reviewed some of the advances in his discussion of the soil system.

Griffin (1963; and in this volume) has pioneered new techniques in soil moisture measurement and together with Smiles (1966) has developed techniques for measuring oxygen concentration and diffusion.

Cook and Papendick (1969) have used a psychrometric method to measure moisture in very small soil masses. Greenwood (1967) has pointed out to us the role of plant roots in transporting oxygen from the atmosphere into the root microhabitat. Continuous measurements of carbon dioxide concentrations have been made (Martin and Pigott, 1965), and attention has been given to other components of the soil atmosphere by Gilbert, Menzies, and Griebel (1968).

Part I of McLaren and Peterson's *Soil Biochemistry* (1967) documents thoroughly the range of compounds that can be found in soil and methods for their detection. We cannot yet ascribe an ecological role for most of these compounds, just as we have not estimated the significance of clay surface activity in the ecology of soil pathogens, but in the near future we may see the significance of these microhabitat factors realized.

CONCLUSIONS.—It would be difficult to overemphasize the importance of techniques in studies on soil-borne fungi, since research must be carried out in field soil in some part of the program if the results are to relate to what actually happens in nature. For example, the development of soil-smear techniques made it possible to follow visually what happens to the fungus in soil in the vicinity of the host, and selective media permit the quantitative measure of the pathogen at all times in the field.

In the past 5 years, special techniques have enabled us to become more and more specific in our study of the ecology of soil-borne plant pathogens. We have called this the microhabital approach.

To us the most significant aspect of the data obtained is the overwhelming importance of substrate and nutrition as a key factor affecting all the events and interactions summarized in Fig. 1. We may conclude that the nutrient stimulus given the pathogen by the host is of such importance as to consider it a principle in plant pathology that the host stimulates the pathogen, which may be in close proximity to it, with the release of nutrients. The stimuli may not only germinate the propagule but orientate the pathogen and also provide necessary energy and enzymes to initiate infection processes and pathogenesis. The nutrients essential to the pathogen may be stored in the resting propagule or food base in which it is lodged, may be provided by the soil solution or undecomposed crop refuse in the soil, or may exude in aqueous or gaseous form from the various organs of the host. Even mycorrhizal fungi must depend to some extent on nutrients external to the host, as has been well demonstrated for nematodes. Further work may show a similar situation to exist with *Streptomyces* and with bacteria which invade below-ground parts of the plants.

FUTURE DEVELOPMENTS.—Having attempted to chronicle and interpret very briefly the multitude of significant research findings of the recent past in an

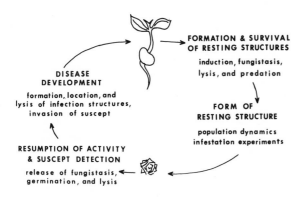

measurement of: nutrients (root and seed exudates etc.),
inhibitors, oxygen, carbon dioxide, other volatiles,
soil water, temperature, pH, and biological spectrum . . .

as they affect critical processes in soil pathogen ecology

FORMATION & SURVIVAL
OF RESTING STRUCTURES

induction, fungistasis,
lysis, and predation

DISEASE
DEVELOPMENT

formation, location, and
lysis of infection structures,
invasion of suscept

FORM OF
RESTING STRUCTURE

population dynamics
infestation experiments

RESUMPTION OF ACTIVITY
& SUSCEPT DETECTION

release of fungistasis,
germination, and lysis

Fig. 1.   Recent Advances in the Study of the Ecology of Soil-Borne Plant Pathogens.

area as complex as the ecology of soil pathogens, the task of prophesying the near future seems less fraught with the dangers of offending omissions and misguided interpretations. At least several years of grace will intervene between these predictions and their relegation to the realm of "false prophecy."

The future should see further definition of microhabitats with the attendant development of techniques capable of making more relevant measurements of biological and physical factors constituting the environment. Quantitization of factors controlling key processes will be emphasized, particularly quantitization of host-derived nutritional factors in the soil whose role will be seen to be even more comprehensive in soil-pathogen ecology. As the specific mechanisms controlling critical processes in soil pathogenesis become known further, more successful attempts will be made to utilize biological control methods for disease control. "The prospects for biological control are brighter than ever before" are the words used by Garrett to conclude his introduction to the First International Symposium. Today our prognosis is the same, but in the meantime we have come a long way.

ACKNOWLEDGMENTS.—The progress we have made, and which I have had the privilege of reviewing here, has been achieved only through the efforts of many scientists located throughout the world. I would like personally to pay tribute to Dennis Garrett, who has continued to provide inspirational leadership in our field, and to express my sincere appreciation to all my graduate students for carrying forward our research at Berkeley. I am also grateful to Andrew Watson for assistance in compiling this manuscript and creating Fig. 1.

LITERATURE CITED

ALEXANDER, J. V., J. A. BOURRET, A. H. GOLD, and W. C. SNYDER. 1966. Induction of chlamydospore formation by *Fusarium solani* in sterile soil extracts. Phytopathology 56:353–354.

BAKER, R. 1968. Mechanisms of biological control of soil-borne pathogens. Ann. Rev. Phytopathol. 6:263–294.

BARTON, R. 1957. Germination of oospores of *Pythium mamillatum* in response to exudates from living seedlings. Nature (London) 180:613–614.

BOURRET, J. A. 1966. Physiology of chlamydospore formation and survival in *Fusarium*. Ph.D. dissertation, University of California, Berkeley. 96p.

BURGES, A. 1967. The soil system, p. 1–13. *In* A. Burges and F. Raw [ed.], Soil biology. Academic Press, London.

BUXTON, E. W. 1962. Root exudates from banana and their relationship to strains of the *Fusarium* causing Panama wilt. Ann. Appl. Biol. 50:269–282.

CARLILE, M. J. 1966. The orientation of zoospores and germ tubes, p. 175–186. *In* M. F. Madelin [ed.], The fungus spore. Butterworths, London.

COLEY-SMITH, J. R. 1960. Studies of the biology of *Sclerotium cepivorum* Berk. Ann. Appl. Biol. 48:8–18.

COOK, R. J., and W. C. SNYDER. 1965. Influence of host exudates on growth and survival of germlings of *Fusarium solani* f. *phaseoli* in soil. Phytopathology 55:1021–1025.

COOK, R. J., and N. T. FLENTJE. 1967. Chlamydospore germination and germling survival of *Fusarium solani* f. *pisi* in soil as affected by soil water and pea seed exudate. Phytopathology 57:178–182.

COOK, R. J., and R. I. PAPENDICK. 1969. Soil water potential as a factor in the ecology of *Fusarium roseum* f. sp. *cerealis* 'Culmorum.' Plant Soil. (In press.)

CUNNINGHAM, J. L., and D. J. HAGEDORN. 1962. Attraction of *Aphanomyces eutiches* zoospores to pea and other plant roots. Phytopathology 52:616–618.

EVANS, G., S. WILHELM, and W. C. SNYDER. 1967. Quantitative studies by plate counts of propagules of the Verticillium wilt fungus in cotton field soils. Phytopathology 57:1250–1255.

FLENTJE, N. T., R. L. DODMAN, and A. KERR. 1963. The mechanisms of host penetration by *Thanatephorus cucumeris*. Austral. J. Biol. Sci. 16:784–799.

FORD, E. J. 1969. Chlamydospore formation in *Fusarium*. Ph.D. dissertation, University of California, Berkeley. 116 p.

GARRAWAY, M. O., and A. R. WEINHOLD. 1968. Period of access to ethanol in relation to carbon utilization and rhizomorph initiation and growth in *Armillaria mellea*. Phytopathology 58:1190–1191.

GARRETT, S. D. 1963. A comparison of cellulose-decomposing ability in five fungi causing cereal root rots. Trans. Br. Mycol. Soc. 46:572–576.

GARRETT, S. D. 1966. Cellulose-decomposing ability of some cereal foot-rot fungi in relation to their saprophytic survival. Trans. Br. Mycol. Soc. 49:57–68.

GERDEMANN, J. W. 1968. Vesicular arbuscular mycorrhiza and plant growth. Ann. Rev. Phytopathol. 6:397–418.

GILBERT, R. G., J. D. MENZIES, and G. E. GRIEBEL. 1968. The influence of volatile substances from alfalfa on growth and survival of *Verticillium dahliae* in soil. Phytopathology 58:1051.

GREENWOOD, D. J. 1967. Studies on the transport of oxygen through stems and roots of vegetable seedlings. New Phytologist 66:337–348.

GRIFFIN, D. M. 1963. Soil moisture and the ecology of soil fungi. Biological Reviews 38:141–166.

HAAS, J. H. 1964. Isolation of *Phytophthora megasperma* var. *sojae* in soil dilution plates. Phytopathology 54:894.

HAYMAN, D. S. 1967. The role of cotton and bean seed

exudate in pre-emergence infection by *Rhizoctonia solani*. Ph.D. dissertation, University of California, Berkeley. 126 p.

HICKMAN, C. J., and H. H. Ho. 1966. Behaviour of zoospores in plant pathogenic phycomycetes. Ann. Rev. Phytopathol. 4:195–220.

JACKSON, R. M. 1957. Fungistasis as a factor in the rhizosphere phenomenon. Nature (London) 180:96–97.

KAMAL, M., and A. R. WEINHOLD. 1967. Virulence of *Rhizoctonia solani* as influenced by age of inoculum in soil. Can. J. Botany 45:1761–1765.

KO, W., and J. L. LOCKWOOD. 1967. Soil fungistasis: Relation to fungal spore nutrition. Phytopathology 57:894–901.

LACY, M. L., and C. E. HORNER. 1965. Verticillium wilt of mint: Interactions of inoculum density and host resistance. Phytopathology 55:1176–1178.

LINDERMAN, R. G. 1967. Behaviour of *Thielaviopsis basicola* in natural soils with emphasis on the effects of products of decomposing plant residues. Ph.D. dissertation, University of California, Berkeley. 175 p.

LINDERMAN, R. G., and T. A. TOUSSOUN. 1968. Predisposition to *Thielaviopsis* root rot of cotton by phytotoxins from decomposing barley residues. Phytopathology 58:1571–1576.

LINGAPPA, B. T., and J. L. LOCKWOOD. 1964. Activation of soil microflora by fungus spores in relation to soil fungistasis. J. Gen. Microbiol. 35:215–227.

LOCKWOOD, J. L. 1964. Soil fungistasis. Ann. Rev. Phytopathol. 2:341–362.

MARTIN, M. H., and C. D. PIGOTT. 1965. A simple method for measuring carbon dioxide in soils. J. of Ecology 53:153–155.

MARX, D. H. 1969. The influence of ectotrophic mycorrhizal fungi on resistance of pine roots to pathogenic infections. I. Antagonism of mycorrhizal fungi to root pathogenic fungi. Phytopathology 59:153–163.

McLAREN, A. D., and G. H. PETERSON. 1967. Soil biochemistry. Marcel Dekker, New York. 509 p.

MENZINGER, W., T. A. TOUSSOUN, and R. S. SMITH. 1966. Reduction of *Fusarium oxysporum* population in soil by aqueous extracts of pine duff. Phytopathology 56:889–890.

MUÑOZ, R. 1967. Biology of the fungus plant pathogen *Ceratocystis fimbriata*. Ph.D. dissertation, University of California, Berkeley. 83 p.

NADAKAVUKAREN, M. J., and C. E. HORNER. 1959. An alcohol agar medium selective for determining *Verticillium* microsclerotia in soil. Phytopathology 49:527–528.

NASH, S. M., T. CHRISTOU, and W. C. SNYDER. 1961. Existence of *Fusarium solani* f. *phaseoli* as chlamydospores in soil. Phytopathology 51:308–312.

NASH, S. M., and W. C. SNYDER. 1962. Quantitative estimations by plate counts of propagules of the bean root rot *Fusarium* in field soils. Phytopathology 52:567–572.

OCAÑA, G., and P. H. TSAO. 1966. A selective agar medium for the direct isolation and enumeration of *Phytophthora* in soil. Phytopathology 56:893.

PARK, D. 1959. Some aspects of the biology of *Fusarium oxysporum* Schl. in soil. Ann. Botany [N.S.] 23:35–49.

PARK, D. 1967. The importance of antibiotics and inhibiting substances, p. 435–447. *In* A. Burges and F. Raw [ed.], Soil biology. Academic Press, London.

ROVIRA, A. D. 1958. Plant root excretions in relation to the rhizosphere effect. I. The nature of root exudates from oats and peas. Plant Soil 7:178–194.

ROYLE, D. J., and C. J. HICKMAN. 1964. Analysis of factors governing in vitro accumulation of zoospores of *Pythium aphanidermatum* on roots. I. Behaviour of zoospores. Can. J. Microbiol. 10:151–162.

SCHROTH, M. N., and W. C. SNYDER. 1961. Effect of host exudates on chlamydospore germination of the bean root rot fungus, *Fusarium solani* f. *phaseoli*. Phytopathology 51:389–393.

SCHROTH, M. N., and D. C. HILDEBRAND. 1964. Influence of plant exudates on root-infecting fungi. Ann. Rev. Phytopathol. 2:101–132.

SCHROTH, M. N., A. R. WEINHOLD, and D. S. HAYMAN. 1966. The effect of temperature on quantitative differences in exudates from germinating seeds of bean, pea, and cotton. Can. J. Botany 44:1429–1432.

SIEGLE, H. 1961. Über mischinfektionen mit *Ophiobolus graminis* und *Didymella exitialis*. Phytopathol. Z. 42:305–348.

SINGH, R. S., and J. E. MITCHELL. 1961. A selective method for isolation and measuring the population of *Pythium* in soil. Phytopathology 51:440–444.

SMILES, D. E., and D. M. GRIFFIN. 1966. The measurement of the diffusion of oxygen in saturated porous media. Australian J. Soil Research 4:87–93.

SNYDER, W. C., M. N. SCHROTH, and T. CHRISTOU. 1959. Effect of plant residues on root rot of bean. Phytopathology 49:755–756.

TOUSSOUN, T. A., S. M. NASH, and W. C. SNYDER. 1960. The effect of nitrogen sources and glucose on the pathogenesis of *Fusarium solani* f. *phaseoli*. Phytopathology 50:137–140.

TSAO, P. H. 1964. Effect of certain fungal isolation agar media on *Thielaviopsis basicola* and on its recovery in soil dilution plates. Phytopathology 54:548–555.

VAARTAJA, O. 1968. *Pythium* and *Mortierella* in soils of Ontario forest nurseries. Can. J. Microbiol. 14:265–269.

VAARTAJA, O., and V. P. AGNIHOTRI. 1967. Inhibition of *Pythium* and *Thanatephorus* (*Rhizoctonia*) by leachates from a nursery soil. Phytopath. Z. 60:63–72.

WALLACE, H. R. 1958. Observations on the emergence from cysts and the orientation of larvae of three species of the genus *Heterodera* in the presence of host plant roots. Nematologica 3:236–243.

WARCUP, J. H. 1967. Fungi in soil, p. 51–110. *In* A. Burges and F. Raw [ed.], Soil biology. Academic Press, London.

WATANABE, TSUNEO. 1967. Populations of microsclerotia of the soil-borne pathogen, *Macrophomina phaseoli*, in relation to stem blight of bean. Ph.D. dissertation, University of California, Berkeley. 97 p.

WEINHOLD, A. R., J. W. OSWALD, T. BOWMAN, J. BISHOP, and D. WRIGHT. 1964. Influence of green manures and crop rotation on common scab of potato. Am. Potato J. 41:265–273.

WEINHOLD, A. R., T. BOWMAN, and R. L. DODMAN. 1969. Virulence of *Rhizoctonia* as affected by nutrition of the pathogen. Phytopathology 59:1601–1605.

WRIGHT, J. N. 1956. The production of antibiotics in soil. IV. Production of antibiotics in coats of seeds sown in soil. Ann. Appl. Biol. 44:561–566.

ZAK, B. 1964. The role of mycorrhizae in root disease. Ann. Rev. Phytopathol. 2:377–392.

ZENTMYER, G. A. 1961. Chemotaxis of zoospores for root exudates. Science 133:1595–1596.

ZENTMYER, G. A. 1963. Biological control of *Phytophthora* root rot of avocado with alfalfa meal. Phytopathology 53:1383–1387.

PART II
◄
POPULATION DYNAMICS OF PATHOGENS IN SOIL

Use of Population Studies in Research on Plant Pathogens in Soil

RALPH BAKER—*Department of Botany and Plant Pathology, Colorado State University, Fort Collins.*

Since the symposium on soil-borne plant pathogens at Berkeley, researchers have intensified efforts to quantitate populations of plant pathogens in soil. For practical purposes, quantitative analysis could locate crop areas where pathogen populations are too high to result in a significant economic crop return (e.g. Leach and Davey, 1938). This is not usually needed since the farmer knows from experience that he has a problem. The more immediate application of quantitative analysis is in the study of the ecological parameters influencing the population levels of plant pathogens in soil and the verification of mathematical systems and models useful for predictions and analyses.

Inoculum theory should be based upon knowledge of the activities of the living organism as it is influenced by the environment. The organism has the potential or energy to perform certain biological functions at a given point in space or time. Theory should attempt to describe such activities quantitatively. This is where models should be useful.

TYPES OF MODELS.—Only a few attempts of this type have been made on the ecology of plant pathogens in soil (excepting nematodes). Sneh, et al. (1966) have applied some of the theory developed by van der Plank (1963) to evaluate methods for determining inoculum potential of *Rhizoctonia* in soil. Dimond and Horsfall (1965) have also reviewed theory concerning inoculum density and multiple infections.

Inoculum in soil exists in three-dimensional space and at times appears to be remarkably uniformly distributed. For example, distribution of the bean root rot *Fusarium* was uniform in cultivated fields at a plow depth of 6–8 in (Nash and Snyder, 1962). Inoculum may not be distributed so evenly in unplowed fields. Trujillo and Snyder (1963) found *Fusarium oxysporum* f. sp. *cubense* associated with cortical tissues of decayed roots of banana but not in decayed tissues of other plants. Its presence in soil outside of tissue was difficult to demonstrate. The principles involved in constructing models in these two cases may be different, but as long as distribution can be determined, there appear to be no insurmountable difficulties.

In describing the interactions between underground plant parts and propagules of pathogens in three-dimensional space, it would seem desirable to explore possibilities of models based on solid geometry. This has been developed for inoculum uniformly distributed in soil (Baker, Maurer, and Maurer, 1967). The relationship between inoculum density and disease severity was the basis for the models: as inoculum density increases there is a corresponding increase in disease severity until a plateau is reached when no increase in symptoms occurs with further increases in inoculum (Dimond and Horsfall, 1965). Four different types of diseases were recognized: (I) nonmotile (fixed) inoculum distributed about a fixed infection court (e.g. below-ground hypocotyl), (II) nonmotile inoculum invaded by a motile infection court (e.g. a moving root tip), (III) motile inoculum about a fixed infection court, and (IV) motile inoculum about a moving infection court. If inoculum is distributed evenly throughout soil, one can apply formulas based on tetrahedrons to determine positions and distances of this inoculum from these various infection courts. Each point of a tetrahedron could be considered the location of a propagule in soil (Baker, 1965). Thus theoretical slopes can be calculated for the models as successful infections increase due to the addition of more and more inoculum to a system. If propagules germinate at a significant distance from the infection court (the rhizosphere effect) in Model I, each increment of increase of inoculum would provide a proportionate increase in the number of successful infections, and the slope or rate of increase should be 1. If only those spores touching the infection court penetrate and infect (a rhizoplane effect) the slope should be .67 (log-log). The relative slopes are shown in Fig. 1-A. For Model II, the slope is also .67 (log-log). In this case the size, number, and rate of growth of the infection courts (root tips) alter the position of the slope but not its value (Fig. 1-B). For motile inoculum (Models III and IV), successful infection was related to distance between motile propagules at any time (Baker, Maurer, and Maurer, 1967). In these cases the slopes are −2. All these, of course, only account for the geometry of the systems and do not correct for multiple infections and other factors associated with actual situations.

APPLICATIONS OF MODELS.—Some limitations in the application of theory to concrete disease situations have been pointed out (Baker, 1965; Baker, Maurer,

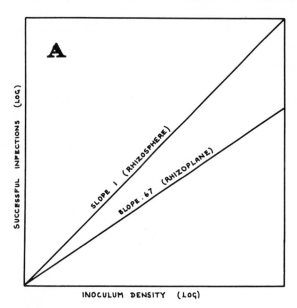

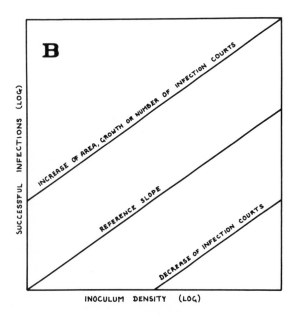

Fig. 1. Theoretical slopes of lines resulting from interaction between successful infections and inoculum density in *A*, Model I, and *B*, Model II.

should correspond directly to a corresponding decrease in disease severity.

Models for biological control.—If nutrients essential for germination and penetration of a pathogen become limiting in the rhizosphere during biological control, the effective volume of influence in the infection court should be reduced to the surface or rhizoplane. The slope then for Model I, for example, should be reduced from one to 0.67 (log-log). If only the rhizoplane significantly affects inoculum, and competition occurs, then the reduction in disease severity would be proportional to pN where p is the probability of propagules germinating and penetrating the infection court and N the number of viable propagules touching the rhizoplane. The ultimate result of this would be similar if antibiosis or lysis occurred, since a proportion of the propagules in this case would be nonoperative.

Predictions involving moving infection courts.—Root tips moving through soil encounter much more inoculum than comparable systems involving fixed infection courts. Models II and IV would predict then very steep slopes (on a non-log-log basis) reaching a plateau at rather low densities for susceptible varieties. To confirm this generalization would require many more experimental examples than are presently available in the literature. Perhaps the examples given by Cook (1968) for Fusarium root and foot rot of cereals (Model I) and Lacy and Horner (1965) for Verticillium wilt of mint (Model II) may illustrate the point, however. Fig. 2 illustrates data taken from their papers. Note that the plateau had not yet been reached for Fusarium root and foot rot at 10^4 propagules/g of soil. In contrast, the plateau was reached at 10^2 propagules/g of soil for Verticillium wilt on a mint variety with low resistance. Certainly these observations may reflect effects of factors other than inoculum density, but one of the implications of this is that biological control may be difficult to attain in Model II types of diseases especially if the mechanism of control involves short-term competition in the rhizosphere. If it does occur this may be due to influence of treatment on the host or large decreases in inoculum density due to lysis or antibiosis.

Mathematical theory should be ultimately supported by experimental observations in concrete situations. First attempts (Baker and Maurer, 1967) were not entirely successful because of the lack of adequate data on the extent of the rhizosphere or rhizoplane influence and difficulty of quantitative analysis of host symptoms (Baker, 1965). Recently, however, Byther (1968) has applied the models to studies of foot rot of wheat caused by *Cercosporella herpotrichoides*. The data relating inoculum density to percentage of lesioned tillers fitted Model I for a rhizoplane (slope 0.67 log-log). The slope obtained when inoculum density was related to saprophytic colonization was less than expected for a theoretical rhizoplane effect, indicating that all conidia in contact with the substrate surface were not capable of colonizing.

and Maurer, 1967); however, applications of the models in analysis of mechanisms of biological control have been suggested recently (Baker, 1968). These include:

Determination of most important mechanisms of biological control.—If control is due to short-term competition in the infection court, no decrease in inoculum density should be noted. If the mechanism is due to lysis, or antibiosis, a decrease in inoculum density should occur. In the latter case, this decrease

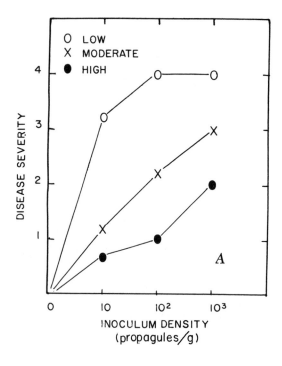

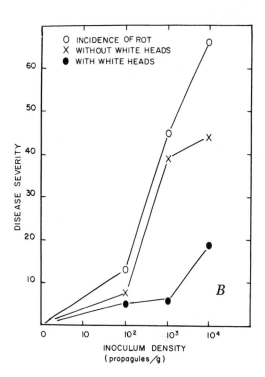

Fig. 2. Relationships of inoculum density to disease severity in *A*, Fusarium root and foot rot of cereals and *B*, Verticillium wilt of mint. Data from Cook (1968) and Lacy and Horner (1965).

Models also may be useful in explaining why and how certain phenomena operate. An example of this application may be found when examining inoculum densities found in nature (Baker and McClintock, 1965). For the particular ecological group in which most soil-borne pathogens are found, reported densities are no more than approximately 3,000 propagules/g of soil. Usually they are less. It is interesting that these densities fall in a range above which large increases in inoculum are required to decrease distance between propagules significantly.

Rao and Rao (1963, 1966a, 1966b) have explored one of the exceptions to the rule that inoculum density is directly proportional to disease severity. Disease development in cotton increased with increases of inoculum of *Fusarium oxysporum* f. sp. *vasinfectum* from 0.5% to 10% of an oatmeal-sand culture in soil. Further increases extended the incubation period. This was probably not due to the inoculum itself, however, as the organic substrates upon which the inoculum was grown also extended the time required for symptoms to appear. This underlines the importance of using inoculum in as "natural" a state as possible. This system should conform to Model II; however, relatively large inoculum potentials applied to the cotton seedlings may have induced somewhat atypical infection of the hypocotyl more characteristic of Model I. Thus the data may be difficult to analyze.

One of the most valuable applications of population studies is related to the ability to detect actual fluctua-

tions of inoculum resulting from various treatments. A good example is the study on populations of pathogens involved in the root rot-Fusarium wilt complex of peas under various crop sequences (Kerr, 1963). Mechanisms of biological control may be detected also by assays of inoculum density before and after treatment. Thus, there was no significant change in inoculum density of *F. solani* f. sp. *phaseoli* following incorporation of cellulose into soil even though control of bean root rot was accomplished (Baker and Nash, 1965; Papavizas, Lewis, and Adams, 1968). This strongly suggests that the most significant mechanism of biological control is competition (Snyder, Schroth, and Christou, 1959; Maurer and Baker, 1965) rather than lysis or antibiosis.

The relation of population levels of plant pathogens in soils to fungicidal dosage also is of interest and more attention to this type of research is needed. An excellent example is that given by Richardson and Munnecke (1963). Their findings confirmed previous work (reviewed in their paper) indicating that fungicide dosage required to control soil pathogens is proportional to the inoculum density. Thus, more attention will have to be given to the importance of standardizing inoculum in soil-fungicide evaluation tests. The relation of this work to mathematical models is treated elsewhere (Baker, 1968).

Ultimately, quantitative analysis of all significant factors contributing to disease severity should have value in achieving a basic understanding of the epi-

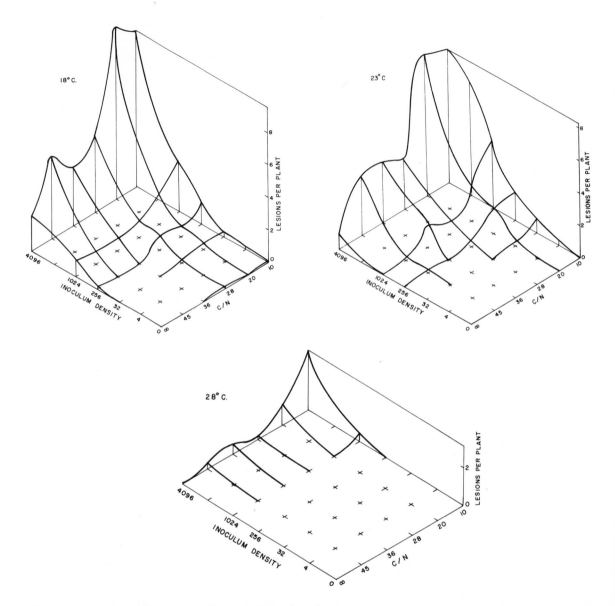

Fig. 3.   Interrelationships of the influence of inoculum density, C:N ratio, and temperature on number of lesions incited by *Fusarium solani* f. sp. *phaseoli* on bean hypocotyls.

demiology of soil-borne pathogens. The equation: disease severity = inoculum potential × disease potential has been suggested as a basis (Baker, 1965). Such a project will require extensive systems analysis and should be developed with better theory than is presently available. A first attempt has been made to interrelate some of the major factors influencing disease severity for bean root rot (Baker and Maurer, 1967). Among these are inoculum density, C:N ratio, and temperature (Fig. 3). Since these experiments were done in growth chambers under controlled conditions, the results cannot be directly applicable to field conditions. They do illustrate, however, a pos-

sible direction for systems analysis applicable to disease situations.

The examples noted above are but a few of the uses of studies concerned with inoculum potential. Other papers in this section will treat the influence of environmental factors, host response, and the significance of inoculum potential in specific pathogen-host interactions.

LITERATURE CITED

BAKER, R.   1965.   The dynamics of inoculum, p. 395–403. *In* K. F. Baker and W. C. Snyder [ed.], Ecology of soil-borne plant pathogens. Univ. of California Press, Berkeley and Los Angeles.

BAKER, R. 1968. Mechanisms of biological control of soil-borne pathogens. Ann. Rev. Phytopathol. 6:263–294.

BAKER, R., and D. L. McCLINTOCK. 1965. Populations of pathogens in soils. Phytopathology 55:495.

BAKER, R., and C. L. MAURER. 1967. Interaction of major factors influencing severity of bean root rot. Phytopathology 57:802. (Abstr.)

BAKER, R., C. L. MAURER, and RUTH A. MAURER. 1967. Ecology of plant pathogens in soil. VII. Mathematical models and inoculum density. Phytopathology 57:662–666.

BAKER, R., and SHIRLEY M. NASH. 1965. Ecology of plant pathogens in soil. VI. Inoculum density of *Fusarium solani* f. sp. *phaseoli* in bean rhizosphere as affected by cellulose and supplemental nitrogen. Phytopathology 55:1381–1382.

BYTHER, R. S. 1968. Etiological studies on foot rot of wheat caused by *Cercosporella herpotrichoides*. Ph.D. thesis, Oregon State University, Corvallis.

COOK, R. J. 1968. Fusarium root and foot rot of cereals in the Pacific Northwest. Phytopathology 58:127–131.

DIMOND, A. E., and J. G. HORSFALL. 1965. The theory of inoculum, p. 404–415. *In* K. F. Baker and W. C. Snyder [ed.], Ecology of soil-borne plant pathogens. Univ. of California Press, Berkeley and Los Angeles.

KERR, A. 1963. The root rot-Fusarium wilt complex of peas. Australian J. of Biol. Sci. 16:55–69.

LACY, M. L., and C. E. HORNER. 1965. Verticillium wilt of mint: interactions of inoculum density and host resistance. Phytopathology 55:1176–1178.

LEACH, L. D., and A. E. DAVEY. 1938. Determining the sclerotial population of *Sclerotium rolfsii* by soil analysis and predicting losses of sugar beet on the basis of these analyses. J. Agr. Res. 56:619–632.

MAURER, C. L., and R. BAKER. 1965. Ecology of plant pathogens in soil. II. Influence of glucose, cellulose, and inorganic nitrogen amendments on development of bean root rot. Phytopathology 55:69–72.

NASH, SHIRLEY M., and W. C. SNYDER. 1962. Quantitative estimations by plate counts of propagules of the bean root rot fusarium in field soils. Phytopathology 52:567–572.

PAPAVIZAS, G. C., J. A. LEWIS, and P. B. ADAMS. 1968. Survival of root-infecting fungi in soil. II. Influence of amendment and soil carbon-to-nitrogen balance on Fusarium root rot of beans. Phytopathology 58:365–372.

PLANK, J. E. VAN DER. 1963. Plant diseases: epidemics and control. Academic Press, New York. 349 p.

RAO, A. S., and M. V. RAO. 1963. Inoculum potential and the Fusarial wilt of cotton. Nature (London) 200:598–599.

RAO, M. V., and A. S. RAO. 1966a. Fusarium wilt of cotton in relation to inoculum potential. Trans. Br. Mycol. Soc. 49:403–409.

RAO, M. V., and A. S. RAO. 1966b. The influence of the inoculum potential of *Fusarium oxysporum* f. *vasinfectum* on its development in cotton roots. Phytopathol. Z. 56:393–397.

RICHARDSON, L. T., and D. E. MUNNECKE. 1964. Effective fungicide dosage in relation to inoculum concentration in soil. Can. J. Botany 42:301–306.

SNEH, B., J. KATAN, Y. HENIS, and I. WALD. 1966. Methods for evaluating inoculum density of *Rhizoctonia* in naturally infested soils. Phytopathology 56:74–78.

SNYDER, W. C., M. N. SCHROTH, and T. CHRISTOU. 1959. Effect of plant residues on root rot of bean. Phytopathology 49:755–756.

TRUJILLO, E. E., and W. C. SNYDER. 1963. Uneven distribution of *Fusarium oxysporum* f. *cubense* in Honduras soils. Phytopathology 53:167–170.

Factors Affecting Plant Pathogen Population in Soil

J. D. MENZIES—*Soil Microbiologist, Soil and Water Conservation Research Division, U.S. Department of Agriculture, Beltsville, Maryland.*

INTRODUCTION.—Soil-borne plant disease, by definition, requires the presence of the pathogen in the soil, but the amount of disease is not clearly related to the size of the pathogen population. Many soil factors influence the final expression of disease, and it is seldom clearly established whether this effect has been upon the amount of the pathogen or upon its pathogenic success.

Because of the indirect relation between disease and amount of inoculum, wide changes in pathogen population will usually be required to produce a significant change in disease damage (Dimond and Horsfall, 1965). Since the major objective of research on soil-borne pathogens is to improve methods of disease control, this discussion will center on those factors most capable of reducing pathogen populations drastically enough to be of practical control value.

The pathogen population is expressed as the mass or number of propagative units of the pathogen per unit of soil. This is the "inoculum density" or "inoculum intensity" of different authors (Baker, Maurer, and Maurer, 1967; Dimond and Horsfall, 1965). It is usually reported as numbers of propagules and should not include consideration of the infective vigor of the inoculum (inoculum capacity). To quantify inoculum density in soil as distinct from disease incidence is not easy because units of inoculum are difficult to define and more difficult to measure directly (Menzies, 1963). The generalizations that follow are based on rather scanty data, a deficiency which may at least stimulate more intensive efforts to evaluate the quantitative features of inoculum in soil.

Factors affecting pathogen populations can best be considered in relation to stages of soil infestation. Inoculum must first be produced by the pathogen, that is, infective units must be increased somewhere in the life cycle. This inoculum must then be dispersed into the soil. There it must survive and possibly increase. The resulting pathogen population is determined by factors influencing all these processes.

INOCULUM PRODUCTION.—The initial inoculum is the high point of population for many soil-borne pathogens. Subsequent events control how fast this population disappears. It seems safe to generalize that soil-borne pathogens characteristically overproduce inoculum. In a natural plant community a pathogen survives by producing and dispersing so much inoculum that somewhere a small residue reaches a new host at just the right time and with just the right conditions to infect. All the rest is lost. The inoculum-producing systems developed by soil-borne pathogens may be barely adequate to cope with natural mixed-plant communities, but they are far over-designed for monoculture. Pathogen populations on the order of millions per cubic foot of top soil are the result (Table 1). Factors associated with the initial production of inoculum are just as important as factors affecting later survival and they may be easier to apply in population control.

A very practical method for preventing initial infestation is the removal and destruction of diseased plant material before it enters the soil or otherwise releases the inoculum. This process of field sanitation is widely recommended but not widely practiced—partly because of expense, partly through lack of

TABLE 1. Some reported populations of soil-borne pathogens in infested field soils

Organism	Propagules per gram of soil	Reference
Fusarium solani f. sp. *phaseoli*	1,000–3,000	Nash and Snyder (1962)
F. roseum f. sp. *culmorum*	1,600–10,000	Cook and Bruehl (1968)
Verticillium dahliae	100–300	Evans, Wilhelm, and Snyder (1967)
Verticillium dahliae	100–1,200	Menzies (unpublished)
Helminthosporium sativum	8–900	Chinn, Ledingham, and Sallans (1960)
Thielaviopsis basicola	3,000–5,000	Papavizas (1964)

100 propagules/gram = 2 million/cu ft ($\pm$)

equipment, and partly through reluctance to destroy organic matter. In the case of a pathogen like *Verticillium dahliae*, initial inoculum production is crucial because this organism survives in soil principally as a population of microsclerotia, all originating in the dead tissue of diseased plants. Potato stems killed by *V. dahliae*, if left on the field, will become filled with sclerotia. A one-inch segment of stem may contain 20,000–50,000 viable sclerotia. Populations of 1,000 per gram of soil have been found by direct assay in fields cropped repeatedly to potatoes. This is enough to insure that for every potato plant the next year there will be something like 50 million sclerotia in the soil mass occupied by its roots.

Practically all of this inoculum could be excluded from the field by removal of the infested plant material before winter. Sclerotial production by *V. dahliae* in dead vines in high-rainfall areas is usually much less than in the dry irrigated areas. Perhaps the numerous saprophytic fungi that grow luxuriously on plant debris in wet regions competitively inhibit production of sclerotia by *V. dahliae*. An indication of this has been observed in culturing the organism on sterile potato stems in flasks. If such a culture becomes contaminated with *Penicillium* or *Aspergillus*, sclerotial formation is prevented in the area overrun by the contaminant. This may be a more effective way to use antagonistic organisms than as biological control agents against the sclerotia in the soil or as inhibitors of infection at the root surface.

Verticillium dahliae has been used as an example where initial inoculum production established the soil population. The same is true for other root pathogens that produce more or less passive propagules—*Fusarium solani*, *Sclerotium rolfsii*, *Thielaviopsis basicola*, and *Sclerotinia* spp., to name only a few. There may be a number of other factors that can be manipulated to influence the amount of inoculum produced by these pathogens, but destruction of the diseased tissue before dispersal is an obvious solution.

INOCULUM DISPERSION.—One way in which dispersion influences the apparent pathogen population is by increasing the frequency of infective sites. Before dispersal occurs, a plant containing a population of pathogen propagules can be considered as a single infective unit. To use *V. dahliae* again as an example, the infected plants may be left standing over winter, and when disced into the soil for seedbed preparation, a certain amount of dispersion may take place. However, the great mass of sclerotia are still embedded in host tissue. These overwintered sclerotia are almost all viable. During the second season the tissue decays and the sclerotia are freed into the soil at that spot. Tillage operations during and after the second season gradually distribute these clusters of sclerotia. So, if one considers the *effective* population of sclerotia from a single infected plant, it is seen that this population increases with tillage over several years as multiple inoculum clumps disperse into separate units.

Dispersion also affects population by exposing the propagules to a wider range of microenvironments in the soil. Although dispersal usually involves the separation of the pathogen propagules from their host tissue—and it is generally true that the pathogen survives better in the host tissue than when free in the soil—the more widely the inoculum is dispersed the greater will be the number of infective units finding a microenvironment suitable for survival.

Pathogens with abilities for competitive saprophytic colonization may be affected by dispersion in a somewhat different way. Bruehl and Lai (1968), in an ingenious study of the wheat root rot organism *Cephalosporium gramineum*, found that when equal amounts of infested and clean straw were mixed in soil they could expect one secondary invasion of healthy straw for every 3 pieces of infested straw. Each infestation of straw, original or secondary, has a life expectancy of 2–3 years or until the straw is decomposed. They calculated that the observed degree of secondary saprophytic colonization did not significantly affect the population survival curve after the first two years. However, we can speculate that a tillage operation like discing would break up both healthy and infested straws and increase the frequency of contact. The net result should be more secondary infection sites with each representing a smaller piece of straw. The apparent population is increased, but the inoculum capacity of each infective unit may be lower because of reduced food base or shorter survival time. Other pathogens with competitive saprophytic ability may also increase this way in the field, but it should be noted that although numerous laboratory studies have shown that *Fusarium roseum* f. sp. *cerealis* 'Culmorum' is one of this type, Cook and Bruehl (1968) were unable to show such saprophytism in the field. Experimental work has not been done, so far as I know, on the role of dispersion by cultivation on the saprophytic increase of this kind of a pathogen.

INOCULUM SURVIVAL.—Given a certain population of pathogen propagules produced on host tissue and then dispersed in the soil, a great many factors, biological or nonbiological, influence the fate of this population.

Probably the first thing to ask about a dispersed population of pathogens is whether it has the capabilities and mechanisms for saprophytic increase. This turns out to be a question of degree. Where careful experiments are done, it is found that even those forms apparently persisting in soil only as dormant propagules may increase in population in unsterile soil if sufficient readily-available energy is added. In our work with *Verticillium dahliae* (Menzies and Griebel, 1967) we have shown that the population can increase 4- to 10-fold in response to additions of sucrose. Papavizas, Lewis, and Adams (1968) have published similar data for *Fusarium solani* showing population increases on addition of soluble carbohydrates. In cases like these the increased population may be conidia arising from sclerotia or chlamydospores. However, there is evidence that in soil, *Verticillium* sclerotia can form small secondary ones (Emmatty and Green, 1967) and it is known that *Fusarium* macroconidia readily convert to new chlamydospores, (Schroth and Hendrix, 1962). Our data on *Verticillium* indicate that the extra population produced by saprophytic multiplication in soil is short-lived and probably is made up of conidia (Menzies and Griebel, 1967). In this case the increase in soil population of propagules may not represent any increase in inoculum potential since the conidia may not be effective agents for infection.

Garrett and his associates at Cambridge have made detailed studies on several important cereal foot-rot fungi that possess various saprophytic abilities but which generally die out before the straw is completely decomposed, some sooner than others. One of their most interesting observations has been that an abundant supply of available nitrogen in the soil prolongs the survival of *Ophiobolus graminis* but shortens that of *Helminthosporium sativum* (Garrett, 1963). Obviously there is more to the story than merely promoting more rapid decay of the straw through additions of nitrogen. Garrett (1966) postulates that the efficiency with which these pathogens utilize the cellulose of the straw for mycelial growth when N is added explains their differential survival (Table 2).

TABLE 2. Effect of nitrogen on saprophytic survival of cereal foot rot fungi as related to cellulosis adequacy index*

Pathogen	Effect of N on survival	Cellulosis Adequacy index†
Ophiobolus graminis	+ +	.59
Fusarium culmorum	+ +	.68
Cercospora herpotrichoides	+	.69
Curvularia ramosa	±	1.25
Helminthosporium sativum	− −	3.50

* Adapted from S. D. Garrett. Trans. Brit. Mycol. Soc. 49:57–68, 1966.
† Index = relative amount of cellulose decomposed per unit of mycelial growth.

The fact that N fertilization inhibits one pathogen but favors another is an unfortunate complication that points up the need for detailed understanding of the consequences of manipulating soil factors to control specific diseases.

In the normal soil situation energy substrates other than host tissue are not usually introduced in sufficient excess to stimulate pathogen populations except by root exudation. There does not seem to be as much specificity in the stimulation effect of root exudates as was once thought. The work reported by Schroth and Hendrix (1962) on the bean root rot *Fusarium* illustrates this point. Chlamydospores of *Fusarium* are stimulated by various amino acids and sugars to germinate and to form new chlamydospores. There are differences in the degree of response from root systems of different crops, but several nonhosts such as lettuce and tomato induce population increases (Table 3).

TABLE 3. Effect on nonhost roots on populations of *Fusarium solani* f. sp. *phaseoli**

Nonhost	Propagules per gram	
Crop	Rhizosphere	Nonrhizosphere
Corn	268	108
Lettuce	140	73
Tomato	247	136
Onion	364	1,680

* From M. N. Schroth and F. F. Hendrix. Phytopathology 52:906–909, 1962.

Populations were apparently reduced in the rhizosphere of onion, but it is possible that this may result from inhibitory substances rather than from the absence of the right kind of amino acid or sugar.

If nonhost root systems can induce germination and growth of dormant pathogen propagules, this seems more likely to result in a mild population increase rather than the hoped-for decrease. The same may be true when other plant residues are incorporated in soil. This may be happening because the stimuli for germination are accompanied by ample substrate for continued growth. In recent work in our laboratory (Menzies and Gilbert, 1967) we have found that plant residues give off volatile compounds that have a remarkable stimulatory effect on germination and respiration of soil organisms. Among the active components are several low-molecular-weight aldehydes and alcohols (Owens et al., 1969) some of which have been shown by others (Sussman, 1965) to stimulate spore germination *in vitro*. By using these volatiles alone instead of with the carbohydrate substrate with which they are usually associated, we may be able to induce pathogen growth in the absence of sufficient exogenous substrate to permit population replenishment. This could lead to decrease of the population through competition and starvation. Initial trials with this approach against *V. dahliae* are quite encouraging (Gilbert and Griebel, 1969).

The gradual decrease in a pathogen population in soil as a result of various adverse factors, biological or environmental, takes time. Historically, not knowing how to speed up these processes, we have withheld planting of host crops until these factors have had time to be practically effective. This is the essence of crop rotation for disease control. There is serious question, however, whether this practice can survive the industrial revolution in agriculture. To the modern farmer time is money, and he does not wish either to wait or to fix his cropping pattern into a long rotation scheme. Since the effectiveness of these inoculum reducing factors is a function of both time and intensity, research emphasis needs to be placed on intensity. For example, the way antibiosis is used in medicine is not by smearing mold cultures on wounds but by isolating the antibiotics and applying them at much higher concentrations than are found in nature.

It is reasonable to expect that the factors effective during crop rotation act by reducing the pathogen population, but the evidence for this is meager. Maloy and Burkholder (1959) used the most-probable-number method for estimating *F. solani* f. sp. *phaseoli* populations in fields following different nonhost crops. They found no great differences in the number of infection units of *F. solani* f. sp. *phaseoli* in field soil following any of the various rotations tested. Incidentally, they also failed to find any direct relation between root rot severity and pathogen populations in these fields.

Nash and Snyder (1962), using a sensitive selective plate count assay for *F. solani* f. sp. *phaseoli* from California fields, found that the propagule population had dropped considerably after a crop of tomatoes, whereas after a crop of barley the population was even higher than after continuous beans. The amount of disease, however, was consistently lower after barley, again emphasizing the poor correlation between pathogen population and disease severity.

There is ample evidence that, *other things being equal*, disease severity correlates rather well with inoculum density (Baker, Maurer, and Maurer, 1967; Dimond and Horsfall, 1965). This evidence comes from controlled experiments where inoculum density was varied by adding the pathogen to soil at different rates. This is entirely different from modifying the system to induce population changes and then measuring subsequent disease. When a pathogen population is increased by a soil treatment, the increased population may differ in kind from the original. For example, sclerotia may produce smaller sclerotia, conidia, or hyphae, and chlamydospores may form conidia or new chlamydospores. The inoculum potential of these new propagules may differ from the original inoculum. Also the soil treatment inducing the population change affects the associated microflora and the host as well, which in turn can modify the disease reaction.

OVERRIDING FACTORS.—As was pointed out earlier in this discussion, factors influencing pathogen populations will be useful for disease control only if they have capabilities for almost complete eradication. Furthermore, such factors should be effective throughout the soil mass, over a wide range of microenvironmental niches, and against a wide range of pathogens. Chemical fumigation is a process that very adequately meets these specifications. Were it not for high cost and perhaps some residue problems, there would be little need to look further. However, in this discussion, we are interested in naturally occurring factors.

Among natural influences, the three most "overriding" factors in pathogen survival are temperature, moisture, and aeration. Temperature effects on root disease have been well studied, but again the direct effect on pathogen populations has seldom been determined. Soil temperature is believed to be important in the geographic distribution of root pathogens as, for example, the restriction of *Phymatotrichum omnivorum* (the cotton root rot organism) to the southern U.S.A., and the restriction of the dark mycelial form of *Verticillium albo-atrum* to more northern regions.

The possibilities of manipulating soil temperature in the field to reduce pathogen populations are limited. Even though lethal temperatures for most plant pathogens are relatively moderate (50°–60°C for 30 minutes) the energy required to heat the soil mass to these temperatures is so great as to be generally impractical at present except on very high value crops (Baker, 1962). Nevertheless, it is possible that atomic power will be so abundant and cheap in the future that heat treatment of soils in general will become feasible.

Inability to use heat for pathogen control more widely is frustrating. It is a truly overriding factor, being able to penetrate all microenvironments much more uniformly than any biologic factor or chemical agent. Heat is effective against all plant pathogenic organisms and against insects and weeds as well. It leaves no residue. We should not be deterred by present costs from doing more research on the field use of various sources of heat as a means of reducing pathogen populations in soil.

The second overriding factor is moisture, more specifically desiccation. (Excess water in soil also affects pathogen populations; this will be considered along with aeration). Large areas of arid lands are farmed under irrigation where it is quite possible to dry soils in the field to very high moisture tensions. A classic case of pathogen control by dry summer fallow is that reported by Elmer (1942), who concluded that *Rhizoctonia solani* was unable to survive summer fallow conditions in Kansas during dry years. The cotton root rot fungus *Phymatotrichum omnivorum* survives in soil by means of sclerotia that appear to be quite sensitive to drying (King, Loomis, and Hope, 1931). Dry cultivation, especially deep tillage, often produces a remarkable control of root rot in cotton (Earl Burnett, personal communication). Desiccation may be an important factor in this disease control. The microsclerotia of *Verticillium dahliae* are quite resistant to drying in the fresh state, but after they have been incubated in soil and induced to go through a period of multiplication they can be killed by a mild air drying of the soil (Menzies and Griebel, 1967). It is probably generally true that if pathogen propagules can be enticed out of dormancy they can then be severely affected by drying. Soils dry by evaporation from the surface, however, and capillary water moves up from lower horizons as this process goes on. To reduce the plow layer of soil to a sufficiently low moisture content will require dry tillage and possibly several weeks of dry fallow.

The last of the three "overriding" factors to be mentioned is aeration. Poor aeration in soil is generally detrimental to pathogen propagules, although some do well in very wet soil. Poor aeration in the field is generally associated with high CO_2 and high water content as well as low O_2. Sewell (1965) reviewed these interrelations recently. It should be emphasized that a shortage of oxygen in soil, especially in the presence of ample decomposable organic material, will favor production of fatty acids, alcohols, and other anaerobic microbial metabolites which can be quite harmful to the obligately aerobic pathogens. In unpublished work in our laboratory we have found that pathogens in culture can be killed by accumulation of toxic metabolites when placed under anaerobiosis if these substances cannot diffuse away. Perhaps a similar killing of pathogens occurs in local sites in soil when green manure is turned under and moisture content is high. Such killing may not be very useful in disease control because of the spotty distribution of such sites. With more research, however, we may find ways of imposing a general anaerobiosis in soil sufficient to eradicate pathogens. Flooding for pathogen control has had some success, but the flood water is oxygenated and downward percolation prevents accumulation of toxic substances. This method might be more dependable if we could find a practical scheme for excluding oxygen from field soil while utilizing organic amendments to impose a severe oxygen stress.

CONCLUSION.—There are a great many factors known to affect pathogen populations in soil. We are interested in those that reduce populations and particularly in those that we can manipulate. Modern agriculture features large acreages of one crop, tends towards monoculture, and uses crop varieties often lacking in root-disease resistance. This has so weighted the balance in favor of massive buildups of pathogen populations in soil that reduction of these populations has to be drastic to be effective in disease control. The most direct way to prevent high pathogen populations is to interfere with production and dispersion of the initial propagules. Practically, this is accomplished by various sanitation practices. Once the propagules are dispersed in soil, effective reduction in population requires the application of an intense pervasive factor such as fumigants, heat, anaerobiosis, or desiccation. The more subtle biologic factors are likely to be specific for certain pathogens, certain soils, and certain microsites. They are more likely to act through modifying the disease expression than by directly reducing the pathogen population. To use these biological controls in the field, especially for population control, we need to find the key factors involved and then devise means of applying them at a much greater intensity than nature provides.

LITERATURE CITED

BAKER, K. F. 1962. Principles of heat treatment of soil and planting material. J. Australian Inst. Agr. Sci. 28:118–126.

BAKER, RALPH. 1968. Mechanisms of biological control of soil-borne pathogens. Ann. Rev. Phytopathol. 6:263–294.

BAKER, R., C. L. MAURER, and R.A. MAURER. 1967. Ecology of plant pathogens in soil. VII. Mathematical models and inoculum density. Phytopathology 57:622–666.

BRUEHL, G. W., and P. LAI. 1968. The probable significance of saprophytic colonization of wheat straw in the field by *Cephalosporium gramineum*. Phytopathology 58:464–466.

CHINN, S. H. F., R. J. LEDINGHAM, and B. J. SALLANS. 1960. Population and viability studies of *Helminthosporium sativum* in field soils. Can. J. Botany 38:533–539.

COOK, R. J., and G. W. BRUEHL. 1968. Relative significance of parasitism versus saprophytism in colonization of wheat straw by *Fusarium roseum* 'Culmorum' in the field. Phytopathology 58:306–308.

DIMOND, A. E., and J. G. HORSFALL. 1965. The theory of inoculum, p. 404–415. *In* K. F. Baker and W. C. Snyder [ed.], Ecology of soil-borne plant pathogens. Univ. of California Press, Berkeley and Los Angeles.

ELMER, O. H. 1942. Effect of environment on the prevalence of soil-borne *Rhizoctonia*. Phytopathology 32:972–977.

EMMATTY, D. A., and R. J. GREEN, JR. 1967. Germination of microsclerotia of *Verticillium albo-atrum* in soil. Phytopathology 57:810–811. (Abstr.).

EVANS, G., S. WILHELM, and W. C. SNYDER. 1967. Quantitative studies by plate counts of propagules of the Verticillium wilt fungus in cotton field soils. Phytopathology 57:1250–1255.

GARRETT, S. D. 1963. A comparison of cellulose-decomposing ability in five fungi causing cereal foot rots. Trans. Brit. Mycol. Soc. 46:572–576.

GARRETT, S. D. 1966. Cellulose-decomposing ability of some cereal foot rot fungi in relation to their saprophytic survival. Trans. Brit. Mycol. Soc. 49:57–68.

GILBERT, R. G., and G. E. GRIEBEL. 1969. The influence of volatile substances from alfalfa on growth and survival of *Verticillium dahliae* in soil. Phytopathology (in press).

GLYNNE, MARY D. 1965. Crop sequence in relation to soil-borne plant pathogens, p. 423–433. *In* K. F. Baker and W. C. Snyder [ed.], Ecology of soil-borne plant pathogens. Univ. of California Press, Berkeley and Los Angeles.

KING, C. J., H. F. LOOMIS, and C. HOPE. 1931. Studies on sclerotia and mycelial strands of the cotton root rot fungus. J. Agr. Res. 42:827–840.

MALOY, O. C., and W. H. BURKHOLDER. 1959. Some effects of crop rotation on the *Fusarium* root rot of bean. Phytopathology 49:583–587.

MENZIES, J. D. 1963. The direct assay of plant pathogen populations in soil. Ann. Rev. Phytopathol. 1:127–142.

MENZIES, J. D., and R. G. GILBERT. 1967. Responses of the soil microflora to volatile compounds in plant residues. Soil Sci. Soc. Amer. Proc. 31:495–496.

MENZIES, J. D., and G. E. GRIEBEL. 1967. Survival and saprophytic growth of *Verticillium dahliae* in uncropped soil. Phytopathology 57:703–709.

NASH, SHIRLEY M., and W. C. SNYDER. 1962. Quantitative estimations by plate counts of propagules of the bean root rot *Fusarium* in field soils. Phytopathology 52:567–572.

OWENS, L. D., R. G. GILBERT, G. E. GRIEBEL, and J. D. MENZIES. 1969. Identification of plant volatiles that stimulate microbial respiration and growth in soil. Phytopathology (in press).

PAPAVIZAS, G. C. 1964. New medium for the isolation of *Thielaviopsis basicola* on dilution plates from soil and rhizosphere. Phytopathology 54:1475–1481.

PAPAVIZAS, G. C., J. A. LEWIS, and P. B. ADAMS. 1968.

Survival of root infecting fungi in soil. II. Influence of amendment and soil carbon-to-nitrogen balance on Fusarium root rot of bean. Phytopathology 58:365–372.

SCHROTH, M. N., and F. F. HENDRIX, JR. 1962. Influence of non-susceptible plants on the survival of *Fusarium solani* f. *phaseoli* in soil. Phytopathology 52: 906–909.

SEWELL, G. W. F. 1965. The effect of altered physical condition of soil on biological control, p. 479–493. *In* K. F. Baker and W. C. Snyder [ed.], Ecology of soil-borne plant pathogens, Univ. of California Press, Berkeley and Los Angeles.

SUSSMAN, A. S. 1965. Dormancy of soil microorganisms in relation to survival, p. 99–109. *In* K. F. Baker and W. C. Snyder [ed.], Ecology of soil-borne plant pathogens. Univ. of California Press, Berkeley and Los Angeles.

Significance of Populations of Major Plant Pathogens in Soil: Bacteria Including Streptomyces

A. R. WEINHOLD—*Department of Plant Pathology, University of California, Berkeley.*

To understand the significance of pathogen population in soil it is necessary to know the relationship between population and disease severity. The key to this is the development of methods for the determination of the population of the pathogen as it exists in the field. The use of host-indexing methods is of limited value because of the lack of information on the number of propagules required for infection. With bacteria including *Streptomyces* the problem is further complicated by the absence of distinctive morphological characteristics with consequent dependence on a pathogenicity test to confirm identification.

The first obstacle is one of numbers. Bacteria and *Streptomyces* spp. constitute a large proportion of the total soil microflora. To isolate a particular plant pathogen which comprises only a very small part of this population, it is first necessary to devise a method whereby a large segment of the nonpathogens can be excluded. This can often be accomplished by soil-dilution techniques and selective media. However, it is extremely difficult to develop a medium on which only the plant pathogenic species in question will grow. There remains, therefore, the question of how best to identify the pathogen following isolation. Approaches to this problem are dictated by the specific organism in question. I wish to present information on the crown gall organism, *Agrobacterium tumefaciens* (M. N. Schroth, personal communication), and the potato scab pathogen, *Streptomyces scabies*.

POPULATION STUDIES ON AGROBACTERIUM TUMEFACIENS.—The studies on *A. tumefaciens* were made possible by the development of a highly selective medium (Schroth, Thompson, and Hildebrand, 1965). This medium is inhibitory to all soil bacteria except *A. radiobacter* and *A. tumefaciens*. Since *A. radiobacter* is not pathogenic, it was necessary to determine the proportion of the crown gall bacterium within this *Agrobacterium* group. This was accomplished by isolating individual colonies and inoculating small tomato plants. Those organisms causing galls were designated *A. tumefaciens*.

The population of *A. tumefaciens* was determined by placing 1g soil in 10 ml of water, shaking the suspension, and placing 0.1 ml on the selective media in petri plates. Colonies were selected from the plates and tested for pathogenicity. In tests where known quantities of the bacteria were added to soil, the recovery efficiency of this method was found to be about 38%. Therefore, population data from field soil were corrected accordingly.

The ratio of *A. radiobacter* to *A. tumefaciens* in most soils averaged 100–500 to 1. Several soils with varying crop histories were assayed. The value for *A. tumefaciens* population in a pasture soil was 42/g, as compared with 200/g for soil from a stone-fruit orchard and 316/g from a fruit-tree nursery soil. Susceptible plants grown on the latter two soils would be expected to have 10–20% crown gall whereas the pasture soil produced a low percentage of diseased plants (Table 1).

POPULATION STUDIES ON STREPTOMYCES SCABIES.— *Streptomyces* spp. are relatively easy to isolate on soil-dilution plates, and the population in soil can be determined, subject to the limitations of this technique.

TABLE 1. Relationship between pathogen population in soil and disease severity for crown gall and potato scab

Pathogen	Source	Population (propagules/g soil)	Disease severity
A. tumefaciens	Pasture	42	Trace to 1[*]
A. tumefaciens	Stone-fruit orchard	200	10–20[*]
A. tumefaciens	Fruit-tree nursery	316	10–20[*]
S. scabies	Rotation plot	6,000–7,000	10–11[†]
S. scabies	Variety testing plot	5,000–6,000	11–21[†]

[*] Percentage of plants with crown gall.
[†] Scab index—average percentage of tuber surface covered by scab lesions.

However, to determine the population of a specific organism such as *S. scabies* is a rather formidable task. To obtain some information on this point we combined two approaches. Several investigators have reported that one of the distinguishing characteristics of *S. scabies* is the production of melanin pigments when grown on suitable media (Menzies and Dade, 1959; Taylor and Decker, 1947). In our studies, 10–15% of the *Streptomyces* spp. present in soil from the central valley of California produced this pigment. By simply transferring colonies from the dilution plates (medium used: 2g agar and 0.1g tyrosine/liter) to peptone agar and selecting melanin-positive isolates, it was possible to eliminate 85–90% of the total population. However, since many nonpathogenic *Streptomyces* spp. also produce melanin pigments, additional selection was needed. For this we turned to serology.

Serological investigations indicated that *S. scabies* belongs to a serologically closely related group which includes some nonpathogens (Bowman and Weinhold, 1963). Fifteen isolates of *S. scabies* and 55 nonpathogens were tested, using the agar-gel double diffusion technique. All of the pathogens and 2 nonpathogens reacted with the pathogen-specific antisera. This suggested that serology could be used to screen *Streptomyces* spp. isolated from soil and single out those most likely to be *S. scabies*, thus reducing time- and space-consuming greenhouse testing.

In an attempt to isolate *S. scabies* directly from soil, dilution plates were prepared. Soil was collected from the rotation plots at the U.S. Cotton Station, Shafter, California. Six hundred colonies were transferred to a potato-dextrose-peptone medium, and those which produced melanin pigment (about 10%) were tested serologically. Two isolates reacted as pathogens, and subsequent pathogenicity tests confirmed their identity as *S. scabies*. Based on the total *Streptomyces* spp. population of this soil, as determined by dilution plates, the population of *S. scabies* was about 6,000 to 7,000/g of soil (Table 1). The soil used in this work was taken from a field which, for the past two years, had produced tubers with a scab index of 10 to 11. This means that the average tuber has 10–11% of its surface covered with scab lesions.

In a second trial to determine the population of *S. scabies*, a field which had a past history of severe scab was selected. This field had been used for many years to test breeding stock for scab resistance and was also located on the U.S. Cotton Research Station in Shafter. A similar process of selection for melanin pigment producing isolates followed by serological testing produced 2 isolates of suspected *S. scabies* from 775 isolates taken from dilution plates. This would represent a population of *S. scabies* of about 5.000–6,000/g of soil. Over a period of several years the scab index for tubers taken from this area ranged from 11 to 21 (Table 1).

The above data present an estimate of the population of *S. scabies* in soil which produces severely scabbed tubers. Because of the relatively small sample, these values are only approximate, but they are not too far out of line with population data on some other soil-borne pathogens (Nash and Snyder, 1965). Other evidence is available which suggests a rather direct relationship between *S. scabies* population and disease severity. In a long-term field experiment to study the effects of cover-crop incorporation on scab buildup, it was revealed that soybean completely prevented an increase in disease severity over a period of 13 years (Weinhold et al., 1964). At the beginning of the test, only a trace of disease (scab index 0.3–0.5) was present on tubers produced in the plots. In the control areas there was a lag phase of 4 years during which the scab index increased only to 2.5. However, during the next 3 years there was a rapid linear increase in disease severity, reaching a peak scab index of 15. In subsequent years, disease severity was relatively constant, with a definite trend toward decreasing severity. In plots on which soybeans were grown and incorporated as green manure, scab index never exceeded a value of 1.0.

There are several lines of evidence which suggest that increase in scab severity reflected an increase in the population of the pathogen. (1) The shape of the curve for disease increase was consistent with that expected for an increase in population; (2) in certain plots soybeans were grown for the first time in the second year of the 3-year period of rapid disease increase. This appeared to halt the increase, but did not result in any decrease in disease severity, suggesting that the effect was on population rather than pathogenic activity; and (3) cessation of the soybean cover crop for a 3-year period after completion of the test did not result in an increase in disease. This indicates that the pathogen population had never increased above the initial low level.

If one assumes on the basis of the above information that there exists a rather direct relationship between population of *S. scabies* and scab severity, it is possible to interpolate from the available population data and arrive at an estimate of the minimum population required for disease occurrence. On this basis, a value of approximately 500 propagules/g of soil would be predicted to result in a scab index of 1.0.

DISCUSSION.—To compare the population of *S. scabies* and *A. tumefaciens* in soil, it is necessary to consider the nature of the diseases caused by these pathogens. With crown gall, a single infection of a seedling can result in an enlarging gall and the plant considered diseased. With potato scab, a single infection results in only one scab lesion. A disease situation where the average tuber had 10% of the surface covered by lesions would require many separate infections for each tuber. Therefore, one might expect that severe scab would require many more infections than that required for a severe crown gall situation and, thus, a higher population of the organisms. This is precisely what our data show. However, such a comparison must be qualified by the realization that we have no information on sporulation of *S. scabies* in scab lesions or the possible importance of secondary infection.

The studies on *S. scabies* and *A. tumefaciens* provide information which indicates a close relationship between disease severity and pathogen population. There are, of course, other factors such as the effects of the physical and biological environment and the vigor and nutritional status of the inoculum in regard to the infection process. The placing of these many variables in proper perspective must await the results of future research. However, it appears certain that pathogen population as regulated by factors affecting buildup and survival is of vital importance in the occurrence and severity of diseases caused by soil-borne bacteria including *Streptomyces*.

LITERATURE CITED

Bowman, Tully, and A. R. Weinhold. 1963. Serological relationships of potato scab organisms and other species of Streptomyces. Nature (London) 200:599–600.

Menzies, J. D., and Caroline E. Dade. 1959. A selective indicator medium for isolation of *Streptomyces scabies* from potato tuber or soil. Phytopathology 49:457–458.

Nash, Shirley M., and W. C. Snyder. 1965. Quantitative and qualitative comparisons of *Fusarium* populations in cultivated fields and noncultivated parent soil. Can. J. Botany 43:939–945.

Schroth, M. N., J. P. Thompson, and D. C. Hildebrand. 1965. Isolation of *Agrobacterium tumefaciens, A. radiobacter* group from soil. Phytopathology 55:645–647.

Taylor, C. F., and P. Decker. 1947. A correlation between pathogenicity and cultural characteristics in the genus *Actinomyces*. Phytopathology 37:49–58.

Weinhold, A. R., J. W. Oswald, Tully Bowman, James Bishop, and David Wright. 1964. Influence of green manures and crop rotation on common scab of potato. Am. Potato J. 41:265–273.

Significance of Populations of Pythium and Phytophthora in Soil

A. F. SCHMITTHENNER—*Plant Pathology Department, Ohio Agricultural Research and Development Center and the Ohio State University, Columbus.*

Most *Pythium* spp. can be quantitatively isolated directly from soil using polyene antibiotics to inhibit nonpythiaceous fungi and wide-spectrum antibiotics to inhibit bacteria and actinomycetes. Combinations of pimaricin, penicillin, and polymyxin (Eckert and Tsao, 1962); endomycin and streptomycin (Schmitthenner, 1962); Mycostatin and streptomycin, (Hendrix and Kuhlman, 1965); and Mycostatin and vancomycin (McCain, Holtzmann, and Trujillo, 1967) have been used successfully. *Pythium graminicola* and most *Phytophthora* spp. have not been quantitatively isolated directly from soil, even though the above antibiotic combinations are suitable for isolating them from diseased tissues. *Phytophthora cinnamomi*, an exception, produces large chlamydospores which germinate readily, and it has been isolated on Mycostatin media flooded with relatively large amounts of suspended soil or sieved organic particles (Hendrix and Kuhlman, 1965; McCain, Holtzmann, and Trujillo, 1967). *Pythium ultimum* (Agnihotri and Vaartaja, 1967a), *Pythium aphanidermatum*, and *Phytophthora parasitica* (Trujillo and Hine, 1965) have been studied with a soil-smear technique. Saprophytic colonization has been used successfully for isolating *Pythium* (Hine and Luna, 1963). Host bioassays are still the major method of detecting most *Phytophthora* species in soil.

RELATIONSHIP BETWEEN POPULATIONS OF PYTHIUM AND PHYTOPHTHORA AND DISEASE SEVERITY.—Vaartaja and co-workers (Agnihotri and Vaartaja, 1967a; Vaartaja and Agnihotri, 1967; Vaartaja and Bumbieris, 1964) have concluded that environmental factors primarily determine severity of seedling root diseases in forest nursery beds, at levels of *Pythium* commonly isolated. In rotation studies I found (unpublished data) that populations of *Pythium ultimum* and incidence of alfalfa seedling blight caused by *P. ultimum* fluctuated independently. Campbell and Hendrix (1967) concluded that *Pythium* spp. are damaging only during or following periods of excessive moisture, but that disease severity may be less dependent upon soil moisture when *Pythium* populations become excessively high. A number of investigators have reported increase in diseases caused by *Pyth-*ium following addition of residues (Trujillo and Hine, 1965), seed (Singh, 1965), or high-carbohydrate amendments (Barton, 1960; Liu and Vaughn, 1966; Yarwood, 1966) to soil. Large amounts of amendments were used and undoubtedly produced unrealistically high *Pythium* populations. For example, Kendrick and Wilbur (1965) concluded that more than 500 propagules *Pythium irregulare*/g soil were needed for severe preemergence kill of lima bean, after building up populations with repeated plantings of lima bean seed. In most studies (Campbell and Hendrix, 1967; Schmitthenner, 1962; Vaartaja and Bumbieris, 1964) populations of pythia/g soil generally were fewer, and pathogenic types (e.g. *Pythium ultimum*, *Pythium irregulare*, and *Pythium aphanidermatum*) generally were less than 100/g soil. High soil moisture apparently is necessary for development of root and seed rots caused by *Pythium* and *Phytophthora* (Griffin, 1963). Soil water, directly or indirectly, may affect the fungus, the host, and the host-pathogen interaction.

EFFECT OF SOIL MOISTURE ON SURVIVAL OF PYTHIUM AND PHYTOPHTHORA.—Survival of *Pythium* definitely is linked to soil moisture. Watson (1966) reported that *Pythium* drastically decreased in soil in summer after a 10-day drought. The fungus could not be recovered following a heavy rain, but built up gradually with the recurrence of wet conditions during the autumn. Apparently *Pythium* became constitutively dormant, *sensu* Sussman and Halvorson (1966), under dry conditions, and was activated after prolonged wetting. A similar type of dormancy was dramatically demonstrated by Hoppe (1966), who stored air-dried soil for 12 years at room temperature. He found high levels of *Pythium* using a corn seedling assay, after 6 years, if the soil was kept wet for 15 days. After 12 years, wetting soil 3 months was required before high levels of *Pythium* were detected. Either sporangia or oospores of *Pythium* survive in soil, depending on the species (Trujillo and Hine, 1965; Vaartaja and Agnihotri, 1967). Mycelium rapidly disappears (Agnihotri and Vaartaja, 1967a; Lockwood, 1960).

All *Phytophthora* structures except chlamydospores

and oospores die under dry conditions (Mircetich and Zentmyer, 1967; Turner, 1965). Sporangia or encysted zoospores or mycelia may persist in moist soils for considerable periods, but under some conditions are rapidly lysed (Mircetich and Zentmyer, 1967). Oospores of all and chlamydospores of some *Phytophthora* spp. may be constitutively dormant. Populations of these *Phytophthora* spp. can be determined only if some method of germinating dormant spores en masse is found.

EFFECT OF SOIL MOISTURE ON SAPROPHYTIC GROWTH OF PYTHIUM AND PHYTOPHTHORA IN SOIL.—All nonconstitutively dormant structures of *Pythium* and *Phytophthora* apparently are exogenously dormant in soil *sensu* Sussman and Halvorson (1966). Germination and growth of these structures are inhibited in soil except when assimilable nutrients are available (Agnihotri and Vaartaja, 1967*a*; Turner, 1964; Vaartaja and Agnihotri, 1967) with one exception (Tsao and Bricker, 1968). Also, there is evidence that soil extracts may be less inhibitory to germination of *Pythium ultimum* when diluted (Agnihotri and Vaartaja, 1967*a*). *Pythium* may grow a considerable distance from a rich food base and colonize many carbohydrate-rich substrates (Barton, 1960; Liu and Vaughn, 1966; Singh, 1965; Yarwood, 1966), but activity of *Phytophthora* in soil is more limited (Trujillo and Hine, 1965; Turner, 1965; Zentmyer and Mircetich, 1966). Neither group can grow in residues already colonized by other microorganisms (Barton, 1961; Hine and Trujillo, 1966). Colonization of residues and amendments is best under high moisture conditions (Barton, 1960; Zentmyer and Mircetich, 1966).

The greatest effects of high soil moisture are on the availability of seed and root exudates (Burstrom, 1965; Kerr, 1964; Schroth and Cook, 1965). Kerr (1964) showed that pea seed exudation was much greater under high soil moisture and that seed rot caused by *Pythium* was correlated with both high moisture and increased seed exudation. Seed genotype and damage, and soil temperature also affect exudation (Kraft and Erwin, 1967; Schroth and Cook, 1964).

The primary effect of high soil moisture is due to decreased oxygen (Burstrom, 1965). Brown and Kennedy (1966) showed that exudation from soybean seed was much greater under anaerobic conditions. *Pythium* seed and root rot of soybeans were severe under low oxygen in naturally infested soils, and under high oxygen only when cornmeal was added. Root exudates may reverse inhibition of spore germination of *Phytophthora* and *Pythium* in soil (Agnihotri and Vaartaja, 1967*b*; Turner, 1963), and the level of nutrients may determine whether mycelia or zoospores are formed upon germination (Aragaki, Mobley, and Hine, 1967; Tsao and Bricker, 1968). Apparently zoospore aggregation on roots is essential for good infection with some fungus-host combinations (Kraft,

Endo, and Erwin, 1967). Water and root exudates are required for formation, movement, and accumulation of zoospores on susceptible roots (Hickman and Ho, 1966).

SOIL MOISTURE EFFECTS ON HOST SUSCEPTIBILITY TO PYTHIUM AND PHYTOPHTHORA.—High soil moisture and accompanying anaerobic conditions may directly affect roots and thus influence root-rot severity and the correlation between soil populations and disease levels. Both *Pythium* and *Phytophthora* grow well in low oxygen that may damage higher plants (Brown and Kennedy, 1966; Klotz, Stolzy, and DeWolfe, 1963). Root growth is retarded under anaerobic conditions, and considerable proliferation of secondary roots may occur (Burstrom, 1965), increasing numbers of immature roots for colonization. Anaerobic conditions also alter root-tissue maturation and suberization and lignification (Burstrom, 1965). Suberized endodermis is the primary barrier to colonization of *Pythium* (Miller et al., 1966). Single zoospore infections caused considerable damage to cotton under high moisture but were reduced under dry conditions (Spencer and Cooper, 1967). Respiratory changes, such as those reported in roots of tomato under low oxygen (Fulton, Erickson, and Tolbert, 1964), might alter the resistance of roots to pathogens. Cruikshank and Perrin, (1967) reported that phaseolin and pisatin production in bean and pea, respectively, were inhibited under low oxygen to such an extent that *Monilinia fructicola*, a nonpathogen, was capable of rotting pod tissue.

CONCLUSIONS.—Pathogenic *Pythium* spp. are widespread in soil and sufficiently abundant to damage crops. Information of *Phytophthora* populations in soil is still meager. Neither genus is capable of causing disease (assuming suitable genotype, temperature, etc.) unless: (1) the propagules are not constitutively dormant, (2) there are sufficient readily assimilable nutrients, and (3) the host is in a susceptible condition. All these requirements are met under high soil moisture (decreased oxygen) conditions. Disease severity, therefore, would be expected to fluctuate independently from population levels under normal moisture conditions. Probably, selective baits used under high moisture conditions would be best for predicting disease severity, but would be valid only if the population were not dormant. *Pythium* and *Phytophthora* might be controlled more easily by manipulating soil moisture, where this is possible, than by attempting to eliminate the pathogens.

LITERATURE CITED

AGNIHOTRI, V. P., and O. VAARTAJA. 1967*a*. Root exudates from red pine seedlings and their effects on *Pythium ultimum*. Can. J. Botany 45:1031–1040.

AGNIHOTRI, V. P., and O. VAARTAJA. 1967*b*. Effects of amendments, soil moisture contents, and temperatures on germination of *Pythium* sporangia under influence of soil mycostasis. Phytopathology 57:1116–1120.

ARAGAKI, M., R. D. MOBLEY, and R. B. HINE. 1967. Sporangial germination of *Phytophthora* from papaya. Mycologia 59:93–102.

BARTON, R. 1960. Saprophytic activity of *Pythium mamillatum* in soils. I. Influence of substrate composition and soil environment. Trans. Brit. Mycol. Soc. 43:529–540.

BARTON, R. 1961. Saprophytic activity of *Pythium mamillatum* in soils. II. Factors restricting *P. mamillatum* to pioneer colonization of substrates. Trans. Brit. Mycol. Soc. 44:105–118.

BROWN, G. E., and B. W. KENNEDY. 1966. Effect of oxygen concentration on *Pythium* seed rot of soybeans. Phytopathology 56:407–411.

BURSTROM, H. G. 1965. The physiology of plant roots, p. 154–169. *In* K. F. Baker and W. C. Snyder [ed.], Ecology of soil-borne plant pathogens. Univ. of California Press, Berkeley and Los Angeles.

CAMPBELL, W. A., and F. F. HENDRIX. 1967. *Pythium* and *Phytophthora* species in forest soils in the southeastern U.S. Plant Disease Reptr. 51:929–932.

CRUIKSHANK, I. A. M., and D. R. PERRIN. 1967. Studies on Phytoalexins. X. Effect of oxygen tension on the biosynthesis of pisatin and phaseolin. Phytopathol. Z. 60:335–342.

ECKERT, J. W., and P. H. TSAO. 1962. A selective medium for isolation of *Phytophthora* and *Pythium* from plant roots. Phytopathology 52:771–777.

FULTON, J. M., A. E. ERICKSON, and N. E. TOLBERT. 1964. Distribution of C14 among metabolites of flooded and aerobically grown tomato plants. Agron. J. 56:527–529.

GRIFFIN, D. M. 1963. Soil moisture and ecology of soil fungi. Biol. Rev. 38:141–166.

HENDRIX, F. F., and E. G. KUHLMAN. 1965. Factors affecting direct recovery of *Phytophthora cinnamomi* from soil. Phytopathology 55:1183–1187.

HICKMAN, C. J., and H. H. HO. 1966. Behavior of zoospores in plant pathogenic Phycomycetes. Ann. Rev. Phytopathol. 4:195–220.

HINE, R. B., and L. V. LUNA. 1963. A technique for isolating *Pythium aphanidermatum* from soil. Phytopathology 53:727–728.

HINE, R. B., and E. E. TRUJILLO. 1966. Monometric studies on residue colonization in soil by *Pythium aphanidermatum* and *Phytophthora parasitica*. Phytopathology 56:334–336.

HOPPE, P. E. 1966. *Pythium* species still viable after 12 years in air-dry muck soil. Phytopathology 56:1411.

KENDRICK, J. B., JR., and W. D. WILBUR. 1965. The relationship of population density of *Pythium irregulare* to pre-emergence death of lima bean seedlings. Phytopathology 55:1064. (Abstr.)

KERR, A. 1964. The influence of soil moisture on infection of peas by *Pythium ultimum*. Australian J. Biol. Sci. 17:676–685.

KLOTZ, L. J., L. H. STOLZY, and T. A. DEWOLFE. 1963. Oxygen requirements of three root-rotting fungi in a liquid media. Phytopathology 53:302–305.

KRAFT, J. M., and D. C. ERWIN. 1967. Stimulation of *Pythium aphanidermatum* by exudates from mung bean seeds. Phytopathology 57:866–868.

KRAFT, J. M., R. M. ENDO, and D. C. ERWIN. 1967. Infection of primary roots of bentgrass by zoospores of *Pythium aphanidermatum*. Phytopathology 57:86–90.

LIU, SHU-YEN, and E. K. VAUGHN. 1966. Control of *Pythium* infection of table beet seedlings by antagonistic microorganisms. Phytopathology 56:986–989.

LOCKWOOD, J. L. 1960. Lysis of mycelium of plant-pathogenic fungi by natural soil. Phytopathology 50:787–789.

MCCAIN, A. H., O. V. HOLTZMANN, and E. E. TRUJILLO. 1967. Concentration of *Phytophthora cinnamomi* chlamydospores by soil sieving. Phytopathology 57:1134–1135.

MILLER, C. R., W. M. DOWLER, D. H. PETERSEN, and R. P. ASHWORTH. 1966. Observations on the mode of infection of *Pythium ultimum* and *Phytophthora cactorum* on young roots of peach. Phytopathology 56:46–49.

MIRCETICH, S. M., and G. A. ZENTMYER. 1967. Production of oospores and chlamydospores of *Phytophthora cinnamomi* in roots and soil. Phytopathology 57:100. (Abstr.)

SCHMITTHENNER, A. F. 1962. Isolation of *Pythium* from soil particles. Phytopathology 52:1133–1138.

SCHROTH, M. N., and R. J. COOK. 1964. Seed exudation and its influence on pre-emergence damping-off of bean. Phytopathology 54:670–673.

SINGH, R. S. 1965. Development of *Pythium ultimum* in soil in relation to presence and germination of seeds of different crops. Mycopathol. Mycol. Appl. 27:155–160.

SPENCER, J. A., and W. E. COOPER. 1967. Pathogenesis of cotton (*Gossypium hirsutum*) by *Pythium* species: zoospore and mycelium attraction and infectivity. Phytopathology 57:1332–1338.

SUSSMAN, A. S., and H. O. HALVORSON. 1966. Spores: their dormancy and germination. Harper and Row, New York and London. 354 p.

TSAO, P. H., and J. L. BRICKER. 1968. Germination of chlamydospores of *Phytophthora parasitica* in soil. Phytopathology 58:1070. (Abstr.)

TRUJILLO, E. E., and R. B. HINE. 1965. The role of papaya residues in papaya root rot caused by *Pythium aphanidermatum* and *Phytophthora parasitica*. Phytopathology 55:1293–1298.

TURNER, P. D. 1963. Influence of root exudates of Cacao and other plants on spore development of *Phytophthora palmivora*. Phytopathology 53:1337–1339.

TURNER, P. D. 1965. Behavior of *Phytophthora palmivora* in soil. Plant Disease Reptr. 49:135–137.

VAARTAJA, O., and V. P. AGNIHOTRI. 1967. Inhibition of *Pythium* and *Thanatephorus* (*Rhizoctonia*) by leachates from a nursery soil. Phytopathol. Z. 60:63–72.

VAARTAJA, O., and M. BUMBIERIS. 1964. Abundance of *Pythium* species in nursery soils in South Australia. Australian J. Biol. Sci. 17:436–445.

WATSON, A. G. 1966. Seasonal variation in inoculum potentials of spermosphere fungi. N. Z. J. Agr. Res. 9:956–963.

YARWOOD, C. E. 1966. Detection of *Pythium* in soil. Plant Disease Reptr. 50:791–792.

ZENTMYER, G. A., and S. M. MIRCETICH. 1966. Saprophytism and persistence in soil by *Phytophthora cinnamomi*. Phytopathology 66:710–712.

The Significance of Populations of Pathogenic Fusaria in Soil

SHIRLEY NASH SMITH—*Ministry of Agriculture, P. B. 757, Marandellas, Rhodesia.*

Commonly pathogenic formae of Fusaria has been found existing in soil as chlamydospores (Bywater, 1959; Cook, 1968; French and Nielson, 1966; Nash, Christou, and Snyder, 1961; Trujillo and Snyder, 1963), and we can assume that the propagules enumerated by plating or "most probable number" techniques usually are these structures—sclerotia, conidia, and hyphal strands normally being of minor or temporary importance. It has also been established that these chlamydospores require an external carbon and nitrogen source for germination and infection (Cook and Schroth, 1965; Maurer and Baker, 1965; Schroth, Toussoun, and Snyder, 1963; Toussoun, Nash, and Snyder, 1960). This makes the nutritional status of the soil itself regarding these elements one of the parameters to be considered, in addition to inoculum level, when disease severity is evaluated in a particular soil (Couch and Bedford, 1966; Maurer and Baker, 1965; Papavizas, Lewis, and Adams, 1968; Toussoun and Snyder, 1961).

DISTRIBUTION.—When a relatively high propagule count of any particular forma specialis of a *Fusarium* species occurs in cultivated soil, the implication is that either a rather severely infected crop has recently grown, or the chlamydospores of the particular forma persist well in soil. Rishbeth (1955) detected as many as 300 propagules of *F. oxysporum* f. sp. *cubense* per gram in soil immediately around banana plants, but this population quickly declined after the plants were removed. Trujillo and Snyder (1963) observed that the propagules of this forma were distributed sparsely and unevenly through the soils of infested banana plantations, existing as chlamydospores inside decaying host-root tissue. Since Stover and Waite (1954) previously reported that this fungus is not a frequent saprophytic colonizer of dead banana tissue in soil, the chlamydospores in the tissue were most probably formed in diseased plants. Wensley and McKeen (1963) obtained populations of *F. oxysporum* f. sp. *melonis* ranging upward to 3,300/g in field soils at sites of wilted muskmelon plants, but mean population in field soils declined steadily during the 9-month interval between crops. At harvest 70% of the *F. oxysporum* isolates at plant sites and 21% from the intersites were pathogenic. Just a few months later only

12–15% of the isolates from random sampling of the infested field soils were pathogenic. French and Neilson (1966) observed that the macroconidia of *F. oxysporum* f. sp. *batatas*, which formed abundantly at ground level on diseased sweet potato stems, converted to chlamydospores quickly in soil. In infested California fields, however, the only place this forma was detected in countable numbers was at diseased plant sites (S. M. Nash, unpublished).

F. solani f. sp. *cucurbitae* (race 1) also forms masses of sporodochia on diseased plants at ground level. These wash into the soil and form chlamydospores which are all relatively short-lived (Nash and Alexander, 1965). Conroy's (1953) effective recommendation of a 3-year rotation and clean seed for control of squash foot rot can be explained on the basis that this fungus is truly a "soil invader" (Garrett, 1956).

PERSISTENCE.—Unlike *F. solani* f. sp. *cucurbitae* chlamydospores and most of those formed by *F. oxysporum* formae, individual chlamydospores of *F. solani* f. sp. *phaseoli* survive for long periods in soils. Eventually numerous clonal types accumulate and can be found distributed evenly throughout a field (Nash and Alexander, 1965; Snyder, Nash, and Trujillo, 1959). Often 1,000–4,000 propagules/g of soil have been found in fields on which beans have frequently grown. Alexander (1964) examined the ultrastructure of the chlamydospore walls of *F. solani* f. sp. *phaseoli* and f. sp. *cucurbitae* as well as a *F. oxysporum* f. sp. *cubense* clone and obtained evidence suggesting that the chlamydospores of the bean root rot *Fusarium* are comparatively the best suited to survival. *F. solani* f. sp. *pisi* appears to survive well in soil (Cook, personal communication).

Although most of the propagules of *F. oxysporum* formae disappear soon after the host crop has been removed, a few may persist in soils for very long periods (Jones and Gilman, 1915). Stover (1962) in Central America and Rishbeth (1955) in Jamaica reported banana wilt quickly recurring in replanted plantations which had been abandoned more than 20 years previously. However, it was not established how the propagules survived—whether in a dormant state or by repeated germination and formation of new propagules. The percentage surviving must have been very small. Rishbeth could not detect 2/g (the smallest detectable population using a comparison technique with hosts and known dilutions of inoculated soils) in soil in which nevertheless only 2% Gros Michel bananas survived until flowering. Waite and Dunlap (1953) found native grass species which harbored the fungus and which may be responsible for some of the persistance of this forma. Similarly Armstrong and Armstrong (1948) and Hendrix and Nielson (1958) found that pathogenic *F. oxysporum* formae could parasitize roots of plants other than susceptible hosts without pathogenesis.

The differences between the progress of cortical rots and vascular wilt diseases in plants has a bearing on how the soil populations influence disease severity. One can parallel this comparison to that which Garrett (1956) made between take-all and vascular wilts. In wilts, once infection has occurred, spread of the fungus is systemic in the vascular tract and is faster in relation to plant growth than in cortical lesions. In cortical rots the extent of infection is proportional to the numbers of individual lesions incited by the numerous propagules as well as subsequent lesion growth. A single propagule of a wilt-causing fungus that succeeds in entering the vascular tract can multiply greatly inside the plant. Furthermore, Baker, Maurer, and Maurer (1967) point out that in *Fusarium* cortical rots the infection court (hypocotyl) is fixed, rather than motile as is the case with the wilts where the infection court is near the root tip. Inoculum level is not likely to be as significant a factor for disease severity when the infection court is motile as when it is fixed. In a crop like bananas, which remain in the soil for a relatively long period, and whose roots search through large volumes of soil, a small undetectable inoculum level in the soil can lead to serious disease.

SAPROPHYTIC ABILITY.—There is no clear-cut evidence for any of the formae specialis being as efficient soil saprophytes as the strictly saprophytic members of the genus. On the other hand, all soil-borne *Fusarium* pathogens must live saprophytically prior to host invasion. The chlamydospore first germinates and forms a small thallus in response to plant exudates (Cook and Snyder, 1965; Papavizas, Lewis, and Adams, 1968; Schroth and Snyder, 1961) or other available nutrients (Toussoun, Patrick, and Snyder, 1963; Toussoun and Snyder, 1961). If it does not succeed in parasitizing a plant, the hypha lyses, but a replacement chlamydospore may first form. Papavizas (1967); and Papavizas, Adams, and Lewis (1968) have reported that populations of soil-borne pathogenic Fusaria may increase in this manner.

Sometimes Fusaria that are ordinarily soil saprophytes are pathogenic to seedlings, mature fruit, or roots in combination with nematodes. Thus 'Gibbosum' and 'Culmorum' cultivars of *F. roseum*—which are nonpathogenic to cereals, and which are aggressive soil saprophytes and colonizers of dead tissue, and

which occur abundantly in cultivated soils the world over—attack maturing cucurbit fruits in the field (Gabrielson, personal communication; Nash and Alexander, 1965). *F. roseum* 'Gibbosum' is also an important causative agent of banana fruit rots (Berg, 1968; Lukesic and Kaiser, 1966). *F. oxysporum* clones that cause sweet potato surface rots (Harter and Weimer, 1919) ordinarily exist as soil saprophytes. *F. solani* damages citrus roots in combination with nematodes (Feder and Feldmesser, 1961; Van Gundy and Tsao, 1963). However, soil-borne Fusaria such as these do not rank as pathogens in the same category with the formae specialis. It is their large numbers in soil or litter as well as the inactivity or senescence of host tissue that allows them to be pathogenic.

In infested California fields *F. oxysporum* f. sp. *vasinfectum* was found a year after the cotton was harvested and some areas contained 1,000 propagules/g of soil (S. M. Nash and F. Dachille, unpublished data). Although counts had declined between crops, the drastic drop reported for the muskmelon wilt *Fusarium* (Wensley and McKeen, 1963) did not occur. *F. oxysporum* f. sp. *vasinfectum* was found to invade bean lesions primarily infected by *F. solani* f. sp. *phaseoli* as frequently as saprophytic *F. oxysporum* in greenhouse tests (Nash and Snyder, 1967). These findings point out that *F. oxysporum* f. sp. *vasinfectum* has a rather high degree of competitive saprophytic ability, agreeing with Garrett's (1956) contention.

F. roseum f. sp. *cerealis* 'Culmorum' is another *Fusarium* that has been found to have strong competitive saprophytic ability (Butler, 1953; Garrett, 1956; Sadasivan, 1939), readily colonizing buried wheat straw. Cook and Bruehl (1968) suggest, however, that this ability may depend on specific conditions. Like several other cortical invaders, this pathogen may require rather high soil populations to do significant crop damage—up to 10,000/g has been reported. Warm, dry climatic conditions are also important to disease development (Cook, 1968; Cook and Bruehl, 1968; Garrett, 1956). The 'Avenaceum' and 'Graminearum' cultivars of *F. roseum* f. sp. *cerealis* can cause cereal foot rots, similar to those produced by 'Culmorum' (Cook, 1968; Oswald, 1950), as well as head blights. However, since neither of these cultivars is capable of producing large numbers of chlamydospores, they are probably soil invaders rather than soil inhabitants. 'Avenaceum' was found to exist at quite high numbers (360/g) in the litter layer of the Rothamsted "wilderness" area, but was a rather infrequent isolate of the field soils of the cereal rotation plots of the station, whereas 'Culmorum' was prevalent in the cultivated soils, particularly those heavily fertilized with nitrogen (Snyder and Nash, 1968).

LITERATURE CITED

ALEXANDER, J. V. 1964. A study of the ultra-structure in the development and germination of chlamydospores of *Fusarium*. Ph.D. dissertation, University of California, Berkeley. 87 p.

ARMSTRONG, G. M., and JOANNE K. ARMSTRONG. 1948. Non-susceptible hosts as carriers of wilt *Fusaria*. Phytopathology 38:808–826.

BAKER, R., C. L. MAURER, and RUTH A. MAURER. 1967. Ecology of plant pathogens in soil. VII. Mathematical models and inoculum density. Phytopathology 57:662–666.

BERG, L. A. 1968. Diamond spot of bananas caused by *Fusarium roseum* 'Gibbosum'. Phytopathology 58:388–389.

BUTLER, F. C. 1953. Saprophytic behavior of some cereal root-rot fungi. I. Saprophytic colonization of wheat straw. Ann. Appl. Biol. 40:284–297.

BYWATER, JOAN. 1959. Infection of peas by *Fusarium solani* var. *martii* forma 2 and the spread of the pathogen. Trans. Brit. Mycol. Soc. 42:201–212.

CONROY, R. J. 1953. *Fusarium* root rot of cucurbits in New South Wales. J. Australian Inst. Agr. Sci. 19:106–108.

COOK, R. J., 1968. *Fusarium* root and foot rot of cereals in the Pacific Northwest. Phytopathology 58:127–131.

COOK, R. J., and G. W. BRUEHL. 1968. Relative significance of parasitism versus saprophytism in colonization of wheat straw by *Fusarium roseum* 'Culmorum' in the field. Phytopathology 58:306–308.

COOK, R. J., and M. N. SCHROTH. 1965. Carbon and nitrogen compounds and germination of chlamydospores of *Fusarium solani* f. *phaseoli*. Phytopathology 55:254–256.

COOK, R. J., and W. C. SNYDER. 1965. Influence of host exudates on growth and survival of germlings of *Fusarium solani* f. *phaseoli* in soil. Phytopathology 55:1021–1025.

COUCH, H. B., and E. R. BEDFORD. 1966. *Fusarium* blight of turf-grasses. Phytopathology 56:781–786.

FEDER, W. A., and J. FELDMESSER. 1961. The spreading decline complex: The separate and combined effects of *Fusarium* spp. and *Radopholus similis* on the growth of Duncan grapefruit seedlings in the greenhouse. Phytopathology 51:724–726.

FRENCH, E. R., and L. W. NIELSON. 1966. Production of macroconidia of *Fusarium oxysporum* f. *batatas* and their conversion to chlamydospores. Phytopathology 56:1322–1323.

GARRETT, S. D. 1956. Biology of root-infecting fungi. Cambridge University Press, London and New York. 293 p.

HARTER, L. L., and J. L. WEIMER. 1919. The surface rot of sweet potatoes. Phytopathology 9:465–470.

HENDRIX, F. F., JR., and L. W. NIELSON. 1958. Invasion and infection of crops other than the forma suscept by *Fusarium oxysporum* f. *batatas* and other formae. Phytopathology 48:224–228.

JONES, L. R., and J. C. GILMAN. 1915. The control of cabbage yellows through disease resistance. Res. Bull. Wis. Agr. Exp. Sta. 38.

LUKESIC, F. L., and W. J. KAISER. 1966. Aerobiology of *Fusarium roseum* 'Gibbosum' associated with crown rot of boxed bananas. Phytopathology 56:545–548.

MAURER, C. L., and R. BAKER. 1965. Ecology of plant pathogens in soil. II. Influence of glucose, cellulose, and inorganic nitrogen amendments on development of bean root rot. Phytopathology 55:69–72.

NASH, SHIRLEY M., and J. V. ALEXANDER. 1965. Comparative survival of *Fusarium solani* f. *cucurbitae* and *F. solani* f. *phaseoli* in soil. Phytopathology 55:963–966.

NASH, SHIRLEY M., T. CHRISTOU, and W. C. SNYDER. 1961. Existence of *Fusarium solani* f. *phaseoli* as chlamydospores in soil. Phytopathology 51:164–168.

NASH, SHIRLEY M., and W. C. SNYDER. 1967. Comparative ability of pathogenic and saprophytic Fusaria to colonize primary lesions. Phytopathology 57:293–296.

OSWALD, J. W. 1950. Etiology of cereal root and foot rots in California. Hilgardia 19:447–462.

PAPAVIZAS, G. C. 1967. Survival of root-infecting fungi in soil. I. A quantitative propagule assay method of observation. Phytopathology 57:1242–1246.

PAPAVIZAS, G. C., P. B. ADAMS, and J. A. LEWIS. 1968. Survival of root-infecting fungi in soil. V. Saprophytic multiplication of *Fusarium solani* f. *phaseoli* in soil. Phytopathology 58:414–420.

PAPAVIZAS, G. C., J. A. LEWIS, and P. B. ADAMS. 1968. Survival of root-infecting fungi in soil. II. Influence of amendment and soil carbon-to-nitrogen balance on *Fusarium* root rot of beans. Phytopathology 58:365–372.

RISHBETH, J. 1955. Fusarium wilt of banana in Jamaica. I. Some observations on the epidemiology of the disease. Ann. Botany 19:293–328.

SADASIVAN, T. S. 1939. Succession of fungi decomposing wheat straw in different soils, with special reference to *Fusarium culmorum*. Ann. Appl. Biol. 26:497–508.

SCHROTH, M. N., and W. C. SNYDER. 1961. Effect of host exudates on chlamydospore germination of the bean root rot fungus, *Fusarium solani* f. *phaseoli*. Phytopathology 51:389–393.

SCHROTH, M. N., T. A. TOUSSOUN, and W. C. SNYDER. 1963. Effect of certain constituents of bean exudate on germination of chlamydospores of *Fusarium solani* f. *phaseoli* in soil. Phytopathology 53:809–812.

SNYDER, W. C., and SHIRLEY M. NASH. 1968. Relative incidence of *Fusarium* pathogens of cereals in rotation plots at Rothamsted. Trans. Brit. Myc. Soc. (In press.)

SNYDER, W. C., SHIRLEY M. NASH, and E. E. TRUJILLO. 1959. Multiple clonal types of *Fusarium solani* f. *phaseoli* in field soil. Phytopathology 49:310–312.

STOVER, R. H. 1962. Fusarial wilt (Panama Disease) of bananas and other *Musa* species. Commonwealth Mycol. Inst. Phytopathological Paper No. 4. 117 p.

STOVER, R. H., and B. H. WAITE. 1954. Colonization of banana roots by *Fusarium oxysporum* f. *cubense* and other soil fungi. Phytopathology 44:689–693.

TOUSSOUN, T. A., SHIRLEY M. NASH, and W. C. SNYDER. 1960. The effect of nitrogen sources and glucose on the pathogenesis of *Fusarium solani* f. *phaseoli*. Phytopathology 50:137–140.

TOUSSOUN, T. A., and W. C. SNYDER. 1961. Germination of chlamydospores of *Fusarium solani* f. *phaseoli* in unsterilized soils. Phytopathology 51:620–623.

TOUSSOUN, T. A., Z. A. PATRICK, and W. C. SNYDER. 1963. Influence of crop residue decomposition products on the germination of *Fusarium solani* f. *phaseoli* chlamydospores in soil. Nature (London) 197:1314–1316.

TRUJILLO, E. E., and W. C. SNYDER. 1963. Uneven distribution of *Fusarium oxysporum* f. *cubense* in Honduras soils. Phytopathology 53:167–170.

VAN GUNDY, S. D., and P. H. TSAO. 1963. Growth reduction of citrus seedlings by *Fusarium solani* as influenced by the citrus nematode and other soil factors. Phytopathology 53:488–489.

WAITE, B. H., and V. C. DUNLAP. 1953. Preliminary host range studies with *Fusarium oxysporum* f. *cubense*. Plant Disease Reptr. 37:79–80.

WENSLEY, R. N., and C. D. McKEEN. 1963. Populations of *Fusarium oxysporum* f. *melonis* and their relations to the wilt potential of two soils. Can. J. Microbiol. 9:237–249.

Significance of Population Level of Verticillium *in Soil*[1]

R. L. POWELSON—*Department of Botany and Plant Pathology, Oregon State University, Corvallis.*

▶

The two major pathogenic types of *Verticillium* are the microsclerotium (MS) and the "dauermycelium" (DM) producers. There is taxonomic controversy on whether the MS type should be called *V. dahliae* or *V. albo-atrum.* This subject has been recently reviewed by Isaac (1967) who has shown that these types can be separated into two distinct ecological groups. The inoculum potential of soil-borne populations of the DM type declines rapidly (2–3 years) in the absence of susceptible hosts (Keyworth, 1942; Robinson, Larson, and Walker, 1957; Sewell and Wilson, 1966), whereas the MS type may maintain a high inoculum potential for many years (Nelson, 1950; Wilhelm, 1955; Purss, 1961; Schreiber and Green, 1962; Martinson and Horner, 1962).

Most investigators classify *Verticillium* as a soil-invading or root-inhabiting fungus, *sensu* Garrett (1956). These fungi are characterized by "an expanding parasitic phase on the living host plants, and by a declining saprophytic phase after its death." According to the host-parasite interactions described by Baker, Maurer, and Maurer (1967), Verticillium wilt diseases would fall into class II "non-motile inoculum invaded by a motile infection court (e.g. a moving root tip)." The amount of root infection increases as new root growth occurs and comes in contact with the inoculum. Thus the number of contacts increases greatly with small increases in inoculum. There is also the implication that when inoculum densities are high, substantial reduction in inoculum would be required to significantly reduce disease severity.

SOURCES OF SOIL-BORNE INOCULUM.—*Verticillium* infects susceptible host plants by penetration of the root cortex followed by systemic invasion of the xylem vessels (Rudolph, 1931; Robinson, Larson, and Walker, 1957; Garber and Huston, 1966). When infected plants become senescent and die, the fungus leaves the xylem and readily permeates the surrounding tissues (Keyworth, 1942; Nelson, 1950). Microsclerotia or dark mycelia are formed within the dead tissue under a moist cool environment or when crop residues are buried in moist ground (Wilhelm, 1950, 1955, and 1956; Evans, Snyder, and Wilhelm, 1966). This is the primary method by which the population level of *Verticillium* increases in soil.

While weed hosts have been implicated in the buildup of *Verticillium* (MS) in soil (Wilhelm and Thomas, 1952; Guthrie, 1960; Oshima, Livingston, and Harrison, 1963), nonpathogenic infection of plant roots may play an important role in the maintenance of populations. Wilhelm (1959) and Baker (1959) have pointed out that the epidermis and cortex of weakened, senescent roots are universally invaded by numerous fungal parasites. This cortical invasion of supposed "nonhosts" allows for survival of these root-infecting fungi. Martinson and Horner (1962) and Lacy and Horner (1966) found that many plants, including members of the Gramineae previously thought to be immune to *Verticillium* (MS) could be infected without showing disease symptoms, and that the fungus could form resting bodies in the roots of these plants.

Wilhelm (1951*b*) reported that *Verticillium* (MS) was incapable of growing saprophytically in unamended soil, and mycelium and conidia were short-lived under natural conditions. Several workers (Isaac, 1946, 1953, 1956; Green, 1960; Schreiber and Green, 1962) have confirmed this in experiments where mycelium and conidial inoculum was noninfective after relatively short incubation periods in soil. Powelson (unpublished), however, found that potato plants became infected when planted in soil infested six months earlier with conidia of *Verticillium* (MS). While conidia appear to be ephemeral in soil, given an available energy source, germination and sporulation commonly occurs (Powelson and Patil, 1963; Powelson, 1966). Sewell (1959) has observed sporulation of *Verticillium* (DM) *in situ* on senescent roots, and Lacy and Horner (1966) found population increases of *Verticillium* (MS) in the rhizosphere of mint roots. Menzies and Griebel (1967); Powelson (1966; unpublished) found that an initial population increase of *Verticillium* (MS) propagules occurred when organic amendments were added to infested soil. Carlstrom (1969) assayed microsclerotia in infected soils under field conditions over a 5-year period and found seasonal sporulation cycles.

[1] Published with the approval of the Director of the Oregon Agricultural Experiment Station as Technical Paper No. 2602.

INOCULUM POTENTIAL.—The probability of a successful infection is determined by the "inoculum potential," a term which has been given several definitions (Garrett, 1960; Dimond and Horsfall, 1960; Martinson, 1963). The mass of inoculum per unit of soil or "inoculum density" has been used to estimate the inoculum potential. The measurement of inoculum potential usually involves the use of host plants and the assignment of quantitative values to disease severity, yield, growth, etc. (Wilhelm, 1950; Tolmsoff and Young, 1947; Schnathorst, 1963; Lacy and Horner, 1965).

Wilhelm (1950) added to soil various amounts of infested stem pieces and diluted infested soil with noninfested soil (1951a) to obtain different inoculum densities. He found that the percentage of tomato plants showing symptoms of Verticillium wilt was directly proportional to the amount of *Verticillium* (MS) in the soil. Tolmsoff (1959), working with potato plants, found that an increase in inoculum density of *Verticillium* (MS) resulted in (1) earlier development of wilt symptoms, (2) more severe symptoms and earlier plant death, and (3) decreased tuber yields. Tolmsoff and Powelson (1964) also found a correlation between inoculum density, number of root infections, and disease severity. Root infection did not occur until after initial tuber formation (approximately 38 days after planting) but did occur at approximately the same time for all inoculum densities. At low inoculum densities, systemic infection occurred, but no recognizable symptoms were produced. Lacy and Horner (1965) grew peppermint in soil infested with 10, 100, and 1,000 microsclerotia per gram of soil. Disease severity increased with increasing inoculum density, but the reduction in mint herb yield was relatively greater at the low inoculum densities.

Both direct and bioassays of field soils indicate that 10–200 propagules of *Verticillium* (MS) per gram of soil may cause severe disease. One of the primary problems in determining populations of soil-borne plant pathogenic fungi is the lack of adequate techniques for isolating and estimating the quantity of propagules present. The relatively low number of *Verticillium* (MS) propagules, which occurs in naturally infested soils, is difficult to detect with direct assay methods (Nadakavukaren and Horner, 1959; Menzies and Griebel, 1967). Recently (Harrison and Livingston, 1966; Evans et al., 1966, 1967) methods have been developed to mechanically concentrate the microsclerotia before plating on selective media.

Martinson and Horner (1962) assayed soil from a potato field with a history of severe Verticillium wilt and found only 10–40 viable *Verticillium* (MS) propagules per gram of soil. Easton (1967) found populations of 40–102 propagules per gram of soil from potato fields in Washington. Large quantities of microsclerotia are produced in cotton infected with Verticillium wilt (Evans, Snyder, and Wilhelm, 1966). Soil assays made by Evans et al. (1966, 1967) shortly after cotton harvest gave counts of 300–400 propagules per gram of soil. Assays made a few months later showed that the population had stabilized at around 40–120 propagules per gram of soil. Populations of *Verticillium* (MS) remain amazingly stable from one year to the next. Carlstrom (1969) found that, while populations stabilized at different densities depending on the soil type and climatic location, they remained nearly constant over a 5-year period.

Crop rotation has been effective in controlling the DM type of *Verticillium* (Keyworth, 1942; Robinson, Larson, and Walker, 1957; Edgington, 1962; Sewell and Wilson, 1966) but not the MS type (Guba, 1934; Nelson, 1950; Wilhelm, 1955; Purss, 1961; Horner, 1962; Powelson, 1965). Sewell and Wilson (1966) found that the hop-infecting DM type of *Verticillium* was eradicated within 3 to 5 years under bare fallow or grass cover conditions, but the MS type was not. Passive survival is affected by soil type, moisture, and temperature. Green (1960) assayed soils 82 weeks after infestation and found the population was reduced 81.7% in a mineral soil and only 19.7% in a muck soil. Flooding has been used to produce anaerobic soil conditions which eliminates *Verticillium* (MS) (Menzies, 1962) and reduces disease severity (Watson, 1964). Nadakavukaren (1961) has shown that high soil moisture and temperature caused a rapid decline in the number of soil-borne Verticillium (MS) propagules. Menzies and Griebel (1967) found that alternate drying and wetting of soil rapidly reduced the number of viable *Verticillium* propagules. They suggested that germinated microsclerotia lose their resistance to desiccation.

The high survival potential of the microsclerotium and the ability of MS strains to infect a wide range of host and nonhost plants makes MS types endemic to soils favorable for survival. Apparently nothing is known about actual population levels of DM strains in soil, but the use of pathogen-free planting material and short rotations with nonhost crops gives good control.

LITERATURE CITED

BAKER, K. F. 1959. Symposium on soil microbiology and root disease fungi, p. 309–379. *In* C. S. Holton et al. [ed.], Plant Pathology—problems and progress 1908–1958. University of Wisconsin Press, Madison.

BAKER, RALPH, CHARLES L. MAURER, and RUTH A. MAURER. 1967. Ecology of plant pathogens in soil. VII. Mathematical models and inoculum density. Phytopathology 57:662–666.

CARLSTROM, R. C. 1969. Survival of *Verticillium dahliae* Kleb. in soils. Ph.D. thesis, Oregon State University.

DIMOND, A. E., and J. G. HORSFALL. 1960. Inoculum and the diseased plant, Vol. 3, p. 1–22. *In* J. G. Horsfall and A. E. Dimond [ed.], Plant Pathology, an advanced treatise. Academic Press, New York.

EASTON, G. D. 1967. The number of *Verticillium* propagules in field soils in Washington. Phytopathology 57:1004. (Abstr.)

EDGINGTON, L. V. 1962. Influence of Connecticut temperatures on the relative pathogenicity of Maine and Connecticut *Verticillium* isolates. Am. Potato J. 39:261–265.

EVANS, G., W. C. SNYDER, and S. WILHELM. 1966. Inoculum increase of the Verticillium wilt fungus in cotton. Phytopathology 56:590–594.

EVANS, G., S. WILHELM, and W. C. SNYDER. 1967. Quantitative studies by plate counts of propagules of the Verticillium wilt fungus in cotton field soils. Phytopathology 57:1250–1255.

GARBER, R. H., and B. R. HOUSTON. 1966. Penetration and development of *Verticillium albo-atrum* in the cotton plant. Phytopathology 56:1121–1126.

GARRETT, S. D. 1956. Biology of root-infecting fungi. Cambridge University Press, London and New York. 292 p.

GARRETT, S. D. 1960. Inoculum potential, Vol. 3, p. 23–56. *In* J. G. Horsfall and A. E. Dimond [ed.], Plant Pathology, an advanced treatise. Academic Press, New York and London.

GREEN, R. J., JR. 1960. The survival of *Verticillium albo-atrum* in muck soils. Phytopathology 50:637. (Abstr.)

GUBA, E. F. 1934. Control of the Verticillium wilt of eggplant. Phytopathology 24:906–915.

GUTHRIE, J. W. 1960. Early dying (Verticillium wilt) of potatoes in Idaho. Idaho Agr. Exp. Sta. Res. Bull. No. 45. 24 p.

HARRISON, M. D., and LIVINGSTON, C. H. 1966. A method for isolating *Verticillium* from field soil. Plant Disease Reptr. 50:897–899.

HORNER, C. E. 1962. Oregon State University mint wilt research project, p. 6–16. *In* Proceedings of the thirteenth annual meeting, Oregon Essential Oil Growers League, Corvallis.

ISAAC, I. 1946. Verticillium wilt of sainfoin. Ann. Appl. Biol. 33:28–34.

ISAAC, I. 1953. Studies in the interaction between species of *Verticillium*. Ann. Appl. Biol. 40:623–629.

ISAAC, I. 1956. Some soil factors affecting Verticillium wilt of antirrhinum. Ann. Appl. Biol. 44:105–112.

ISAAC, I. 1967. Speciation in *Verticillium*. Ann. Rev. Phytopathol. 5:201–222.

KEYWORTH, W. G. 1942. Verticillium wilt of the hop (*Humulus lupulus*). Ann. Appl. Biol. 29:346–357.

LACY, M. L., and C. E. HORNER. 1965. Verticillium wilt of mint: Interaction of inoculum density and host resistance. Phytopathology 55:1176–1178.

LACY, M. L., and C. E. HORNER. 1966. Behaviour of *Verticillium dahliae* in the rhizosphere and on roots of hosts and non-hosts. Phytopathology 56:427–430.

MARTINSON, C. A. 1963. Inoculum potential relationships of *Rhizoctonia solani* measured with soil microbiological sampling tubes. Phytopathology 53:634–638.

MARTINSON, C. A., and C. E. HORNER. 1962. Importance of nonhosts in maintaining the inoculum potential of *Verticillium*. Phytopathology 52:742. (Abstr.)

MENZIES, J. D. 1962. Effect of anaerobic fermentation in soil on survival of sclerotia of *Verticillium dahliae*. Phytopathology 52:743. (Abstr.)

MENZIES, J. D., and G. E. GRIEBEL. 1967. Survival and saprophytic growth of *Verticillium dahliae* in uncropped soil. Phytopathology 57:703–709.

NADAKAVUKAREN, M. J. 1961. Influence of soil moisture and temperature on survival of *Verticillium* microsclerotia. Phytopathology 51:66. (Abstr.)

NADAKAVUKAREN, M. J., and C. E. HORNER. 1959. An alcohol agar medium selective for determining *Verticillium* microsclerotia in soil. Phytopathology 49:527–528.

NELSON, R. 1950. Verticillium wilt of peppermint. Michigan Agr. Exp. Sta. Tech. Bull. No. 221, 259 p.

OSHIMA, N., C. H. LIVINGSTON, and M. D. HARRISON. 1963. Weeds as carriers of two potato pathogens in Colorado. Plant Disease Reptr. 47:466–469.

POWELSON, R. L. 1964. Studies on Verticillium wilt of potatoes in Oregon. Am. Potato J. 41:303. (Abstr.)

POWELSON, R. L. 1965. Verticillium wilt of potatoes in arid regions of the Pacific Northwest, pp. 14–15. Proceedings, 4th Annual Washington State Potato Conference.

POWELSON, R. L. 1966. Availability of diffusable nutrients for germination and growth of *Verticillium dahliae* in soils amended with oat and alfalfa residues. Phytopathology 56:895. (Abstr.)

POWELSON, R. L., and S. S. PATIL. 1963. Influence of oat and alfalfa residues on soil respiration, fungistasis, and survival of *Verticillium albo-atrum*. Phytopathology 53:1141. (Abstr.)

PURSS, G. S. 1961. Wilt of peanut (*Arachis hypogea* L.) in Queensland, with particular reference to Verticillium wilt. Qd. J. Agr. Sci. 18:453–462.

ROBINSON, D. B., R. H. LARSON, and J. C. WALKER. 1957. Verticillium wilt of potato in relation to symptoms, epidemiology and variability of the pathogen. Madison, Wisconsin. Agr. Expt. Sta. Res. Bull. No. 202, 49 p.

RUDOLPH, B. A. 1931. Verticillium hadromycosis. Hilgardia 5:197–361.

SCHNATHORST, W. C. 1963. Theoretical relationships between inoculum potential and disease severity based on a study of the variation in virulence among isolates of *Verticillium albo-atrum*. Phytopathology 54:906. (Abstr.)

SCHREIBER, L. R., and R. J. GREEN, JR. 1962. Comparative survival of mycelium, conidia, and microsclerotia of *Verticillium albo-atrum* in mineral soil. Phytopathology 52:288–289.

SEWELL, G. W. F. 1959. Direct observation of *Verticillium albo-atrum* in soil. Trans. Brit. Mycol. Soc. 42:312–321.

SEWELL, G. W. F., and J. F. WILSON. 1966. Verticillium wilt of the hop: the survival of *V. albo-atrum* in soil. Ann. Appl. Biol. 58:241–249.

TOLMSOFF, W. J. 1959. The influence of quantity of inoculum on the severity of Verticillium wilt. M.Sc. thesis, Oregon State University.

TOLMSOFF, W. J., and ROY A. YOUNG. 1947. Relation of inoculum potential of *Verticillium albo-atrum* to development and severity of wilt in potatoes. Phytopathology 47:536. (Abstr.)

WATSON, R. D. 1964. Eradication of soil fungi by a combination of crop residue, flooding and anaerobic fermentation. Phytopathology 54:1437. (Abstr.)

WILHELM, S. 1950. Vertical distribution of *Verticillium albo-atrum* in soils. Phytopathology 40:368–376.

WILHELM, S. 1951a. Effect of various soil amendments on the inoculum potential of the Verticillium wilt fungus. Phytopathology 41:684–690.

WILHELM, S. 1951b. Is *Verticillium albo-atrum* a soil invader or soil inhabitant? Phytopathology 41:944–945. (Abstr.)

WILHELM, S. 1955. Longevity of the Verticillium wilt fungus in the laboratory and field. Phytopathology 45:180–181.

WILHELM, S. 1956. A sand culture technique for the isolation of fungi associated with roots. Phytopathology 46:293–295.

WILHELM, S. 1959. Parasitism and pathogenesis of root-disease fungi, p. 356–366. *In* C. S. Holton et al. [ed.], Plant pathology—Problems and Progress 1908–1958. Univ. of Wisconsin Press, Madison.

WILHELM, S., and H. E. THOMAS. 1952. *Solanum sarachoides* an important weed host to *Verticillium albo-atrum*. Phytopathology 42:519–520. (Abstr.)

Significance of Population Level of Rhizoctonia solani in Soil

Y. HENIS—*The Hebrew University of Jerusalem, Faculty of Agriculture, Rehovot, Israel.*

▶

The significance of the population level (i.e. inoculum density) of *Rhizoctonia solani* in soil to disease severity is determined by the host and by the particular strain of *R. solani*, as well as the environment (Baker, 1965; Baker, Maurer, and Maurer, 1967; Baker and Martinson, 1969; Papavizas, 1969). The problem may be examined from two points of view: first, the relative importance of the population level of *R. solani* to disease severity, and second, significance of the quantitative estimation of population level of *R. solani* as a means to predict disease.

IMPORTANCE OF POPULATION LEVEL TO DISEASE SEVERITY.—A relationship between disease severity and the abundance of *R. solani* in the soil has been observed in both field and laboratory experiments. Thus, Boosalis and Scharen (1959) found 6,800 out of 80,000 plant debris to be infested with *R. solani* in 100 g of soil taken from a field with a history of high incidence of crown rot, as compared with 600 infested particles out of 30,500 found in a field with a low disease incidence. According to Martinson (1963), 8,000 ppm of sand-cornmeal-grown inoculum of *R. solani* mixed with *Rhizoctonia*-free soil were required in order to cause preemergence damping-off of radish seedlings in the greenhouse.

Henis, Ben-Yephet, and Katan (unpublished) compared the distribution and the severity of damping-off disease in bean seedlings, using various fractions of propagules of uniform size and avoiding the addition of nutrients to the soil along with the inoculum. Both propagule number and size were of importance to disease incidence and severity. Thus, propagules 0.15–0.25 mm in diameter did not incite disease when added to soil at a concentration of 4/g (dry weight) soil, whereas with propagules 0.25–1.0 mm in diameter about 10% of the host population showed disease symptoms when the soil contained as little as 1 propagule/g; no decrease in disease severity could be observed at high concentration of inoculum. The lower effectiveness of heavy inocula reported by Jacks (1951), and the decrease in pathogenicity at high inoculum concentrations observed by Sanford (1941) and Das and Western (1959), is probably due to the stimulation of the soil microflora by the organic matter which is often added to the soil along with the inoculum, or to self-inhibition of *R. solani* propagules as a result of CO_2 production (Papavizas, 1969) which is also stimulated by the added organic matter.

THE QUANTITATIVE ESTIMATION OF RHIZOCTONIA TO PREDICT DISEASE.—Papavizas (1969) has summarized reports on the ability of *R. solani* to proliferate saprophytically in the soil. Perhaps we shall not be too far from the real situation if we assume that at the time of planting a host, most of the population of *R. solani* exists in the soil in the form of multicellular propagules of heterogeneous size, composed of sclerotia and of hyphal cells at various stages of development, embedded in and associated with soil organic matter (Boosalis and Scharen, 1959). The relative importance of this fraction of the *R. solani* population in soil was recently emphasized by Papavizas (1968). These propagules may be regarded as "resting" bodies in which death and multiplication rates have reached a certain dynamic equilibrium. We have to date no method of measuring the total number and size of propagules of *R. solani* in the soil. The abundance of the plant debris which harbor the fungus can be estimated by plating the isolated debris on a suitable agar medium and examining for *R. solani* (Boosalis and Scharen, 1959). Other methods, such as saprophytic colonization (plant segment colonization, soil microbiological sampling tube) and host infection, also estimate population level indirectly (Menzies, 1963; Martinson, 1963; Sneh et al., 1966; Papavizas, 1969). The use of these methods make it possible to find a relationship between inoculum density and disease in artificially and naturally infested soil (Baker and Martinson, 1968). As suggested by Baker (1965), disease severity depends on inoculum potential *sensu* Garrett and disease potential *sensu* Granger. Other factors being constant, the population level of *R. solani* in soil should reflect disease severity. On the other hand, similar population levels of strains differing in inoculum and disease potential may affect the host differently. Although a direct relationship between saprophytism and pathogenesis of *R. solani* has been demonstrated by several authors (Martinson, 1963; Papavizas and Ayers, 1965; Sneh et al., 1966), this has not been done with many strains of *R. solani*. Furthermore, according to Sanford (1941), such a correlation may not exist.

Whatever assay method is used, it should be based on the assumptions that (a) during the period elapsing from assay to planting, the population of *R. solani* in the field remains constant; (b) the period during which the host is susceptible to *R. solani* is relatively short; and (c) multiplication of the pathogen during the assay procedure is similar to that occurring later in the field during the period of susceptibility. A typical example for such a case is damping-off. The situation may be quite different in the case of root rot, which requires a relatively long time to develop (Baker and Martinson, 1968). Here, population level at time of assay does not necessarily reflect the level at infection time. *R. solani* is capable of building up in the soil; moreover, pathogenicity tests may not help in predicting disease severity, as new strains may develop during the growing season and adapt themselves to the host (Papavizas and Davey, 1962; Daniels, 1963). Nevertheless, any nonselective method used for the population estimation should suffice for a rough prediction of disease severity.

MATHEMATICAL MODELS.—It appears that both colonization and disease are not linearly correlated to inoculum density at high population levels (van der Plank, 1963; Baker and Martinson, 1968; Dimond and Horsfall, 1965; Sneh et al., 1966). Several models have been proposed which explain the behaviour of soil-borne plant pathogens in the soil. In the model proposed by Baker (1965) and further developed by Baker, Maurer, and Maurer (1967), equations are derived which describe the correlation between inoculum density and the number of successful infections of a host population, assuming a uniform spatial distribution of the pathogen in the soil and a possible mutual growth of propagules and host roots toward each other.

It may be expected that the number of successful infections will be positively related to disease severity of the total population, as measured by the infection index method. However, distribution of disease severity within the host population is not considered in this model. Moreover, no methods are available at present which estimate quantitatively the number of infections caused by *R. solani* as well as the relationship between disease severity and infection.

Other models (van der Plank, 1963; Dimond and Horsfall, 1965) deal with the correlation between inoculum density and the proportion of diseased vs healthy host plants, assuming a Poisson distribution of the pathogen in the soil and taking into account the occurrence of multiple infections.

Martinson (1963) obtained a nearly linear increase in colonization of tubes (SMST) and preemergence damping-off disease of radish seedlings with the logarithmic increase of inoculum of *R. solani* to 8,000 ppm. This might imply that Baker's model is valid here. On the other hand, Sneh et al. (1966) used Poisson equations to explain the relationship between inoculum density and disease caused in bean seedlings by *R. solani*, and interpreted their results on the basis of the "multiple infection" theory. More studies are required in order to assess the usefulness of these models for the quantitative estimation and understanding of the relationship between inoculum and disease caused by *R. solani*. Whatever the model may be, it should express the different correlations at various concentrations of inoculum, between inoculum density and colonization or disease incidence, as well as the increase in the proportions of severely diseased host plants observed at increased inoculum density (Martinson, 1963; Sneh et al., 1966). The slope of the straight line obtained by any mathematical treatment of the experimental data (Martinson, 1963; Sneh et al., 1966; Baker, Maurer, and Maurer, 1967) will depend upon the soil type and microflora, temperature and moisture conditions, nature of the infected host, age of host, the particular strain of *R. solani* involved, and on its external and internal nutriment (Papavizas, 1969; Baker and Martinson, 1969; Baker, Maurer, and Maurer, 1967).

CONCLUSIONS.—Population level of *R. solani* influences disease incidence and severity at infection time, depending on the particular strain, on propagule number, distribution, and size, on the host, and on the environment. Other factors being constant, at low propagule concentrations the correlation between population level and disease incidence and severity is close to linearity, and any reduction in this level is reflected in concomitant decrease in disease incidence and severity. At high population levels, however, a relatively greater change in this level is required in order to obtain a decrease in disease incidence and severity. At present, the methods available for the quantitative estimation of *R. solani* in soil can be used only for a rough prediction of disease severity. A possible practical solution to the prediction of severity of damping-off disease may involve the random collection of soil samples from the infested field, testing the particular host in these samples by the infection index method, and predicting the probability of disease severity in the whole field using precalculated statistical tables. More defined systems which take into account inoculum potential, disease potential, and field conditions need to be developed, and mathematical models should be used in order to obtain a better prediction and an understanding of the mechanisms involved.

Does the positive correlation between disease severity and inoculum density mean that the *Rhizoctonia* population in soil should be reduced in order to obtain disease control (Baker and Martinson, 1969), or would the arrest of its saprophytic activity be sufficient? The control of *R. solani* with PCNB which acts as a fungistat at low concentrations (Kendrick and Middleton, 1954) indicates that eradication is not obligatory. Mild means, such as organic amendments (Papavizas, 1965) could be useful in control in those cases in which host susceptibility is of short duration.

ACKNOWLEDGEMENTS.—I wish to express my gratitude to G. C. Papavizas, J. D. Menzies, and J. Katan for most fruitful discussions and comments on the topics covered in this paper.

LITERATURE CITED

BAKER, RALPH. 1965. The dynamics of inoculum, p. 395–403. *In* K. F. Baker and W. C. Snyder [ed.], Ecology of soil-borne plant pathogens. Univ. of California Press, Berkeley and Los Angeles.

BAKER, R., C. L. MAURER, and RUTH A. MAURER. 1967. Ecology of plant pathogens in soil. VII. Mathematical models and inoculum density. Phytopathology 57:662–666.

BAKER, R., and C. A. MARTINSON. 1969. Epidemiology of diseases caused by *Rhizoctonia solani*. APS Symposium on *Rhizoctonia solani* and related forms. Miami, Florida, 3–7 October, 1965. (In press.)

BOOSALIS, M. G., and A. L. SCHAREN. 1959. Methods for microscopic detection of *Aphanomyces euteiches* and *Rhizoctonia solani* and for isolation of *Rhizoctonia solani* associated with plant debris. Phytopathology 49:192–198.

DANIELS, JOAN. 1963. Saprophytic and parasitic activities of some isolates of *Corticium solani*. Trans. Brit. Mycol. Soc. 46:485–502.

DAS, A. C., and J. H. WESTERN. 1959. The effect of inorganic manures, moisture and inoculum on the incidence of root disease caused by *Rhizoctonia solani* Kuhn in cultivated soil. Ann. Appl. Biol. 47:37–38.

DIMOND, A. E., and J. G. HORSFALL. 1965. The theory of inoculum, p. 404–415. *In* K. F. Baker and W. C. Snyder [ed.], Ecology of soil-borne plant pathogens. Univ. of California Press, Berkeley and Los Angeles.

JACKS, H. 1951. The efficiency of chemical treatments of vegetable seeds against seed-borne and soil-borne organisms. Ann. Appl. Biol. 38:135–160.

KENDRICK, J. B., JR., and J. T. MIDDLETON. 1954. The efficacy of certain chemicals as fungicides for a variety of fruit, root and vascular pathogens. Plant Disease Reptr. 38:350–353.

MARTINSON, C. A. 1963. Inoculum potential relationships of *Rhizoctonia solani* measured with soil microbiological sampling tubes. Phytopathology 53:634–638.

MENZIES, J. D. 1963. The direct assay of plant pathogen propagules in soil. Ann. Rev. Phytopathol. 1:127–142.

PAPAVIZAS, G. C. 1968. Survival of root-infecting fungi in soil. VIII. Distribution of *Rhizoctonia solani* in various physical fractions of naturally and artificially infested soils. Phytopathology 58:746–751.

PAPAVIZAS, G. C. 1969. Colonization and growth of *Rhizoctonia solani* and related forms. APS Symposium on *Rhizoctonia solani* and related forms. Miami, Florida, 3–7 October, 1965. (In press.)

PAPAVIZAS, G. C., and C. B. DAVEY. 1962. Isolation and pathogenicity of *Rhizoctonia* saprophytically existing in soil. Phytopathology 52:834–840.

PAPAVIZAS, G. C., and W. A. AYERS. 1965. Virulence, host range and pectolytic enzymes of single-basidiospore isolates of *Rhizoctonia praticola* and *Rhizoctonia solani*. Phytopathology 55:111–116.

PLANK, J. E. VAN DER. 1963. Plant Diseases: epidemics and control. Academic Press, New York. 349 p.

SANFORD, G. B. 1941. Studies on *Rhizoctonia solani* Kuhn. V. Virulence in steam sterilized and natural soil. Can. J. Res. (C) 19:1–8.

SNEH, B., J. KATAN, Y. HENIS, and I. WAHL. 1966. Methods for evaluating inoculum density of *Rhizoctonia* in naturally infested soil. Phytopathology 56:74–78.

Measurement of Host Reactions to Soil-borne Pathogens

J. G. BALD—Department of Plant Pathology, University of California, Riverside.

Symptoms are the qualitative changes that distinguish diseased from healthy plants. They may be expressed in numbers by counting, measuring, or estimating; and almost any characteristic that is consistently altered by infection may thus be made to serve as an index of disease. The farmer's index is usually the reduction in yield and quality of the product he harvests from an infected crop, and this has been much used in agricultural research. In experimental work, however, yield data generally give less information than measurements taken during the development of a disease.

Aspects of the general subject of host reactions to soil-borne infection are taken to include leaf areas or growth rates of diseased and healthy plants in soil as a substitute for terminal measurements of yield; numbers of diseased plants in relation to numbers of pathogen propagules; and the direct effects of infection, in terms of (a) single lesions and (b) distribution and level of lesion populations. Here the emphasis will be on quantal estimates of plant reactions to infections with soil-borne pathogens. Other aspects are discussed elsewhere (Bald, 1969).

QUANTAL DATA.—*The binomial distribution.*—In research on plant diseases many attributes of diseased and healthy plants are estimated by whole numbers, e.g., numbers of diseased plants in a sample of plants or in a field, or numbers of lesions on single plants. The binomial distribution is the basis for the analysis of quantal data of this type. However, the binomial itself is less often applicable than the derivative Poisson and normal distributions, which are appropriate under certain limiting conditions.

Use of the binomial series is illustrated by data from an experiment on damping-off of aster seedlings by *Pythium ultimum* (Bald and Munnecke, 1962). The infested soil was diluted to various concentrations with sterilized soil and planted in 20 pots on 3 successive occasions with aster seed. The data were tabulated as occurrence of symptoms in each pot on 0, 1, 2, or 3 occasions (Table 1), and an estimate of the probability of infection, p, was given by the proportion of the readings in which damping-off was recorded. The proportion of readings without recorded infections was q, $p + q = 1$, and the terms of the expansion

$$(q = p)^3 \times 20 = 20(q^3 + 3q^2p + 3qp^2 + p^3)$$

gave the numbers of pots expected to yield 0, 1, 2, or 3 positive readings.

The calculated and observed values in Table 1 are very similar. Numbers of infections were related to inoculum concentration, but apart from this, damping-off in individual pots seemed to be largely governed by chance. The common picture of susceptible roots penetrating infested soil and meeting propagules or hyphae of pathogens by chance fits these results.

The Poisson series.—The Poisson series is the limiting case of the binomial series where p, the probability of an event, is very small. According to Fisher (1932) "it may be shown theoretically that if the probability of an event is exceedingly small, but if a sufficiently large number of independent cases are taken to obtain a number of occurrences, then this number will be distributed in the Poisson series." In a volume of soil large enough to contain the root system of a plant, a single pathogenic fungal or bacterial propagule is unlikely to meet and infect a single susceptible cell; but if there is a large number of propagules distributed at random in the soil and it contains a susceptible root system consisting of millions of cells, the probability

TABLE 1. Numbers of pots among 20 per treatment in which preemergence damping-off and/or postemergence symptoms occurred. Results are fitted with binomials of the form $y = (q + p)^3$

Dilution	Not inf.	q^3 calc.	Inf. 1 time	$3pq^2$ calc.	Inf. 2 times	$3p^2q$ calc.	Inf. 3 times	p^3 calc.
1	0	0.2	1	1.9	10	7.7	9	10.2
10	2	1.6	6	6.4	8	8.4	4	3.6
100	10	9.6	7	8.0	3	2.2	0	0.2

of infecting a cell somewhere on that root system may be high. If the original infection readily expands to a visible lesion involving many cells, the probability of a lesion occurring may also be high. If y = the probability of finding one lesion or more on a root system, the proportion of infected root systems in a group of N susceptible plants growing in the same infested soil, will be given by an equation of the form

$$y = 1 - e^{-m} \text{ (Equation 1)}$$

Here e is the base of natural logarithms, and m is the number of lesions on the N plants, divided by N. If lesions become confluent, m is the number of original infections expanding to form lesions.

Agreement with the Poisson series may be illustrated by fitting data of Martinson (1963). He compared samples of biologically buffered soil containing different concentrations of *Rhizoctonia* inoculum. He measured simultaneously the capacity of the *Rhizoctonia* to enter 16 holes in a soil-sampling tube, and to cause damping-off in radish seedlings. Damping-off reached 100% at the highest concentrations of inoculum, but the maximum number of holes penetrated was around 13, i.e., an average of 3 holes in a tube seemed blocked to penetration by the *Rhizoctonia*. Poisson series were calculated and imposed on a tracing of Martinson's Fig. 2 (Fig. 1). The agreement was good, indicating that the Poisson series was applicable under the conditions of these experiments both to infection of seedlings and to penetration of soil-sampling tubes by *Rhizoctonia*.

An example of departure from a Poisson distribution is a soil-dilution series with *Verticillium* and tomato seedlings by Wilhelm (1951). His results cannot be fitted directly, but if it is assumed that for some reason about 20% of inoculated seedlings were protected from infection, a fit can be obtained. This assumption might suggest that Wilhelm's *Verticillium* experiment was analogous to Martinson's experiments on the penetration of soil-sampling tubes by *Rhizoctonia*. Infection might be blocked, for example, by a strongly competitive microflora on the seedling roots. By revealing possibilities of this kind, deviation from a Poisson distribution may in itself be a valuable indication of influences at work in the pathogen-host interaction.

The first requirement for agreement with the Poisson series is that single propagules shall be capable of causing infection. Other requirements often stated without qualification are, in fact, subject to qualification. They include uniform virulence of propagules, uniform susceptibility of the host, and random distribution of the propagules over all host plants in an experiment or in a field. If these requirements are not absolute, the question arises how much variability can there be without serious departure from a Poisson distribution? In the absence of a mathematical definition of allowable heterogeneity, one can obtain a partial answer by examining hypothetical examples given specific arithmetical values. When such tests were made, they indicated agreement below 50% infection between the numbers of plants infected in 2 contrasting situations, (a) when the m of equation 1 was constant and (b) when m varied within a set of samples and the chances of infecting variable plants were near-normally distributed. As the range of normally distributed variation increased, values for numbers of infected plants fell below the calculated values in the range above 55 or 60%, but the divergence was less than might have been expected. If in practice such divergence is extreme, a skewed or bimodal distribution of variability or some other cause for the deviation should be considered, as in the example drawn from Wilhelm's data.

The normal distribution.—As the sample size of a binomial distribution increases, the readings, which we shall assume to be numbers of plants infected, tend more and more to fall evenly on either side of a mean value. The distribution is discontinuous, i.e., it is confined to whole numbers. When the numbers of plants, n, in a sample rises to 10, and p, the probability of infection, ranges from 0.2 to 0.8, it is possible to draw a normal curve through the plot of binomial values without radical deviations. As the sample size increases, curves for the discontinuous variable with limits 0 and n approach normal curves for a continuous random variable with limits − α and + α, provided p is not too small nor too close to 1. The normal distribution can be used for quantal data where such mathematical requirements are satisfied, and the host-pathogen system may result in an approximately normal distribution of infective power at points of penetration. For example, there are relationships in which a single propagule may infect, but if one enters, others follow. Some species of nematodes and Phycomycetes act this way. In such instances variations in infectivity

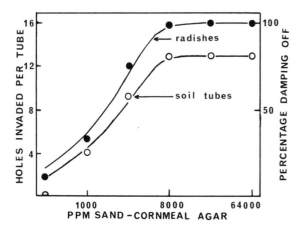

Fig. 1. Data of Martinson (1963) (dots or circles) fitted with Poisson series (curves). Percentage of preemergence damping-off of radish seedlings by *Rhizoctonia solani*, and holes invaded per microbiological sampling tube by *R. solani*, at different inoculum densities.

might be considered equivalent to a post-factum dosage effect. The normal distribution could very well apply to them, and also to examples in which there was extreme variability in virulence and susceptibility or in distribution of inoculum. Many such pathogen-host combinations need examination directly for their statistical relations (see below). In the meantime it would be advisable, wherever the statistical relations are important, to test first for agreement with the Poisson distribution. If there is significant disagreement, the type and extent of divergence will most likely give useful information about the system under study.

TRANSFORMATIONS.—*Probits versus the multiple-infection transformation.*—Leach and Davey (1938) sampled sclerotia of *Sclerotium rolfsii* from the soil of sugar beet fields before planting. Within the subsequent crop in each sampled field they counted numbers of infected plants, and related sclerotia per unit of soil before planting with percentages of infected beets. There was a positive association which enabled them to predict from the counts of sclerotia whether planting would be safe in a particular field. There were departures from this association (a) when extreme localization of inoculum was observed; (b) when high soil nitrogen or a high water table affected the host plants. The data from such fields were discarded, and percentages of infection (1) as probits, the transformation appropriate to a normal distribution, and (2) as log multiple-infection transformations, appropriate to a Poisson distribution, were plotted for the remaining 33 fields against log numbers of sclerotia. The discarded data, when included, increased variability but did not affect the comparison between transformations.

The plot of log multiple-infection transformation against log number of sclerotia (Fig. 2) was well fitted by a straight line of slope 1, as required by agreement with the Poisson series (see next section). Any slight systematic disagreement was concentrated in a tendency for low infection values at higher concentrations of sclerotia. Probits seemed to fall more on an upward curve than on a straight line.

The opposing tendencies for departure from straight-line relationships may provide a rule of thumb for use of the 2 transformations; when probit values plotted against log numbers of propagules fall on a rising curve, use the multiple-infection transformation; when the multiple-infection transformation gives values falling away markedly in the higher ranges, use probits.

Other useful transformations.—The standard transformation for quantal data with variation distributed according to the binomial series is the angular transformation (Fisher and Yates, 1963). If infections are in agreement with the Poisson series, the mean number of infections per plant, m, should be directly related to the amount of inoculum in the soil. Log m, calculated from percentage of infection and plotted

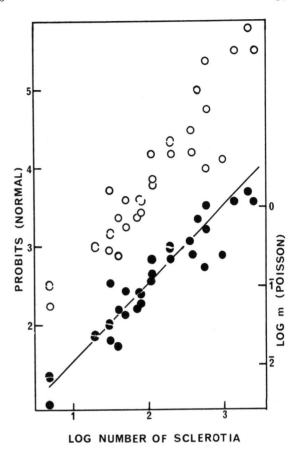

Fig. 2. Data of Leach and Davey (1938). Log number of *Sclerotium rolfsii* sclerotia sampled before planting from soils in 33 sugar beet fields, plotted against the probit transformation (white circles) and log m transformation (black dots) of subsequent percentage of infection with *S. rolfsii.*

against log number of propagules in the soil, should give a straight line of slope 1. If infection is less than 90 or 95%, m can be read off from Gregory's (1948) multiple-infection table or van der Plank's (1963) table of $\log_e \frac{1}{1-x}$. Either m or log m can be analyzed according to the way in which the data are to be used.

When the numbers of infections per plant, root system, or standard length of root average more than 3, a square-root transformation of lesion numbers may give the best data for analysis. If infections are distributed at random, about 95% of plants will be infected when m = 3. If there are zero values, the transformation, $\lambda + 1$, where λ = number of lesions, may be used. According to Fisher and Yates (1953) the square-root transformation has the same properties as the angular transformation at either end of the scale when the observations on each plot (of an experiment or field trial) are distributed according to a Poisson distribution: for this reason the angular transformation is sometimes used for analysis of data agreeing with the Poisson as well as the binomial distribution.

The probit transformation is most familiar in its application to toxicity tests, where dosage (= log dose) of a substance is the experimental variable, and variation of the test subjects, organisms or tissues, is not subject to control. The main difference between this and the usual infection system is that dosage becomes almost a continuous variable, and a minimum active dose may consist of a large number of identical molecules. Usually, in a host-pathogen system, propagules are units, any one of which may cause infection. It is a matter of chance which of many active propagules does cause infection. When a "dose" is applied, serious departures from a normal distribution in the susceptibility of organisms or in the chances of a toxic substance reaching individual organisms may make the use of probits inaccurate or unreliable (Bald and Jefferson, 1956); when unitary propagules are applied, variability in virulence, susceptibility, and the chances of inoculum reaching the roots of individual plants, and particularly the demand that two or more propagules acting together are necessary for infection—all these situations can make the use of multiple-infection transformation inaccurate or misleading. However, the binomial, Poisson, and normal distributions can stand a lot of abuse from natural systems without responding badly; if this were not so, the statistical analysis of many forms of data would be very difficult. In practice, one or other of the transformations mentioned above will generally reduce infection data to a usable form when arithmetic values are inadequate.

Total Area of Lesions.—This section will discuss the basis for measuring the surface area of tissue covered or affected by a population of lesions. There are 2 general cases: (1) lesions developing on a continuous surface, i.e. stems, tubers, bulbs, rhizomes, etc., and (2) lesions developing on a branching, finely divided surface, i.e. on a root system. A continuous surface is very small and compact per unit of substrate or mass compared with the divided surface of a branching root system. Multiple lesions on a stem or tuber may rapidly cover the surface; the same number on a root system may remain discrete. The type of interference in the former instance is direct—lesions become confluent or repress each other; in the latter, indirect—lesions may develop distally to one another on the same section of the root system. A lesion near the base of a root causing serious damage will partly or wholly nullify the effects of distal lesions. For a first analysis assume the 2 kinds of interference to be merely statistical; lesions develop independently of other lesions until they meet on a continuous surface, or they may arise serially and at random along a section of root system. The proportion of a continuous tissue surface covered by lesions, or of a root system damaged by lesions, should be related to the concentration of inoculum in the soil by a Poisson series, which may be reduced to the form of equation 1.

These cases were formulated by the writer, and a similar formulation was given independently by van der Plank (1963, p. 115) for the leaf area of crops affected by "local lesion" diseases. Van der Plank wrote, "One interprets $\frac{ak}{b}$ (= m of equation 1) as the area of diseased tissue, relative to total leaf area, that one would expect if there were no overlapping of lesions." Essentially the same assumption applied to a continuous surface was tested by a detailed series of trials with a model system (Bald and Tsao, unpublished), and applied to data for infection of carrot discs from soil infested with *Thielaviopsis basicola* (Tsao, 1962). The model gave results in agreement with a Poisson series, but the results with *Thielaviopsis* were widely divergent. If *Thielaviopsis* lesions had been distributed at random and had expanded without mutual interference, the percentages of carrot disc surfaces covered by lesions should have been calculable from the concentration of inoculum, simply by fitting Poisson series with values of m proportional to the dilution of the original infested soil. The most probable explanation for the departure from theory was that lesions developing from crowded infection foci tended to suppress each other.

The most-probable-number method (Maloy and Alexander, 1958), which has been tested by Tsao and Canetta (1964) and others for estimating the inoculum concentration in soil, is based on agreement with the Poisson distribution (Halvorson and Ziegler, 1933). In our results there was agreement at concentrations of inoculum where the most-probable-number method was applicable. In Tsao and Canetta's estimates of soil-borne propagules and probably in many others this essential condition was satisfied. The most-probable-number method is useful both in itself and in providing a basis for testing less laborious methods of measurement. The serial-dilution endpoint is a less accurate method meant to achieve a somewhat similar end.

Conclusion.—Some methods previously outlined for estimating amounts or intensity of inoculum in the environment by tissue or plant reactions (see summaries by Large, 1966; Dimond and Horsfall, 1965) now appear sound and farsighted; and the principles elaborated here may be considered an effort to continue some of the earlier work toward the point where the subject can be submitted to mathematicians for adequate treatment. The essential features of the discussion are contained in the examples. Among the quantal data some agree with simple probabilistic models, while others are widely divergent. Agreement indicates that events occurred or disease propagules were distributed more or less at random; divergence indicates a variable environment or nonrandom events, often unforeseen, or definitely nonrandom distributions of propagules. Qualitative methods or summation and comparison of simple numbers often fail to reveal either agreement with, or significant differences from, theoretical models. The revelation of disagreement with theory can sometimes be more useful than evidence of agreement. Deviations from simple mathematical relations like those discussed above tend to open up new ideas and new sources of information.

LITERATURE CITED

BALD, J. G., and R. N. JEFFERSON. 1956. Interpretation of results from a soil fumigation trial. Plant Disease Reptr. 40:840–846.

BALD, J. G., and D. E. MUNNECKE. 1962. Determination of inoculum potential of infested soil. Phytopathology 52:724 (Abstr.)

BALD, J. G. 1969. Estimation of leaf area and lesion sizes for studies on soil-borne pathogens. Phytopathology. (In press.)

DIMOND, A. E., and J. G. HORSFALL. 1965. Theory of inoculum, p. 404–415. *In* K. F. Baker and W. C. Snyder [ed.], Ecology of soil-borne plant pathogens. Univ. of California Press, Berkeley and Los Angeles.

FISHER, R. A. 1932. Statistical methods for research workers. Oliver and Boyd, Edinburgh. 307 p.

FISHER, R. A., and F. YATES. 1963. Statistical tables for biological, agricultural and medical research. Oliver and Boyd, Edinburgh. 146 p.

GREGORY, P. H. 1948. The multiple-infection transformation. Ann. Appl. Biol. 35:412–417.

HALVORSON, H. O., and N. R. ZIEGLER. 1933. Application of statistics to problems in bacteriology. I. A means of determining bacterial populations by the dilution method. J. Bacteriol. 25:101–121.

LARGE, E. C. 1966. Measuring plant disease. Ann. Rev. Phytopathol. 4:9–25.

LEACH, L. D., and A. E. DAVEY. 1938. Determining the sclerotial population of *Sclerotium rolfsii* by soil analysis and predicting losses of sugar beets on the basis of these analyses. J. Agr. Res. 56:619–631.

MALOY, O. C., and M. ALEXANDER. 1958. The "most probable number" method for estimating populations of plant pathogenic organisms in the soil. Phytopathology 48:126–128.

MARTINSON, C. A. 1963. Inoculum potential relationships of *Rhizoctonia solani* measured with soil microbiological sampling tubes. Phytopathology 53:634–638.

PLANK, J. E. VAN DER. 1963. Plant diseases. Epidemics and control. Academic Press, New York and London. 349 p.

TSAO, P. H. 1962. A quantitative technique for estimating the degree of soil infestation by *Thielaviopsis basicola*. Phytopathology 52:366.

TSAO, P. H., and ALICE C. CANETTA. 1964. Comparative studies of quantitative methods used for estimating the population of *Thielaviopsis basicola* in soil. Phytopathology 54:633–635.

WILHELM, S. 1951. Effect of various soil amendments on the inoculum potential of the Verticillium wilt fungus. Phytopathology 41:684–690.

PART III

◄

GENETICAL ASPECTS OF PATHOGENIC AND SAPROPHYTIC BEHAVIOR IN ROOT-INFECTING FUNGI

Genetical Aspects of Pathogenic and Saprophytic Behaviour of Soil-borne Fungi: Basidiomycetes with Special Reference to Thanatephorus cucumeris

N. T. FLENTJE—*Department of Plant Pathology, Waite Agricultural Research Institute, University of Adelaide, South Australia.*

INTRODUCTION.—There are relatively few Basidiomycetes which are soil-borne and which cause serious plant diseases. *Armillaria mellea, Fomes* spp., *Helicobasidium purpureum, Poria* spp., *Marasmius* spp., and *Thanatephorus cucumeris* are among the most important.

The plant pathologist is interested mainly in two aspects of these fungi: first, their ability to attack plants, and second, their ability to survive and grow in soil. In the past he has frequently either overlooked the fact that each fungus is capable of considerable variation, or catalogued variants without recognizing their dynamic character and potential value as tools for investigating their pathogenic and saprophytic behaviour. In recent years, however, there has been much greater interest in the application of genetical studies to the investigation of soil-borne fungi, and an accompanying realization that a detailed background of genetical information is required for each fungus to be investigated. In the past ten years preliminary genetical studies have been carried out on *T. cucumeris* and closely related species in several laboratories. Relatively little genetical work has been carried out on the other fungi mentioned.

There are, however, a number of other Basidiomycetes which, although soil borne, do not cause diseases in higher plants, but which have been subjected to detailed genetical analysis for certain characters. These fungi include *Agaricus* spp., *Coprinus* spp. and *Schizophyllum commune*. The genetical analyses have been concerned mostly with mating type behaviour and compatability, characters which, although not directly influencing pathogenic and saprophytic behaviour, will influence mechanisms of variation and are likely to be important in the soil-borne pathogens.

Finally the rust and smut fungi, although not soil borne, are important pathogens, and the extensive genetical analyses of pathogenicity carried out with these fungi may help in investigating the pathogenicity of the soil-borne fungi.

MECHANISMS OF VARIATION.—It is generally accepted that genetic variation in a range of characters in fungi can occur by (a) mutation, (b) heterocaryosis, (c) parasexual recombination, (d) sexual recombination. Variation may also occur by adaptation, but this does not involve genetic change. Most of these mechanisms have been demonstrated to occur among the Basidiomycetes, although parasexual recombination has been demonstrated in only a few cases, e.g. *Ustilago maydis* and *Schizophyllum commune*, and the extent to which it occurs is unknown. Adaptation as a mechanism of variation is still somewhat controversial and was reviewed by Buxton (1960). The major problem here is to distinguish between nongenetical changes on one hand and mutation followed by selection on the other hand. There is strong evidence for the stimulation of enzyme production in response to suitable substrates, so it seems likely that both pathogenic and saprophytic abilities of a fungus could be adaptively influenced, but much more critical investigation is needed in this field.

PATHOGENIC BEHAVIOUR.—Pathogenic behaviour may be considered in two parts, virulence and host range. Virulence is a measure of the severity of attack of a fungus on a host, whereas host range is the measure of the different types of host—whether varieties, species, genera, or families—which are attacked.

Virulence.—In both rusts and smuts virulence to a particular host variety appears to be determined by a small number of genes, and changes in virulence occur by a range of mechanisms including mutation. Avirulence generally appears to be dominant. It should be pointed out, however, that because of our present inability to grow the rusts and the dicaryon of the smuts on laboratory media, the changes in virulence investigated have been restricted to those changes which affect the growth of the fungus within the host tissue. There is little information on changes which might affect the initial penetration of the host. It is quite likely, then, that there is a much larger number of genes affecting virulence than is known at present. As the rust and smut fungi grow extensively through living host cells, they must achieve a close physiological balance with the host cells and it is therefore quite feasible that small genetic changes in the fungus could upset this balance, leading to either rapid death of the host cells or inability to provide essential nutrients and consequently limited lesion or avirulent reactions. There is still little information, however, on the physiological changes influencing virulence. Holliday (1961) has attempted to investigate the influence of various amino acids on virulence by using auxotrophic mutants of *U. maydis*, but the results, like those of Kline, Boone, and Keitt (1958) with *Venturia*, are inconclusive.

A range of studies with *T. cucumeris* has provided further genetical information on variation in virulence. *T. cucumeris* isolates studied so far are homothallic (Stretton, Flentje, and McKenzie, 1967), but various workers have shown that they generally occur in the field as heterocaryons. When maintained in vegetative culture in the laboratory, they show little if any variation in virulence. However, single basidiospore cultures derived from one parent show quite a wide variation in virulence. But as with the obligate parasites, there is little information on physiological changes affecting virulence. Papavizas and Ayers (1965) found no correlation between virulence and production of various pectinase enzymes. Flentje, Stretton, and McKenzie (1967) and McKenzie et al. (1969), using a homocaryon single-spore line of an isolate of *T. cucumeris* pathogenic to radish, obtained a range of 16 avirulent mutants which occurred spontaneously or in response to ultraviolet irradiation. The process of infection of radish by a virulent isolate is known in detail and occurs as a series of stages, namely (a) attachment of hyphae to the stem, (b) growth along the stems following the lines of junction of epidermal cells, (c) multiple short branching of side branches to form infection cushions, (d) penetration from the base of the cushions, and (e) intercellular and intracellular development of hyphae, causing spreading lesions. The avirulent mutants were tested to determine at which stage of the infection process they failed. Of 16 mutants tested, 8 failed at stages (a) and (b), 3 failed at stage (c), 2 failed at stage (d), and 3 failed at stage (e). Heterocaryons formed from pairs of the mutants were as virulent as the original parent. Thus it appears, on present evidence at least, that in *T. cucumeris* virulence is dominant and is controlled by a large number of genes most of which affect prepenetration growth and only a small number of which, as in the rusts and smuts, affect growth within the host tissue. The fact that virulence is dominant and that vegetative cells are multinucleate would explain why virulence is maintained during laboratory culture.

One aspect of avirulence, namely the hypersensitive reaction, is of particular interest. With obligate parasites it has been suggested (Hare, 1966) that the hypersensitive reaction occurs in response to some toxic material produced by the fungus-host interaction which causes the death first of host and then of fungus cells. Some avirulent mutants of *T. cucumeris* produce a hypersensitive reaction in the host, but if these mutants are allowed to form a heterocaryon, the latter is now virulent and does not produce a hypersensitive reaction. These results in themselves give little further light on the nature of the hypersensitive reaction, but the avirulent mutants of a facultative pathogen such as *T. cucumeris* may be very useful tools for investigating the mechanism.

Isolates of the same pathogenic strain of *T. cucumeris* obtained from different geographic locations appear to be closely similar in the genetic factors controlling virulence (Fig. 1, 2). Avirulent mutants from

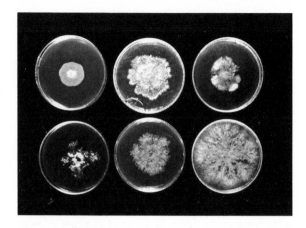

Fig. 1. Virulent and avirulent mutants from a "crucifer strain" isolate of *T. cucumeris* from Clare, South Australia. Top: avirulent, virulent, virulent. Bottom, avirulent, avirulent, avirulent.

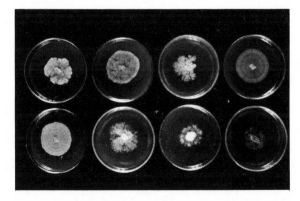

Fig. 2. Avirulent mutants from a "crucifer strain" isolate of *T. cucumeris* from the Waite Institute, South Australia.

the different isolates (Fig. 3) complement each other for virulence in the heterocaryon, and recombination occurs in the sexual stage to yield virulent wild type progeny which do not vary markedly in cultural appearance (Flentje and Stretton, unpublished).

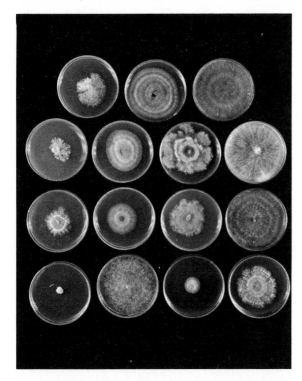

Fig. 3. Range in variation in recombinants from crossing two mutants obtained from two different "crucifer strain" isolates of *T. cucumeris*. Top line, mutant 1, mutant 2, heterocaryon between 1 and 2. Remainder, single basidiospore cultures from the heterocaryon.

Host range.—With the rust and smut fungi attention has been largely concentrated on varietal host range with regard to resistance breeding. There is evidence that an increase in varietal host range can occur by mutation or by heterocaryon formation followed by recombination. This latter mechanism has led to surprisingly wide increases in varietal susceptibility in some studies which are not explained by our present knowledge. Changes in generic host range have also been demonstrated by several workers (Johnson, 1960; Day, 1966). Watson and Luig (1962) showed that *Puccinia graminis* var. *secalis* when inbred on barberry gave rise to strains which readily attacked wheat as well as rye. There is limited capacity, however, for interspecific crossing between the smuts or the rusts because, in general, the different varieties or species are incompatible (Johnson, 1960).

In *T. cucumeris* one of the outstanding features is the apparent lack of varietal resistance, the only record being that of Kernkamp et al. (1952) in potato varieties. We have tested several field isolates and a large number of homocaryotic single-spore cultures derived from them, against different varieties of susceptible hosts and have found no real indication of varietal resistance. Also, a number of the avirulent mutants of *T. cucumeris* described above were tested and found to be avirulent on a range of radish varieties

and related cruciferous hosts. Thus lack of varietal resistance to field isolates of *T. cucumeris* is apparently not due to their heterocaryotic condition. It might be suggested that the mechanisms of pathogenesis for an organism like *T. cucumeris*, resulting in the death of invaded host cells, are much less delicately balanced with the host cells than in the rusts and smuts and that therefore much greater differences in host-cell composition would be required for varietal resistance than with the rust and smut fungi where the gene-for-gene hypothesis of pathogenicity suggests that small differences are required for resistance. However, the fact that mutants of *T. cucumeris* can be obtained which either fail to enter the host or produce a hypersensitive reaction in the host cells, and on the other hand, the fact that some rusts can attack several host genera, would throw doubt on this argument. Much more thorough investigation of varietal resistance to *T. cucumeris* is, in my opinion, warranted.

In an attempt to investigate factors determining pathogenic specialization Flentje and co-workers (unpublished) have attempted to form heterocaryons between large numbers of mutants derived from isolates of *T. cucumeris* pathogenic to different hosts. From some 200 attempts none has been successful. Microscopic study of opposing hyphae from the paired mutants has shown that heterocaryon formation fails because the hyphae react in one or other of the following ways: (a) they are repelled and do not make contact; (b) they make contact but do not fuse; (c) the cell walls fuse but there is no cytoplasmic connection; (d) cytoplasmic connection is established, but the participating cells together with several neighbouring cells on either side are killed because of an incompatibility reaction. There is indication from studies with *Neurospora* (Holloway, 1955) that these failures may be controlled by several different genes. The incompatibility reaction is similar to the vegetative incompatibility described by Esser and Keunen (1965), but it has no apparent relation with the sexual stage. McKenzie (1966) showed that it occurred, apparently with different degrees of severity, between mutants from one parent as well as between mutants from different parents. In mutants from one parent, however, it was not severe enough to prevent heterocaryon formation, therefore some modification of the nuclear behaviour apparently occurs when different nuclei are associated in common cytoplasm. An attempt was made by Garza-Chapa and Anderson (1966) to obtain further genetical information on factors determining heterocaryon formation. They produced some evidence that it was determined by more than 2 alleles, but because of the lack of specific markers for the identification of recovered progeny, their evidence cannot be regarded as satisfactory.

Parmeter and Sherwood (personal communication) have recently carried out a study of anastomosis with a wide range of field isolates and have shown that the isolates could be separated into several groups on their ability to anastomose with each other. This appears to be a very promising line of investigation which could greatly assist an understanding of genetical relationships.

Several workers have shown that anastomosis can be influenced by environmental factors such as culture media (Bouchier, 1957; Cabral, 1951) and temperature (McKenzie, 1966), and it is well known that the monocaryon lines of smuts, while readily forming the dicaryon in the host, will not do so on culture media.

Anastomosis and particularly the incompatibility reaction in *T. cucumeris* warrant detailed study. Incompatibility would lead to biological isolation and, with subsequent mutation followed by selection, could lead to the development of quite separate strains. The fact that incompatibility occurs within single-spore isolates from one parent would facilitate such divergent development.

As vegetative incompatibility appears at present to be a serious obstacle to investigating factors that determine pathogenicity to different hosts, an alternative possibility would be through the study of segregation in single-spore cultures and mutants from a parent isolate which has a wide host range. Papavizas and Ayers (1965) tested the pathogenicity of single-spore cultures obtained from an isolate which attacked five different hosts, namely cabbage, cotton, lettuce, radish, and sugar beet. Six of the eight single-spore cultures failed to attack cotton but were still pathogenic on the remaining hosts. Flentje and Stretton (unpublished) have obtained similar results with single-spore isolates from two different parent cultures which had a wide host range.

Thus it would appear that in isolates which have a wide host range, at least some genes may determine pathogenicity to specific hosts and can be inherited independently of genes determining pathogenicity for other hosts. This information, although tentative, would indicate that this might be a profitable field of investigation.

SAPROPHYTIC BEHAVIOUR.—The saprophytic growth and survival in soil of many different fungi have been compared, but there are usually so many differences between the fungi that it has proved impossible to define the factors responsible for their different behaviour. Presence or absence of so-called survival structures such as sclerotia, ability to utilize different substrates, and tolerance or resistance to antibiotics have all been suggested as important factors.

Baker et al. (1967), Olsen, Flentje, and Baker (1967), and Papavizas (1964) have made extensive analysis of the survival of different field isolates and single basidiospore cultures of *T. cucumeris* in several soils. Single basidiospore cultures obtained from one parent show widely different survival ability in soil, some of the single-spore cultures surviving even better than the parent isolates. Olsen, Flentje, and Baker (1967) showed that ability to grow through soil immediately after inoculation was not necessarily an indication that an isolate would survive well, as some

single-spore isolates which grew prolifically survived poorly. Apparently then, we have to make more careful distinction between growth and survival in soil. Olsen, Flentje, and Baker (1967) also showed that the single-spore cultures which survived best apparently did not produce sclerotia in soil.

In attempts to better understand growth and survival in soil, various studies have been made of the growth of single basidiospore cultures in the laboratory. Many such cultures obtained from the one parent grew as well on complete laboratory media as did the parent, but they exhibited a wide variation in cultural appearance (Whitney and Parmeter, 1963; Flentje and Stretton, 1964; Papavizas, 1964), in their ability to utilize different forms of nitrogen (Whitney and Parmeter, 1963), in enzyme production, especially pectinases (Papavizas and Ayers, 1965), in tolerance to CO_2, to antibiotics and other toxic chemicals, and in ability to survive in previously colonized substrates (Papavizas, 1964). All of these characters might be expected to influence saprophytic growth and survival, but as yet no detailed genetical analysis of them has been carried out. Whitney and Parmeter (1963) obtained some preliminary evidence that ability to use different forms of nitrogen may be simply inherited. Olsen, Flentje, and Baker (1967) showed that single-spore cultures from one parent differing in survival ability would anastomose readily, and McKenzie et al. (1969) have shown that recombination between the same isolates used by Olsen, Flentje, and Baker (1967) can be readily achieved. It appears then that useful tools are now available for genetical study of the physiological factors affecting growth and survival, at least of *T. cucumeris*, in soils.

GENERAL DISCUSSION.—With respect to pathogenicity there appear to be many points of similarity between the obligate airborne pathogens like the rusts and smuts and the facultative soil-borne pathogens. The major differences are in the abundance of varieties resistant to the obligate pathogens and the almost complete absence of varietal resistance to the facultative pathogens.

I feel that, in the future, because of the similarities that exist, much more use could be made of the genetical information about the obligate pathogens in investigations of soil-borne fungi.

With regard to the soil-borne fungi, we are beginning to accumulate substantial genetical information about *T. cucumeris*. It accumulates in the soil as different pathogenic strains which also differ in saprophytic ability. Many of these strains apparently are identical in the basidial stage, and it might be argued that they had a common origin. If this were so, variants would have arisen initially by mutation and would have become segregated either by the formation of basidiospores or by fragmentation of vegetative hyphae. Although single basidiospore cultures may survive well when inoculated into soil, there is some doubt (Flentje and Stretton, 1964) as to whether the spores themselves, if introduced into soil, can establish colonies. If we assume they can, there is now ample evidence that they would give rise to cultures which vary in virulence and saprophytic ability. Cultures of similar host range could anastomose, and the various factors determining survival in soil would cause selection of the types best adapted for the particular soil. Transfer of the fungus to another soil environment with different survival factors would result, however, in selection of a different type from the genetic pool available.

There is no clear evidence as to how isolates differing in host range arise. From the evidence available at present it might be argued that host range is a far more stable character than virulence or saprophytic ability. A more likely explanation is that a change in host range involves a series of mutations subject to continuous selection in one or other direction. The types of mutation involved might well concern the various stages, outlined earlier, which appear to be necessary for penetration. For example, a change in host range may require first a mutation which will cause attachment to the stem of a new host, then a mutation that will allow a response to the exudate of a new host to form infection structures, and so on. If such a succession of changes is required, it will be difficult to investigate in *T. cucumeris*, as there is at present no rapid method of obtaining and screening large numbers of mutants for such changes. It is also a matter of conjecture how often such changes would occur in the field unless the basidial stage is freely formed. The number of different pathogenic races occurring naturally suggests it must happen with reasonable frequency.

Investigation of these mechanisms of pathogenicity and the possibility of obtaining varietal resistance to the soil-borne fungi is of urgent importance, for resistance would offer valuable possibilities for control.

LITERATURE CITED

BAKER, K. F., N. T. FLENTJE, C. M. OLSEN, and HELENA M. STRETTON. 1967. Effect of antagonists on growth and survival of *Rhizoctonia solani* in soil. Phytopathology 57:591–597.

BOUCHIER, R. J. 1957. Variation in cultural conditions and its effect on hyphal fusion in *Corticium vittercum*. Mycologia 49:20–28.

BUXTON, E. W. 1960. Heterokaryosis, saltation and adaptation, Vol. 2, p. 359–405. *In* J. G. Horsfall and A. E. Dimond [ed.], Plant pathology, an advanced treatise. Academic Press, New York.

CABRAL, R. V. DE G. 1951. Anastomosis meceliais. Bol. Soc. Broteriana Ser. [2] 25:291–362.

DAY, P. R. 1966. Recent developments in the genetics of the host parasite system. Ann. Rev. Phytopathol. 4: 245–268.

ESSER, K., and R. KEUNEN. 1965. Genetik der Pilze. Springer, Berlin. 497 p.

FLENTJE, N. T., and HELENA M. STRETTON. 1964. Mechanisms of variation in *Thanatephorus cucumeris* and *T. praticolus*. Australian J. Biol. Sci. 17:686–704.

FLENTJE, N. T., HELENA M. STRETTON, and A. R. McKENZIE. 1967. Mutation in *Thanatephorus cucumeris*. Australian J. Biol. Sci. 20:1173–1180.

GARZA-CHAPA, R., and N. A. ANDERSON. 1966. Behaviour of single spore isolates and heterocaryons of *Rhizoctonia solani* from flax. Phytopathology 56:1260–1268.

HARE, R. C. 1966. Physiology of resistance to fungal diseases in plants. Bot. Rev. 32:95–137.

HOLLIDAY, R. 1961. The genetics of *Ustilago maydis*. Genet. Res. 2:204–230.

HOLLOWAY, B. W. 1955. Genetic control of heterocaryosis in *Neurospora crassa*. Genetics 40:117–129.

JOHNSON, T. 1960. Genetics of pathogenicity, Vol. 2, p. 407–509. *In* J. G. Horsfall and A. E. Dimond [ed.], Plant pathology, an advanced treatise. Academic Press, New York.

KERNKAMP, M. F., D. J. DE ZEEUW, S. M. CHEN, B. C. ORTEGA, C. T. TSIANG, and A. M. KHAN. 1952. Investigations on physiologic specialization and parasitism of *Rhizoctonia solani*. Tech. Bull. Minn. Agr. Expt. Sta., No. 200, p. 1–36.

KLINE, D. M., D. M. BOONE, and G. W. KEITT. 1958. *Venturia inaequalis* (CRC) Wint. XIV. Nutritional control of pathogenicity of certain induced biochemical mutants. Am. J. Botany 44:797–803.

MCKENZIE, A. R. 1966. Studies on genetically controlled variation in *T. cucumeris*. Ph.D. thesis, University of Adelaide.

MCKENZIE, A. R., N. T. FLENTJE, HELENA M. STRETTON, and M. JEAN MAYO. 1969. Heterocaryon formation and genetic recombination in *Thanatephorus cucumeris*. Australian J. Biol. Sci. 22. (In press.)

OLSEN, C. M., N. T. FLENTJE, and K. F. BAKER. 1967. Comparative survival of monobasidial cultures of *Thanatephorus cucumeris* in soil. Phytopathology 57:598–601.

PAPAVIZAS, G. C. 1964. Survival of single basidiospore isolates of *Rhizoctonia practicola* and *Rhizoctonia solani*. Can. J. Microbiol. 10:739–746.

PAPAVIZAS, G. C., and W. A. AYERS. 1965. Virulence, host range and pectolytic enzymes of single basidiospore isolates of *Rhizoctonia practicola* and *Rhizoctonia solani*. Phytopathology 55:111–116.

STRETTON, HELENA M., N. T. FLENTJE, and A. R. McKENZIE. 1967. Homothallism in *Thanatephorus cucumeris*. Australian J. Biol. Sci. 20:113–120.

WATSON, I. A., and N. H. LUIG. 1962. Selecting for virulence on wheat while inbreeding *Puccinia graminis* var. *secalis*. Proc. Linn. Soc. of N.S.W. 87:39–44.

WHITNEY, H. S., and J. R. PARMETER. 1963. Synthesis of heterocaryons in *Rhizoctonia solani* Kuhn. Can. J. Botany 41:879–886.

▶

Genetical Aspects of Pathogenic and Saprophytic Behavior of the Phycomycetes with Special Reference to Phytophthora[1]

M. E. GALLEGLY—*Department of Plant Pathology and Bacteriology, West Virginia University, Morgantown.*

▶

The important soil-borne plant pathogens among the Phycomycetes include species of *Aphanomyces, Pythium,* and *Phytophthora.* Common to all these species is the inability to grow, compete, and survive as a saprophyte in soil. Long-term survival depends on resting structures such as oospores and chlamydospores. The genetical information on these Oomycetes has been obtained mostly from studies of variation in artificial culture and variation in pathogenicity. Most of the genetical studies have been with species of *Phytophthora.* Recent notable exceptions are the cytological studies of species of *Pythium* by Sansome (1961, 1963), and the observation of heterothallism in *Pythium sylvaticum* (Campbell and Hendrix, 1967).

Prior to this discovery, all species of *Pythium* were considered to be homothallic (Middleton, 1943). *Pythium sylvaticum* produced oospores on hemp-seed agar only when certain cultures were paired. Among five isolates, three produced only antheridia, and two produced only oogonia (Papa, Campbell, and Hendrix, 1967). This indicates that the two mating types differ in morphological sex rather than in compatibility factors as noted below for species of *Phytophthora.*

The cytological studies of Sansome (1963, 1965) indicate that certain species of *Pythium* and *Phytophthora* are diploid in their vegetative stages, and that meiosis occurs just prior to fertilization. This concept, if proven to be correct, materially alters interpretations of genetical data obtained from these organisms heretofore considered to be haploid. This concept will be discussed later.

Sexual Patterns in Species of Phytophthora.—The recent studies by Savage et al. (1968) have eliminated much of the confusion and controversy regarding sexuality in *Phytophthora.* The literature was reviewed by them and will not be included here. They presented the following groupings.

Homothallic with predominantly paragynous antheridia: *P. cactorum* (Leb. and Cohn) Schroet., *P.*

[1] Published with the approval of the Director of the West Virginia University Agricultural Experiment Station as Scientific Paper No. 1037. The unpublished studies reported were supported in part by National Science Foundation Grant GB-3858.

citricola Saw., *P. lateralis* Tucker and Milb., *P. megasperma* Drechs., *P. porri* Foister, *P. sojae* Kaufmann and Gerdemann, *P. syringae* (Kleb.) Kleb.

Homothallic with amphigynous antheridia: *P. boehmeriae* Saw., *P. erythroseptica* Pethyb., *P. fragariae* Hickman, *P. heveae* Thompson, *P. hibernalis* Crane, *P. ilicis* Buddenhagen and Young, *P. phaseoli* Thaxt., *P. richardiae* Buis.

Heterothallic with compatibility types A[1] and A[2] and amphigynous antheridia: *P. arecae* (Colem.) Pethyb., *P. cambivora* (Petri.) Buis., *P. capsici* Leonian, *P. cinnamomi* Rands, *P. citrophthora* (R. E. Sm. and E. H. Sm.), *P. colocasiae* Rac., *P. cryptogea* Pethyb. and Laff., *P. drechsleri* Tucker, *P. infestans* (Mont.) de By., *P. meadii* McRae, *P. mexicana* Hotson and Hartge, *P. palmivora* (Butl.) Butl., *P. parasitica* Dast., *P. parasitica* (Dast.) var. *nicotianae* Tucker.

Perhaps the most interesting discovery was that the compatibility types A[1] and A[2], previously described in *P. infestans* by Galindo and Gallegly (1960), were present in 13 species and 1 variety of the genus. There was no evidence of additional compatibility factors.

Equally interesting was the simultaneous discovery of relatively free interspecific mating between the heterothallic species, but only when an A[1] isolate was paired with an A[2] isolate. Proof of true interspecific hybridization awaits the establishment of single-zoospore cultures from single oospores obtained from interspecific matings. However, the direct observation of the fusion of gametangia of *P. capsici* and *P. infestans* by Savage et al. (1968), with the formation of mature oospores, is strong indication that interspecific hybridization occurs in some species.

In considering the implications of interspecific hybridization in relation to development of root-rot diseases, it must be remembered that these fungi do not grow for long in the soil. Thus, there probably would be little chance for mating to occur, other than in the infected host tissue. Host-specific pathogenicity among species would further limit interspecific hybridization, as would the type of tissue normally invaded (e.g. foliage vs roots). However, the root and other tissues of some hosts are susceptible to several species. For instance, buckeye rot of tomato may be caused by *P. parasitica, P. capsici,* and *P. drechsleri;*

isolations from citrus roots have yielded *P. citroph-thora* and *P. parasitica.* Similarly *P. palmivora, P. meadii,* and *P. arecae* frequently occur on the same host. Simultaneous infections of the same host by two or more species of opposite compatibility types could provide a means for variation through interspecific sexual recombination. Perhaps taxonomic difficulties encountered among certain species could be attributed to such hybridization.

SEXUAL MECHANISMS.—The antheridia of all species of *Aphanomyces* and *Pythium* are paragynous, whereas both paragyny and amphigyny occur among species of *Phytophthora.* When paragyny occurs the fertilization tube from the antheridium enters the oogonium directly through the oogonial wall. In species with amphigynous antheridia the oogonial hypha first penetrates the antheridium and then passes through to form the oogonium. A fertilization tube penetrates the part of the oogonial stalk within the antheridium, or, as suggested by Galindo and Zentmyer (1967), the antheridial contents are discharged into the oogonium through a pore in the oogonial stalk. Gallegly (1968) has described the fertilization process for *Phytophthora infestans.* Following fertilization the protoplasm in the oogonium rounds into an oospore which becomes thick-walled.

Light stimulates oospore germination (Berg, 1966) which may occur *in situ* in agar cultures at 20° C. Germination is first marked by a swelling of the oospore (rehydration) followed by the emergence of a germ tube from the oospore wall. The germ tube usually is terminated by a single germ-sporangium, but branching of the germ tube and continued hyphal growth sometimes occurs. The terminal sporangium (germ-sporangium) usually liberates zoospores when held in water at low temperatures (12° C for *P. infestans*). In *P. infestans* 16 zoospores are usually produced, but as many as 38 have been liberated by a single germ-sporangium (Laviola, 1968).

The current controversy concerning the sexual mechanism centers on where meiosis occurs. It has generally been assumed that meiosis occurs in the oospore, following fertilization, and that the single-nucleate zoospores in the germ sporangium are haploid. The studies of Sansome (1966), and Galindo and Zentmyer (1967) suggest that meiosis occurs in the gametangia prior to fertilization, and that upon germination the zoospores in the germ sporangium are diploid.

CYTOLOGY.—Sansome (1961, 1963, 1965, 1966) has provided valuable information concerning the cytology of the sexual stages of certain members of the Oomycetes, including species of *Pythium* and *Phytophthora*; Galindo and Zentmyer (1967) presented information on the cytology of the sexual stages of *Phytophthora drechsleri*; and Marks (1965) has studied the cytology of the asexual stages of *Phytophthora infestans.*

Sansome (1963) observed an association of four chromosomes in dividing nuclei in the antheridium and oogonium of *Pythium debaryanum.* This observation coupled with simultaneous division of the nuclei in one oogonium, the occurrence of a distinct metaphase stage in dividing oogonial nuclei, and the absence of metaphases in the vegetative hyphae and sporangia, indicated that meiosis occurs in the gametangia prior to fertilization, and that the vegetative stages are diploid.

Sansome (1965) reached similar conclusions in studies with *Phytophthora cactorum* and *P. erythroseptica.* Two nuclear divisions occurred in the gametangia, with the first division having a long prophase. The nuclei were about half the size of vegetative nuclei following the second division. A bridge and fragment were observed at anaphase in *P. cactorum,* and multivalents were observed in nuclei considered to be polyploid following treatment of gametangia of *P. cactorum,* and *P. erythroseptica* with camphor. These observations were regarded as critical evidence that the divisions in the gametangia were meiotic. Sansome has suggested that *Phytophthora* species (n = 9 ca.) and *Pythium* species (n = 18 ca.) belong to a polyploid series.

The observations by Marks (1965) in cytological studies of the asexual stages of *Phytophthora infestans,* a heterothallic species, do not support Sansome's conclusions for homothallic species. Marks noted that mitotic stages in the hyphal tips of *P. infestans* resembled those in the oogonia and antheridia of *Pythium debaryanum* regarded by Sansome as meiotic stages. Gallegly (1968) speculated that the heterothallic species may be haploid and the homothallic ones diploid. However, the cytological observations by Galindo and Zentmyer (1967) with *Phytophthora drechsleri,* a heterothallic species, were similar to those of Sansome. It is obvious that additional cytological studies are needed to resolve the question of ploidy in species of *Phytophthora* and *Pythium.*

INHERITANCE OF PATHOGENICITY AND OTHER CHARACTERS.—The first evidence that genetic recombination occurs via the sexual stage of species of *Phytophthora* was presented by Gough (1957) and Smoot et al. (1958) with *P. infestans,* but only two single-oospore cultures were established. Savage and Gallegly (1960) established two additional oospore cultures which also were recombinants. More recently, Galindo and Zentmyer (1967) with *Phytophthora drechsleri,* Satour and Butler (1968) with *P. capsici,* Romero (1967) with *P. infestans,* and Laviola (1968) with *P. infestans,* have studied progenies with larger numbers of individuals.

Phytophthora drechsleri.—Recombination was observed for compatibility type and several other genetic markers among 173 single-oospore colonies established by Galindo and Zentmyer (1967) from oospores and single-zoospore cultures. However, only one phenotype was obtained from each oospore, even in the six cases where colonies were established from different zoospores from the same germ sporangium. Phenotypic ratios among over 100 single-oospore cultures derived from one cross were: A^1 to A^2 compatibility type, 1:1; repulsion to stimulation, 2:1; and white to gray colonies, 2:1. Ratios among 17 single-oospore cultures from one backcross were: A^1 to A^2 type, 1:1; arbuscular to thread branching of mycelia, 1:1; and brown-pigmented to colorless colony, 1:1.

Phytophthora capsici.—Satour and Butler (1968) observed recombination for several genetic markers in the F_1 of A^1 and A^2 crosses of *P. capsici*. The parent A^1 isolate produced star-appearing colonies, its sporangia liberated zoospores, and it was pathogenic to tomato and pepper. The parent A^2 isolate was nonstar in colony appearance, its sporangia liberated zoospores, and it was pathogenic to tomato and pepper. Thus, among these markers the parent isolates differed only in compatibility type and colony appearance. However, some of the isolates in the F_1 were nonpathogenic to tomato and pepper, and the sporangia of some failed to liberate zoospores. Similarly, star patterns different from the pattern in the parent appeared in the F_1. Segregation ratios among 51 single-oospore cultures from one cross were as follows: star to nonstar colonies, 2:1; zoospore to nonzoospore formation, 2:1; A^1 to A^2 compatibility type, 2:1; pathogenic to nonpathogenic to tomato, 2:1; and pathogenic to nonpathogenic to pepper, 1:2. Only one phenotype was detected in each oospore culture. Cultures from single zoospores liberated by germ sporangia were not established, but single-zoospore and hyphal-tip cultures obtained from previously established single-oospore cultures were phenotypically the same.

Phytophthora infestans.—In potato, nine R-genes (R_1, R_2, R_3, etc.) have been identified (Gallegly, 1968), and each is inherited in a monogenic dominant manner to confer resistance expressed as a hypersensitive reaction following inoculation with an apathogenic race of *P. infestans*. In the fungus, pathogenic races have been identified and labeled according to the R-gene hosts which they will attack. For example, race *0* is pathogenic only on plants without R-genes (recessive), race *1* is pathogenic only on recessive and R_1 plants, race *2* is pathogenic only on the recessive and R_2 plants, race *1,3* is pathogenic on recessive R_1, R_3, and R_1R_3 plants, and race *1,2,3,4* is pathogenic on the 16 host genotypes possible with genes R_1, R_2, R_3, and R_4.

Romero and Erwin (1968) found recombinations of pathogenic race characters and compatibility types among the progenies from three crosses of different races of *P. infestans*. Segregation ratios for the presence to absence of the four race characters carried by one of the parent isolates was 10:10 for race *1*, 3:17 for race *2*, 9:11 for race *3*, and 10:10 for the race *4* character. Cultures from single zoospores liberated by germ sporangia were not analyzed by Romero and Erwin to determine whether segregation occurred during oospore germination. However, an-

alysis of single-zoospore cultures from previously established single-oospore cultures showed that each germinating oospore gave rise to only one phenotype. Romero (1967) also noted that four cultures established from bodies considered to be parthenogenic oospores, which occasionally appeared in single cultures of their isolate 445, were of the same phenotype as the parent 445.

Laviola (1968) has studied the pathogenic race and compatibility type characteristics of individuals in the F_1 from four crosses of *P. infestans*. Twenty-nine cultures were established from single oospores producing germ tubes directly from germ sporangia, and 194 cultures were established from single zoospores liberated by the germ sporangia of 59 oospores.

With one exception, the F_1 from a cross between weakly pathogenic isolates of race *3*, compatibility type A^1 × race *1, 2*, compatibility type A^2 were nonpathogenic on all the differential hosts including recessive plants. The one pathogenic isolate was race *4*. The predominance of nonpathogenicity among the offspring was attributed to a predominance of nonpathogenic nuclei in the heterocaryotic mycelium of the parents. The appearance of race *4* in one isolate was not unexpected since Gallegly and Eichenmuller (1959) observed that this character frequently appeared spontaneously in the isolates studied.

Among 44 F_1 cultures of cross *3* A^1 × *0* A^2 (8 single-oospore and 36 single-zoospore cultures from 24 oospores), seven were nonpathogenic. The cultures were either race *3* or race *0* except for the appearance of the race *4* character in a few isolates to give races *3,4* or *4*. Recombination for compatibility type and pathogenic race was evident among the progeny. Segregation for compatibility type was in a ratio slightly less than 3:1.

All F^1 cultures from cross *1,2,3,4* A^1 × *1,2* A^2 (three single-oospore and 14 single-zoospore cultures from five oospores) were of compatibility type A^1, and all but one carried pathogenic race characters *1* and *2*. Segregation for the race *3* and *4* characters occurred, with recombination for compatibility type and pathogenic race being evident among the offspring, e.g. race *1,2* of type A^1.

The cultures established from cross *1,2,3,4* A^1 × *0* A^2 included 7 single-oospore and 140 single-zoospore cultures from 26 germinated oospores. Segregation for pathogenic race and compatibility type occurred, with all but three of the possible 16 races appearing among the F_1; races *2*, *2,4*, and *1,2,4* were not detected. The ratio for A^1:A^2 was approximately 3:1. Among the phenotypes in the progeny, segregation ratios for the presence to absence of race characters *1*, *3*, and *4* was about 1:1; for the race *2* character the ratio was 6:33.

In six instances Laviola (1968) found more than one phenotype among the cultures established from zoospores liberated by the germ sporangium produced by a single oospore. Thus, recombination for pathogenic race was evident among the individuals of a single oospore.

Abundant oospores were produced in single culture by two single-oospore isolates and 28 single-zoospore isolates established by Laviola (1968). Only three isolates lost this selfing ability upon serial transfer. Further studies are necessary to determine whether any of these are true homothallic isolates, whether they consist of intermingled hyphae of A^1 and A^2 types, or whether the A^1 and A^2 characters are present in the same hypha in a heterocaryotic condition.

ASEXUAL VARIATION.—Asexual variation in the fungi is generally attributed to mutation, heterocaryosis, parasexuality, physiological adaptation, and cytoplasmic control. The extent of variation in the genus *Phytophthora* has been reviewed thoroughly by Erwin et al. (1963). An excellent discussion of asexual variation in *Phytophthora infestans* has been presented by Caten and Jinks (1967). Asexual variation in regard to pathogenicity of *P. infestans* has been reviewed recently by Gallegly (1968).

It has been assumed that new pathogenic-race characters of *P. infestans* arise through mutation (Gallegly, 1968; Gallegly and Niederhauser, 1959) because the new R-gene-specific races are stable and do not revert to race 0 when cultured on a recessive host, and because such races first appear in nature only under severe epiphytotic conditions when the inoculum level is high. However, in the laboratory, new pathogenic races appear regularly after serial passage of an apathogenic race through senescent or juvenile leaf tissue of resistant potato hosts (Graham, Dionne, and Hodgson, 1961). In contrast, efforts to artificially induce pathogenic race mutants through the use of conventional mutagenic agents have met with little success (Wilde, 1961). The relative ease in securing new races through host passage may mean that the host-passage technique simply provides a method of working with large populations of the pathogen on resistant tissue where a specific mutant will grow when it appears.

Heterocaryosis is a possible mechanism for asexual variation in these Phycomycetes. However, anastomosis seems to be rare among them, especially in species of *Phytophthora*. It is possible for heterocaryons to originate from zoospore fusions (Hiddema and Kole, 1954), but the most likely way would be from mutations in certain nuclei which would be perpetuated along with the original nuclei. Gallegly and Eichenmuller (1959) explained the frequent appearance of the race 4 character in many isolates of *P. infestans* on this basis.

Parasexuality is not known in species of *Phytophthora* and *Pythium*. However, Sansome (1966) suggested that the asexual stages may be diploid. If so, the mitotic crossing over and nondisjunction might be involved in variation in these organisms. Sansome (1965) suggested that the continuous variation observed by Buddenhagen (1958) in single-zoospore cultures of *Phytophthora cactorum* could be explained by the diploid hypothesis.

Caten and Jinks (1967) concluded that the continuous variation in cultural characteristics among single-zoospore cultures of *P. infestans* could be explained best on the basis of cytoplasmic control. The results of Buddenhagen (1958) may perhaps also be explained on the basis of cytoplasmic control. Examination of the inheritance data presented above suggests that control of compatibility type may be cytoplasmic. Galindo and Gallegly (1960) pointed out that isolates of *P. infestans* varied in relative sexual strength. Some isolates were strong males, others strong females, and others intermediate in relative sexual strength. If compatibility type is under cytoplasmic control, presumably the cytoplasm of the oogonium, rather than that contributed by the antheridium during fertilization, would determine the type of cultures established from single oospores. Ratios of segregation for compatibility type would then be determined by the relative sexual strength of the parent isolates.

DISCUSSION OF INHERITANCE STUDIES.—Galindo and Gallegly (1960) assumed that *P. infestans* was haploid in its vegetative stages and therefore considered the compatibility types A^1 and A^2 to be allelic. The occurrence of the two types in Mexico (Gallegly and Galindo, 1958) in a ratio of 1:1 was cited as supportive evidence since the expected ratio of segregation in haploid organisms of monogenically controlled characters is 1:1. However, the data of Romero (1967) and Laviola (1968) with *P. infestans* showed a predominance of A^1 types in the progeny of A$^1 \times$ A^2 crosses. Galindo and Zentmeyer (1967) obtained the expected 1:1 ratio in progeny from crosses of *P. drechsleri*, whereas Satour and Butler (1968) obtained a 2:1 ratio in progeny of *P. capsici*. It is possible that compatibility type may be cytoplasmically controlled with individuals among the progeny having the same type as that of the parent which produced the oogonium. If compatibility type is under intranuclear control, another explanation for the predominance of the A^1 type must be found.

If species of *Phytophthora* prove to be diploid in their vegetative stages, as suggested by the cytological studies of Sansome (1965) and Galindo and Zentmyer (1967), a different interpretation of genetic control of compatibility type would be needed; if intranuclear and monogenically controlled, perhaps dominance and recessiveness might be involved. One type could be *aa* and the other *Aa*; *aa* × *Aa* should yield segregation ratios of 1:1 in the F$_1$.

Whereas the cytological studies indicate that species of *Phytophthora* may be diploid, the results from the inheritance studies are more indicative of haploidy in these species. If they were diploid, it seems only logical that one of the parents would be homozygous for at least one of the genetic markers studied. However, the segregation data showed recombination in the F$_1$ for every character. The data of Laviola (1968) particularly support the premise that these heterothallic species are haploid in their asexual stages.

In addition to observing recombination in the F$_1$,

and segregation ratios of 1:1 for the presence to absence of pathogenic race characters, Laviola (1968) obtained more than one phenotype from one oospore in six cases. His data strongly suggest that meiosis occurs following fertilization, and not before as suggested by Sansome (1965). Laviola's data further indicate that segregation may occur in both the first and second division. However, Laviola found that most commonly a single germinated oospore gave rise to only one phenotype, as found in every instance by previous workers who studied recombination in progeny from A $^1\times$ A^2 crosses of species of *Phytophthora*. The finding of only one phenotype per oospore has been used as evidence to support the diploid hypothesis. Although Galindo and Zentmyer's (1967) cytological observations supported the diploid hypothesis, their genetic data did not. They proposed that all but one of the meiotic nuclei possibly degenerated during oospore germination so that only one genotype was present among the zoospores of the germ sporangium of an oospore. Laviola's (1968) results certainly support this hypothesis as the usual behavior, but point out that in about one of ten cases more than one genotype can be detected among the individual cultures established from zoospores released by a germ sporangium. Thus, if degeneration occurs, it is apparent that occasionally not all of the products of meiosis degenerate, and that more than one meiotic nucleus may appear in the germ sporangium. If meiosis occurred prior to fertilization, and the resulting diploid nuclei divided mitotically, an explanation of more than one genotype being present in a germ sporangium would involve more than one fertilized nucleus per oospore.

Additional discussions of the inheritance studies of species of *Phytophthora*, and other aspects of the genetics of pathogenicity of *P. infestans*, have been presented by Gallegly (1968).

LITERATURE CITED

BERG, L. A. 1966. Influence of light and other factors on oospore germination in species of *Phytophthora*. Ph.D. dissertation, West Virginia Univ., Morgantown. 106 p.

BUDDENHAGEN, I. W. 1958. Induced mutations and variability in *Phytophthora cactorum*. Am. J. Botany 45:355–365.

CAMPBELL, W. A., and F. F. HENDRIX, JR. 1967. A new heterothallic *Pythium* from southeastern United States. Mycologia 59:274–278.

CATEN, C. E., and J. L. JINKS. 1967. Spontaneous variability of single isolates of *Phytophthora infestans*. I. Cultural variation. Can. J. Botany 46:329–348.

ERWIN, D. C., G. A. ZENTMYER, J. GALINDO, and J. S. NIEDERHAUSER. 1963. Variation in the genus *Phytophthora*. Ann. Rev. Phytopathol. 1:375–396.

GALINDO-A., J., and M. E. GALLEGLY. 1960. The nature of sexuality in *Phytophthora infestans*. Phytopathology 50:123–128.

GALINDO-A., J., and G. A. ZENTMYER. 1967. Genetical and cytological studies of *Phytophthora* strains pathogenic to pepper plants. Phytopathology 57:1300–1304.

GALLEGLY, M. E. 1968. Genetics of pathogenicity of *Phytophthora infestans*. Ann. Rev. Phytopathol. 6:375–396.

GALLEGLY, M. E., and J. J. EICHENMULLER. 1959. The spontaneous appearance of the potato race 4 character in cultures of *Phytophthora infestans*. Am. Potato J. 36:45–51.

GALLEGLY, M. E., and J. GALINDO. 1958. Mating types and oospores of *Phytophthora infestans* in nature in Mexico. Phytopathology 48:274–277.

GALLEGLY, M. E., and J. S. NIEDERHAUSER. 1959. Genetic controls of host-parasite interactions in the Phytophthora late blight disease, p. 168–182. *In* C. S. Holton et al. (ed.) Plant pathology–problems and progress 1908–1958. Univ. of Wisconsin Press, Madison, 588 p.

GOUGH, F. J. 1957. The sexual stage of *Phytophthora infestans*, the cause of potato and tomato late blight. Ph.D. dissertation, West Virginia Univ., Morganton. 46 p.

GRAHAM, K. M., L. A. DIONNE, and W. A. HODGSON. 1961. Mutability of *Phytophthora infestans* on blight-resistant selections of potato and tomato. Phytopathology 51:264–265.

HIDDEMA, J., and A. P. KOLE. 1954. Enkele waarnemingen over versmelten van zoosporen bis *Phytophthora infestans* (Mont.) de Bary. Tijdschr. Plantenziekten 60:138–139. (English Summary.)

LAVIOLA, C. 1968. Studies on the genetics of *Phytophthora infestans*. Ph.D. dissertation, West Virginia Univ., Morgantown, 74 p.

MARKS, G. E. 1965. The cytology of *Phytophthora infestans*. Chromosoma 16:681–692.

MIDDLETON, J. T. 1943. The taxonomy, host range and geographic distributions of the genus *Pythium*. Mem. Torrey Botan. Club 20:1–171.

PAPA, K. E., W. A. CAMPBELL, and F. F. HENDRIX, JR. 1967. Sexuality in *Pythium sylvaticum*: heterothallism. Mycologia 59:589–595.

ROMERO-C., S. 1967. Effects of genetic recombination on the pathogenicity of *Phytophthora infestans*. Ph.D. dissertation, Univ. of California, Riverside. 68 p.

ROMERO, S., and D. C. ERWIN. 1969. Variation in pathogenicity of progeny from germinated oospores of *Phytophthora infestans* (Mont.) de Bary. Phytopathology. (In press.)

SANSOME, EVA. 1961. Meiosis in the oogonium and antheridium of *Pythium debaryanum* Hesse. Nature (London) 191:827–828.

SANSOME, EVA. 1963. Meiosis in *Pythium debaryanum* Hesse and its significance in the life history of the biflagellatae. Trans. Brit. Mycol. Soc. 46:63–72.

SANSOME, EVA. 1965. Meiosis in diploid and polyploid sex organs of *Phytophthora* and *Achlya*. Cytologia 30:103–117.

SANSOME, EVA. 1966. Meiosis in the sex organs of the Oomycetes, Vol. I, p. 77–83. *In* C. D. Darlington and K. R. Lewis [ed.], Chromosomes today. Oliver and Boyd, Edinburgh, Scotland.

SATOUR, M. M., and E. E. BUTLER. 1968. Comparative morphological and physiological studies of the progenies from intraspecific matings of *Phytophthora capsici*. Phytopathology 58:183–192.

SAVAGE, E. J., C. W. CLAYTON, J. H. HUNTER, J. A. BRENNEMAN, C. LAVIOLA, and M. E. GALLEGLY. 1968. Homothallism, heterothallism, and interspecific hybridization in the genus *Phytophthora*. Phytopathology 58:1004–1021.

SAVAGE, E. J., and M. E. GALLEGLY. 1960. Problems with germination of oospores of *Phytophthora infestans*. Phytopathology 50:573. (Abstr.)

SMOOT, J. J., F. J. GOUGH, H. A. LAMEY, J. J. EICHENMULLER, and M. E. GALLEGLY. 1958. Production and germination of oospores of *Phytophthora infestans*. Phytopathology 48:165–171.

WILDE, P. 1961. Ein Beitrag zur Kenntnis der Variabilität von *Phytophthora infestans* (Mont.) de Bary. Arch. Microbiol. 40:163–195.

The Genetics of Asexual Phytopathogenic Fungi with Special Reference to Verticillium

A. C. HASTIE—*Botany Department, The University, Dundee, Scotland.*

Prospects for genetic analysis in asexual fungi were first reviewed by Pontecorvo (1956) where he suggested that the parasexual cycle could open the way to genetic studies with asexual plant pathogenic fungi. Ten years after Pontecorvo's original review, Roper (1966) concluded that nothing was known of the role, or even the existence, of the parasexual cycle in wild fungal populations. Parmeter, Snyder, and Reichle (1963) reached essentially the same conclusion regarding heterokaryosis in plant pathogenic fungi. Progress toward an appraisal of the role of heterokaryosis and parasexuality in the biology of asexual plant pathogenic fungi has been slow. This paper briefly reviews the limited progress made recently and proposes that an examination of the techniques used to study parasexual recombination in these species is necessary. The information which parasexual genetic analysis can provide may make little direct contribution to the control of plant diseases, but it is nevertheless vital to an appreciation of the general biology of the fungi responsible. This information will certainly not be provided by applying present techniques in the laboratory to an ever increasing range of species—a superficial treatment which has been in vogue over the past ten years.

PARASEXUAL RECOMBINATION IN ASPERGILLUS AND VERTICILLIUM.—The possibility of conducting genetic analysis through the parasexual cycle was first explored and developed with *Aspergillus nidulans* (Pontecorvo, 1956). The techniques developed with that species have since been widely adopted for studies with asexual plant pathogens. The essential steps in demonstrating the parasexual cycle are (1) isolation of complementary auxotrophic mutants; (2) synthesis and selection of a heterokaryon using these auxotrophs; (3) selection of heterozygous diploid strains from the heterokaryon; (4) recovery of diploid and haploid segregant genotypes from the heterozygous strain.

The subsequent analysis is based upon the information provided by the types and proportions of the various novel segegant genotypes (Pontecorvo, 1959). Most diploid segregants of *Aspergillus* are formed by mitotic recombination, which requires crossing over and therefore gives information about the rela-

tive positions of genes located in one linkage group. Formation of haploid segregants is by haploidisation, involving mitotic nondisjunction and therefore reassortment of whole chromosomes. It follows that haploid segregants give the information necessary for assigning particular genes to their linkage groups.

Both mitotic recombination and haploidisation are rather infrequent in *Aspergillus* and selective devices are employed to recover the segregants formed. Pontecorvo (1959) gave estimates of 1% and 0.1% respectively for the proportions of diploid and haploid segregants which are recovered in random samples of conidia from segregating *Aspergillus* diploids. More recent estimates by Kafer (1961) indicate that either mitotic recombination or mitotic nondisjunction occurs in about 1 in 25 vegetative nuclear divisions of *A. nidulans*. Certainly the two processes are sufficiently infrequent to make their coincident occurence in one nuclear lineage very rare during the growth of a colony.

The parasexual cycle of *Verticillium albo-atrum* has been described in outline by Hastie (1964). These studies together with more recent investigations show some differences in detail from the *Aspergillus* model. Heterokaryons of *Verticillium* are much less stable than those of *Aspergillus* when both are investigated at temperatures giving optimum vegetative growth. *Verticillium* hyphal cells are usually uninucleate (Caroselli, 1957) while those of *Aspergillus* are multinucleate (Clutterbuck and Roper, 1966), and this difference could obviously account for the more frequent segregation of *Verticillium* heterokaryons. In my experience the easiest method of demonstrating that heterokaryosis has occured in laboratory cultures of *Verticillium* is simply to search for heterozygous diploids. Their recovery implies at least a transient heterokaryotic phase.

Heterozygous diploids of *V. albo-atrum* are also less stable than those of *A. nidulans*. This relative instability has been detected using haploids derived from hops (Hastie, 1962) and from potato and tomato (Hastie, 1964). Isolates of *V. albo-atrum* from different geographical localities have also consistently shown this instability. All the *Verticillium* heterozygotes investigated form homozygous diploid segregants very frequently, and most of these are formed by mitotic recombination. These diploid segregants are easily found in random samples of conidia from segregating cultures (Fig. 1) and their formation can

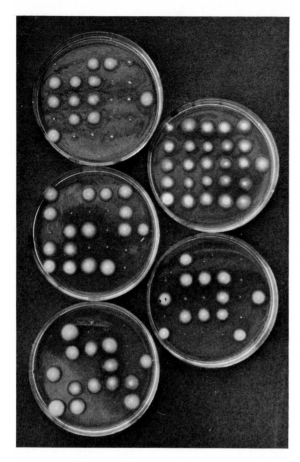

Fig. 1. Replicates of 26 random monoconidial isolates from a heterzygous diploid of *V. albo-atrum.* Top right, replica on complete medium. One of the four possible requirements is omitted from each of the other plates.

be followed in families of conidia borne on single conidiophore phialides (Hastie, 1967). The cytological events in phialides have been observed by MacGarvie and Isaac (1966) and by Hastie (1967). Each phialide is initially uninucleate. Nuclear division in a phialide is immediately followed by the formation of a uninucleate conidium so that the phialide again becomes uninucleate and the sequence is repeated. It follows that there is a simple arithmetical relationship between the number of nuclear divisions in a phialide and the number of conidia it bears. The conidia from single phialides (phialide families) can be isolated by micromanipulation, and segregation of genes because of mitotic recombination in the formation of the families can be detected. The information obtained makes chromosome mapping possible and gives a direct estimate of the frequency of mitotic recombination per nuclear division. Hastie (1967) has estimated a frequency of 0.021 for a gene affecting uracil syn-

thesis *(ur-1).* It has been roughly estimated that the general frequency of mitotic recombination affecting some part of the *V. albo-atrum* genome may be at least 0.2 per nuclear division. These frequencies refer to the mitotic recombination at nuclear divisions in phialides, and there is now some indication that the corresponding frequencies for nuclear divisions at conidial germination may be lower (Hastie, 1968).

Most diploids of *Verticillium* synthesised in the laboratory to date also appear to show higher frequency of mitotic nondisjunction than that found in *A. nidulans.* Random samples of about 100 conidia from a segregating two-week-old monoconidial culture contain few diploids (Hastie, 1964). The scarcity of diploids in these samples is due partly to frequent mitotic nondisjunction and partly to more rapid growth and prolific conidiation of haploid mycelium. Mitotic nondisjunction, like mitotic recombination, can also be detected in phialide families, but there are unresolved complications in estimating its frequency in these less stable *Verticillium* diploids.

Other *Verticillium* diploids synthesised by Ingram (1968) show very infrequent mitotic nondisjunction and therefore haploidisation. In that respect they resemble *Aspergillus* diploids, but mitotic recombination in them is frequent. These stable *Verticillium* diploids all have at least one parent derived from *V. dahliae* var. *longisporum,* and the significance of this will be discussed later.

In summary the main difference between the parasexual systems of *A. nidulans* and *V. albo-atrum* is in the frequencies with which new genotypes are formed. Pontecorvo (1959) has estimated, in general terms and under laboratory conditions, that the sexual cycle of *A. nidulans* is about five hundred times more active than the parasexual cycle in forming recombinant genotypes. The much higher frequency with which recombinants are formed parasexually in *V. albo-atrum* brings it closer to the activity of a sexual system. Little information is available on the frequency of segregational events in diploids of other imperfect plant pathogens, but there are certainly indications that the *Fusarium* system may be nearer to that of *V. albo-atrum.* Buxton (1956) found segregant genotypes in random asexual progeny of *F. oxysporum* f. sp. *pisi,* and Hoffmann (1966a) has found them in *F. oxysporum* f. sp. *callistephi.* Similarly, Sanderson and Srb (1965), and Tuveson and Coy (1961), did not require selective techniques to detect segregation in heterozygous diploids of *Ascochyta imperfecta* and *Cephalosporium mycophilum* respectively. These observations are in accord with the generalisation made by Pontecorvo (1959) that parasexual segregation seems to be more frequent in those fungi which apparently do not have a sexual stage.

THE PARASEXUAL CYCLE IN WILD POPULATIONS.— Until very recently nothing was known of the occurrence of the parasexual cycle in wild fungal populations. We have progressed only slightly from this position to the recognition of all the essential phases of the system in wild-type isolates of *Verticillium*. There is now no doubt that hyphal anastomosis occurs on natural substrates. Schreiber and Green (1966) have demonstrated hyphal fusions between germinating conidia of *V. albo-atrum* in contact with soil. This confirms the fact that *Verticillium* has at least the potential for transient heterokaryosis in natural environs—the minimum requirements for the formation of heterozygous diploids in these conditions.

Stark (1961) isolated from diseased horseradish a wild variant of *V. dahliae* with large uninucleate conidia. He named it *V. dahliae* var. *longisporum*. Relatively large uninucleate conidia are precisely what is expected of diploid strains, and Ingram (1968) has now given convincing indirect evidence that *V. dahliae* var. *longisporum* is a stable diploid. Her observations are indirect only in the sense that chromosome counts have so far proved impossible to obtain. Three observations support the hypothesis that it is diploid. First, Ingram (1968) has shown that it forms small-spored variants when grown on media containing p-fluorophenylalanine. This is a standard technique used to induce haploids of *A. nidulans* and other species (Lhoas, 1961), and the results with *V. dahliae* var. *longisporum* are clearly analogous. A second indirect test of diploidy was provided by a search for auxotrophic mutants from both type *V. dahliae* var. *longisporum* and also from the small-spored variants derived from it. Auxotrophic mutants were detected only in the latter. Experience with all fungi suggests that auxotrophic mutants are always recessive, and attempts to obtain dominant auxotrophs in diploid yeast have failed (Snow, as quoted by Day, 1966). This clearly agrees with the hypothesis that *V. dahliae* var. *longisporum* is diploid and auxotrophic mutations induced in it are concealed by their nonmutant dominant alleles. Further support for this hypothesis could be obtained by progeny-testing mutagen-treated *"longisporum"* conidia. This is essentially the procedure used by Snow (1966) in demonstrating that recessive auxotrophic mutants were induced in diploid *Saccharomyces cereviseae*. Finally Ingram (1968) has used complementary auxotrophs from *V. dahliae* var. *longisporum*, obtained indirectly in small-spored variants, to synthesise diploids which are morphologically indistinguishable from Stark's type culture. These diploids produce frequent mitotic recombinants, but are stable in that spontaneous haploids from them have not been detected.

Ingram (unpublished) has also succeeded in selecting heterozygotes composed of a haploid genome from *"longisporum"* and a haploid derived from either *V. dahliae* or *V. albo-atrum*. Diploids of both these origins behave as if heterozygous for genetic factors controlling the frequency of haploidisation. The precise genetic basis of this variation in stability is not yet known, but clearly the fact that it occurs may be of considerable importance in regard to the genetic structure and potentialities of wild populations. Kafer (1963) has shown how mitotic recombination in *Aspergillus* diploids heterozygous for chromosomal rearrangements may lead to the formation of unstable clones. Haploid segregants formed in these clones are at a considerable selective advantage, and quickly grow out. This situation resembles that seen with the relatively unstable *Verticillium* diploids, but linkage studies have not revealed the existence of translocations in our stocks. The currently available information on the diverse stability of *Verticillium* diploids does not, however, rigorously exclude their occurrence. It is possible that the diverse stability is a consequence of chromosomal rearrangements caused by the mutagen treatments used to obtain auxotrophs rather than a reflection of genetic variation in wild strains.

In Stark's (1961) original studies with *V. dahliae* var. *longisporum* he isolated it along with typical *V. dahliae*, and in recent correspondence he has pointed out that the two were sometimes obtained from the same root. It is tempting to speculate that Stark was here observing the occurence of spontaneous haploidisation in the wild diploid. Clearly this is a hypothesis which is easily tested by inoculation experiments. At least it can now be claimed there is minimal evidence for the operation of the parasexual cycle in wild *Verticillium* populations. Future work will show whether it is prevalent.

V. dahliae var. *longisporum* was detected as a large-spored variant because it is relatively stable. Other wild *Verticillium* diploids may be much less stable and therefore more difficult to find. The large-spored variant with multinucleate conidia reported by Roth and Brandt (1964) is the only other appropriate reference I have found in the literature, but reports of bimodal frequency distributions of conidial lengths do occur (e.g. Van den Ende, 1958) and this is the pattern of conidial dimensions one expects in samples of conidia from rapidly segregating *Verticillium* colonies. It would be naïve to pretend that ploidy is the only factor influencing conidial size; but it is the most convenient criterion of ploidy, and searches for large-spored variants would be a simple way of assessing the prevalence of the parasexual cycle in field conditions. Most mycologists and plant pathologists undoubtedly work with *Verticillium* isolates from diseased plants. Perhaps searches elsewhere, and also with other fungi, would be more fruitful. One can expect cytological evidence of parasexual phenomena at the nuclear level to be difficult to obtain in view of the size of fungal chromosomes and the uncertainty regarding the details of regular nuclear divisions in vegetative mycelium (Robinow and Bakerspigal, 1965). Bakerspigal (1965) could find no evidence in the many species he examined. I suggest that a comparison of *V. dahliae* var. *longisporum* and *V. dahliae* might yield interesting results.

GENETICS OF SURVIVAL AND SAPROPHYTIC GROWTH. —Survival of a potential pathogen, from a practical viewpoint, is the persistence of the organism over the period intervening between planting successive susceptible crops. This may involve either resting structures or a prolonged period of saprophytic growth.

Both these factors may operate in many species of root-infecting fungi, and Park (1965) has pointed out that fungi which persist by active growth in soil have greater opportunities for variation than those which rely entirely upon resting structures. Mutations may occur in resting structures, but other mechanisms like parasexual phenomena will be restricted to actively growing mycelium.

The capacity of stable heterokaryons to shelter recessive variation in culture led to speculation that heterokaryons may function in this way in the biology of soil fungi (Buxton, 1959). It is indeed conceivable that genotypes adapted to either a saprophytic or a parasitic habit may temporarily or permanently coexist in the same cytoplasm, the nuclear ratio changing in response to prevailing environmental factors. Parmeter, Snyder, and Reichle (1963) emphasised the lack of convincing evidence for the existence of heterokaryons in wild populations of phytopathogenic fungi, and they urged a critical reexamination of the issue. Limited progress in this direction can now be claimed although many recent papers on heterokaryosis in these fungi, like the earlier reports, refer to experiments in laboratory conditions with auxotrophic mutants. Stephan (1967) has demonstrated heterokaryosis in *Colletotrichum gloeosporioides* using wild morphological variants; Menzinger (1966) also used morphological traits to investigate heterokaryons of *Botrytis* spp.; Ming, Lin, and Yu (1966) similarly demonstrated heterokaryons in *Fusarium fujikuroi*, and Hoffmann (1966*a*) used auxotrophs to study heterokaryons in *Fusarium oxysporum* f. sp. *callistephi*.

Stable heterokaryons are clearly necessary if heterokaryosis is to offer a functional alternative to heterozygosity as a store of recessive variability. On any definition of heterokaryosis, stability is primarily dependent upon the number of nuclei per hyphal cell and the frequency of hyphal fusions. Hoffmann (1964), working with *F. oxysporum* f. sp. *callistephi* has shown that environmental factors such as temperature, pH and C/N ratio of the culture medium may influence the number of nuclei in hyphal cells. Generally, in temperatures promoting optimum vegetative growth, he found that hyphal tip cells were multinucleate (average 5.8 nuclei/cell) but most other cells uninucleate. At suboptimal temperatures there were more multinucleate cells. Hoffmann (1966*b*) has also examined the frequency of hyphal fusions in the same fungus on artificial media and has concluded that conditions for maintaining heterokaryons were far from ideal in these circumstances. We are all aware of the dangers in extrapolating from results obtained in pure culture, and there are additional hazards in attempting to predict the behaviour of heterokaryons from results obtained with homokaryons, but Hoffmann's contributions indicate that we may reasonably expect greater heterokaryon stability on some natural substrates than others.

There are probably many species of soil mycoflora of which certain genotypes persist as heterokaryons in albeit restricted environmental conditions and which

are therefore not detectable as heterokaryons in culture. The standard technique for proving heterokaryosis by isolating single hyphal tips has the limitations discussed by Davis (1966). Any appraisal of its importance in nature is also biased by the selection of species which have been studied. It is a reasonable expectation that the capacity of a fungus to form heterokaryotic conidia has an adaptive value for the species. In neo-Darwinian terms one expects the value to arise from the success of such fungi to disperse as heterokaryons and therefore, by extension, that heterokaryosis has particular significance in the biology of these fungi. This is a tenuous argument. I offer it merely to emphasise that, with two exceptions, the fungi used in the laboratory studies of heterokaryosis and parasexuality are not of this type. The exceptions known to me are *Helminthosporium sativum* (Tinline, 1962) and *Botrytis* spp. (Stephan, 1967). Fungi which form uninucleate (or homokaryotic) conidia are more convenient to use in genetic studies, but there is surely greater likelihood of finding an important role for heterokaryosis in fungi which form heterokaryotic conidia.

The report by Ming, Lin, and Yu (1966) of heterokaryosis in *Fusarium fujikuroi* is particularly interesting, for the wild heterokaryotic isolates they describe probably represent precisely the sort of model about which we have probably all speculated at some time or other, (e.g. Buxton, 1959). They describe three distinctly coloured strains—purple, red, and white—which propagate true to type from single uninucleate microconidia. These three forms differ significantly in the total amounts of gibberellin which they produce in culture. No attempt was made to distinguish the types of gibberellins involved. The white form produced very little gibberellin, the purple form at least five hundred times as much, and the red form was intermediate. Gibberellin production was closely correlated with virulence, and indeed the white form was practically avirulent. Special interest is attached to their observation that these three genotypes occur regularly as components of wild heterokaryotic isolates in China. Most heterokaryons they found contained only two of the three genotypes in a single hyphal tip. It is not stated by Ming, Lin, and Yu (1966) whether the original heterokaryotic isolates were from soil or from infected plants, and it would obviously be of further interest to know the frequency of heterokaryotic isolates from these different sources.

Spector and Phinney (1968), using the sexual form *Gibberella fujikuroi*, have demonstrated two genes controlling gibberellin synthesis, and there can be little doubt that the variation observed by Ming, Lin, and Yu (1966) had a nuclear basis. It is one of the few references to heterokaryosis with respect to natural variation in virulence in a largely asexual fungus. That heterokaryosis may be important in the same connection in other fungi awaits investigation.

There is no direct evidence of stable *Verticillium* heterokaryons in field conditions. Heterokaryons synthesised in the laboratory are unstable (Hastie, 1962,

Heale, 1966), but in view of Hoffmann's studies perhaps we should be cautious about generalising (Hoffmann, 1964, 1966b). Even if wild heterokaryons are unstable, they can nevertheless provide the necessary opportunity for nuclear fusions so that transient heterokaryosis may meet the requirements for genetic recombination.

Survival of *Verticillium* spp. in soil is a function of both resting structures and saprophytic growth. The former are especially important in regard to short-term survival, and the latter for survival over prolonged periods. Isaac (1967) has reviewed the evidence for this. Sewell and Wilson (1966) have found *V. albo-atrum* associated with dicotyledonous weeds which therefore promote long-term persistence of the fungus. However, a grass cover over 3–5 years, and associated with the suppression of dicot weeds, effectively reduced the population of *V. albo-atrum* although it was ineffective against *V. dahliae*.

There is no knowledge of the genetic basis of these species differences, and there is indeed disagreement over whether microsclerotial and dark mycelial types should be ranked as different species (Isaac, 1967). The evidence for differences in the potential survival of these forms in various environments is now compelling, and it is surely wrong to ignore the pleas of Sewell and Wilson (1966) and Isaac (1967) that they should be distinguished for obvious practical reasons. In the context of disease control, elementary practical considerations must outweigh the fact that hyphal fusions and genetic recombination can occur between the two forms (Fordyce and Green, 1964).

The dark resting mycelium of *V. albo-atrum* maintains its viability in soil for at least nine months under the conditions used by Heale and Isaac (1963). Hyaline variants, forming no resting mycelium, are rarely isolated from infected plants although they occur frequently in laboratory cultures. Isaac (1967) takes this, together with their reduced virulence on inoculation, as evidence of the important role of the dark mycelial in the biology of the fungus. The cultural variability of the trait has been studied frequently, and it is widely recognised that the expression of the trait is greatly influenced by cultural conditions (Robinson, Larson, and Walker, 1957; Heale and Isaac, 1965). My experience with auxotrophic mutants is that these also may influence its expression. The genetic factors responsible for the formation of the dark torulose hyphae are apparently complex, and extranuclear factors are involved. Information implicating extranuclear factors comes from both heterokaryons (Hastie, 1962; Heale, 1966) and heterozygous diploids (Hastie, 1962). The evidence from heterokaryosis follows the rationale developed by Jinks (1963) in connection with the almost unique opportunities which heterokaryosis offers for studying extranuclear inheritance. Essentially, if any given pair of mutually exclusive characters is entirely determined by nuclear material, then uninucleate (monoconidial) haploid segregants from an appropriate heterokaryon will exhibit either one or the other trait. This is consistently observed with nutri-

tional mutants of *Verticillium* but not for the presence and absence of dark torulose hyphae. Heterokaryons composed of complementary black and white auxotrophs show some reassortment of mycelial colour relative to the auxotrophic nuclear markers. Segregation from a heterozygous diploid selected from such a heterokaryon shows that mycelial colour is not inherited as a simple monogenic trait (Hastie, 1962). The prototrophic diploid had black hyphae, and the auxotrophic markers, which certainly had chromosomal loci, segregated regularly and frequently among asexual progeny. None of the segregants had exclusively white mycelium, and there was great variation amongst them, both regarding the time required for black hyphae to form, and also in the amount of dark mycelium each ultimately produced.

Certain laboratory mutants of *V. albo-atrum* show that pigmentation can be influenced independently of the morphogenetic requirements for the formation of torulose hyphae. These I refer to as *sooty* mutants to distinguish them from the wild black phenotype (Hastie, in press). There is not yet compelling biochemical evidence to prove that the pigment in sooty mutants is identical to that in wild type, but the fact that pigmentation is influenced by the same environmental factors in both types is suggestive. Apart from the absence of torulose hyphae in sooty strains, the main difference between sooty and black wild types is the time required for pigment to develop and the area of the colony which becomes pigmented (Fig. 2). Sooty mutants form completely black colonies in four or five days, and the pigmentation extends to all the

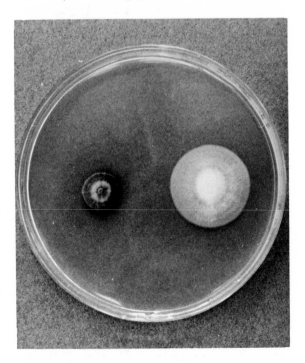

Fig. 2.   Wild type and sooty (black) mutant of *V. albo-atrum* after six days growth at 23°C.

hyphae, and finally to the hyphal tips when growth of the colony is arrested. This sooty character segregates as a monogenic trait, and its genetic linkage to an arginineless mutant is established (Hastie, in press).

PROSPECTS FOR THE GENETIC ANALYSIS OF ECOLOGICALLY IMPORTANT TRAITS.—Apart from the probable occurence and significance of the parasexual cycle in wild populations, it offers the possibility of conducting formal genetic analysis of ecologically important traits. In *Verticillium* these include variations in host range, virulence, survival in soil, and the *albo-atrum* dimorphism (or polymorphism). It is this possibility which has motivated most investigations into the genetics of asexual plant pathogens, and it has always been assumed that such analyses are possible with the techniques borrowed from studies with *Aspergillus* (Pontecorvo, 1959). I should like to think they are, but there are considerable obstacles to be overcome, at least with *Verticillium*, and there is evidence of their wider application.

The first and essential requirement for conducting a genetic analysis is getting heterozygous diploids, and it is at this point that I feel our techniques are perhaps unsuitable. The standard techniques, as already outlined, rely upon the use of induced auxotrophic mutants for the selection and identification of heterozygotes. It has long been known that these mutations often interfere with the development of a disease syndrome and therefore with the assessment

Fig. 3. Left and center, tomato plants infected with wild type *V. albo-atrum* and a methionine requiring mutant respectively. The plant on the right was not inoculated.

of pathogenicity and virulence as determined by other genes (Keitt and Boone, 1956; and see Fig. 3.). Furthermore, and vitally important, this interference cannot always be alleviated by applications of the appropriate metabolites to the host plant (Beraha, Garber, and Stromnaes, 1964). It was precisely this sort of interference which confused my attempts to

find segregants with new host specificities from synthetic diploids of *V. albo-atrum*. That this interference, apparently arising from pleiotrophic effects of auxotrophic mutations, is more generally applicable is evident from similar interference with resting structure formation in *Verticillium* and also with penicillin production in *Penicillium chrysogenum* (MacDonald, Hutchison, and Gillett, 1963). More than half the auxotrophs isolated in *Penicillium* give reduced penicillin yields when compared with the original strain, and as with virulence in other fungi, this yield is often not restored by supplying the required metabolites in quantities exceeding those necessary to give wild type growth.

In theory it is of course possible to avoid this interference by selecting out the auxotrophic alleles and scoring segregation of the ecological traits in the progeny of the then twice-selected diploids. Potential hosts or particular soil conditions could be utilised for this selection. No doubt such methods could be readily devised, but in practice they will be at best tedious, and with fungi like *Verticillium* exhibiting a high frequency of mitotic recombination, they may often be fruitless because of homozygosity established before the second selection is applied. They will be less tedious, and more fruitful either with fungi which have a lower frequency of mitotic recombination or in species with which smaller quantities of inoculum can be conveniently used in pathogenicity tests.

Wild populations offer an alternative source of diploids which do not incorporate auxotrophic mutants. Given their availability, they are likely to be homozygous in fungi which recombine freely. Ingram (unpublished) has found no evidence of heterozygosity in *V. dahliae* var. *longisporum*. It is of course possible that there are genes in wild *Verticillium* which influence the frequency of mitotic recombination, just as there appear to be genetic factors influencing the frequency of nondisjunction. Finding such a variant would encourage optimism, and it is worth noting that this in itself is an ecologically important trait.

My views on this aspect of my subject are deliberately pessimistic and are intended to provoke discussion of the techniques employed relative to the sort of information wanted. The ideal system for studying the segregation of wild-type traits would be one in which conidia could be hybridized without first obtaining auxotrophic mutants. These mutants may have some value in studying host-parasite interactions at the biochemical level, but despite optimistic hopes, they have not yet provided anything more than a microbiological assay of the readily available metabolites in the host. It is difficult to imagine what could replace them as selective markers for identifying heterozygotes in parasexual studies. Perhaps intensive biochemical studies will reveal natural variation which can be used in this way. Protoplasts prepared from some fungi tend to aggregate in suspension (Lopez-Belmonte, Garcia-Acha, and Villanueva, 1966). It would be interesting to see how these behave in hybridization experiments. Given that such a system yields heterozygous diploids, it would still be necessary to have a device for recognising them. Modifications of present techniques certainly seem desirable. The problem is perhaps analogous to that encountered in applying parasexual techniques to mammalian somatic cell cultures; it is essentially a matter of finding the right markers (Pontecorvo, 1962).

LITERATURE CITED

BAKERSPIGAL, A. 1965. Cytological investigation of the parasexual cycle in fungi. I. Nuclear fusion. Mycopathol. Mycol. Appl. 26:233–240.

BERAHA, L., E. D. GARBER, and O. STROMNAES. 1964. Genetics of phytopathogenic fungi. Virulence of colour and nutritionally deficient mutants of *Penicillium italicum* and *Penicillium digitatum*. Can. J. Botany 42:429–436.

BUXTON, E. W. 1956. Heterokaryosis and parasexual recombination in pathogenic strains of *Fusarium oxysporum*. J. Gen. Microbiol. 15:133–139.

BUXTON, E. W. 1959. Mechanisms of variation in *Fusarium oxysporum* in relation to host parasite interactions, p. 183–191. *In* C. S. Holton et al. [ed.] Plant Pathology—problems and progress, 1908–1958. Univ. of Wisconsin Press, Madison.

CAROSELLI, N. E. 1957. Verticillium wilt of Maple. Rhode Island Agr. Exp. Sta. Bull. 335.

CLUTTERBUCK, A. J., and J. A. ROPER. 1966. A direct determination of nuclear distribution in heterokaryons of *Aspergillus nidulans*. Genet. Res. 7:185–194.

DAVIS, R. H. 1966. Heterokaryosis, Vol. II, p. 567–586. *In* G. C. Ainsworth and A. S. Sussman [ed.], The fungi. Academic Press, New York.

DAY, P. R. 1966. Recent developments in the genetics of the host-parasite system. Ann. Rev. Phytopathol. 4:245–268.

FORDYCE, C., and R. J. GREEN. 1964. Mechanisms of variation in *Verticillium albo-atrum*. Phytopathology 54:795–798.

HASTIE, A. C. 1962. Genetic recombination in the hop wilt fungus *Verticillium albo-atrum*. J. Gen. Microbiol. 27:373–382.

HASTIE, A. C. 1964. The parasexual cycle in *Verticillium albo-atrum*. Genet. Res. 5:305–315.

HASTIE, A. C. 1967. Mitotic recombination in conidiophores of *Verticillium albo-atrum*. Nature (London) 214:249–252.

HASTIE, A. C. 1968. Phialide analysis of mitotic recombination in *Verticillium*. Molec. Gen. Genetics 102:232–240.

HEALE, J. B. 1966. Heterokaryon synthesis and morphogenesis in *Verticillium*. J. Gen. Microbiol. 45:419–427.

HEALE, J. B., and I. ISAAC. 1963. Wilt of lucerne caused by species of *Verticillium*. IV. Pathogenicity of *V. albo-atrum* and *V. dahliae* to lucerne and other crops; spread and survival of *V. albo-atrum* in soil and weeds; effect upon lucerne production. Ann. Appl. Biol. 52:439–451.

HEALE, J. B. and I. ISAAC. 1965. Environmental factors in the production of dark resting structures in *Verticillium albo-atrum*, *V. dahliae* and *V. Tricorpus*. Trans. Brit. Mycol. Soc. 48:39–50.

HOFFMANN, G. M. 1964. Untersuchungen über die Kernverhaltnisse bei *Fusarium oxysporum* f. *callistephi*. Arch. Mikrobiol. 49:51–63.

HOFFMANN, G. M. 1966a. Untersuchungen über die heterokaryosebildung und den parasexualcyclus bei *Fusarium oxysporum*. III. Paarungsversuche mit auxotrophen mutanten von *Fusarium oxysporum* f. *callistephi*. Arch. Mikrobiol. 56:40–59.

HOFFMANN, G. M. 1966b. Untersuchungen über die heterokaryosebildung und den parasexualcyclus bei *Fusarium oxysporum*. I. Anastomosenbildung im mycel

und kernverhaltnisse bei der conidienentwicklung. Arch. Mikrobiol. 53:331–347.

INGRAM, RUTH. 1968. *Verticillium dahliae* Kleb. var. *longisporum* Stark: a stable diploid. Trans. Brit. Mycol. Soc. 51:339–341.

ISAAC, I. 1967. Speciation in *Verticillium*. Ann. Rev. Phytopathol. 5:201–222.

JINKS, J. 1963. Cytoplasmic inheritance in fungi, p. 325–354. *In* W. J. Burdette [ed.], Methodology in basic genetics. Holden-Day, Inc.

KAFER, ETTA. 1961. The processes of spontaneous recombination in vegetative nuclei of *Aspergillus nidulans*. Genetics 46:1581–1609.

KAFER, ETTA. 1963. Radiation effects and mitotic recombination in diploids of *Aspergillus nidulans*. Genetics 48:27–45.

KEITT, G. W., and D. M. BOONE. 1956. *In* Genetics and plant breeding. Brookhaven Symp. Biol. No. 9:209–225.

LHOAS, P. 1961. Mitotic haploidisation by treatment of *Aspergillus niger* with para-fluorophenylalanine. Nature (London) 190:744.

LOPEZ-BELMONTE, F., I. GARCIA-ACHA, and J. R. VILLANUEVA. 1966. Observations on the protoplasts of *Fusarium culmorum* and their fusion. J. Gen. Microbiol. 45:127–134.

MacDONALD, K. D., J. M. HUTCHISON, and W. A. GILLETT. 1963. Isolation of auxotrophs of *Penicillium chrysogenum* and their penicillin yields. J. Gen. Microbiol. 33:365–374.

MacGARVIE, Q. D., and I. ISAAC. 1966. Structure and behaviour of the nuclei of *Verticillium* spp. Trans. Brit. Mycol. Soc. 49:687–693.

MENZINGER, W. 1966. Zur variabilität und Taxonomie von arten und formen der gattung *Botrytis* Mich. II. Untersuchungen zur variabilität des Kulturtyps under konstanten kulturbedingungen. Zentralbl. Bakteriol., Parasitenk., Infektionskrankh.-u. Hyg. II. Abt. 120:179–196.

MING, Y. N., P. C. LIN, and T. F. YU. 1966. Heterokaryosis in *Fusarium fujikuroi*. (Sacc.) Wr. Scientia Sinica XV:371–378.

PARK, D. 1965. Survival of microorganisms in soil, p. 82–98. *In* K. F. Baker and W. C. Snyder [ed.], Ecology of soil-borne plant pathogens. Univ. of California Press, Berkeley and Los Angeles.

PARMETER, J. R., W. C. SNYDER, and R. E. REICHLE. 1963. Heterokaryosis and variability in plant pathogenic fungi. Ann. Rev. Phytopathol. 1:51–76.

PONTECORVO, G. 1956. The parasexual cycle in fungi. Ann. Rev. Microbiol. 10:343–400.

PONTECORVO, G. 1959. Trends in genetic analysis. Columbia University Press, New York.

PONTECORVO, G. 1962. Methods in microbial genetics in an approach to human genetics. Brit. Med. Bull. 18:81–84.

ROBINOW, C. F., and A. BAKERSPIGAL. 1965. Somatic nuclei and forms of mitosis in fungi, Vol. I, p. 119–142. *In* G. C. Ainsworth and A. S. Sussman [ed.], The fungi. Academic Press, New York.

ROBINSON, D. B., R. M. LARSON, and J. C. WALKER. 1957. Verticillium wilt of potato in relation to epidemiology, symptoms and variability of the pathogen. Univ. of Wisconsin Res. Bull. 202.

ROPER, J. A. 1966. The parasexual cycle, Vol. II, p. 589–617. *In* G. C. Ainsworth and A. S. Sussman [ed.], The fungi. Academic Press, New York.

ROTH, J. N., and W. H. BRANDT. 1964. Nuclei in spores and mycelium of *Verticillium*. Phytopathology 54:363–364.

SANDERSON, K. E., and A. M. SRB. 1965. Heterokaryosis and parasexuality in the fungus *Ascochyta imperfecta*. Am. J. Botany 52:72–81.

SCHRIEBER, L. R., and R. J. GREEN. 1966. Anastomosis

in *Verticillium albo-atrum* in soil. Phytopathology 56: 110.

SEWELL, G. W. F., and J. F. WILSON. 1966. *Verticillium* wilt of hop: The survival of *Verticillium albo-atrum* in soil. Ann. Appl. Biol. 58:241–249.

SNOW, R. 1966. An enrichment method for auxotrophic yeast mutants using the antibiotic nystatin. Nature (London) 211:206–207.

SPECTOR, C., and B. O. PHINNEY. 1968. Gibberellin biosynthesis: Genetic studies in *Gibberella fujikuroi*. Physiol. Plant. 21:127–136.

STARK, C. 1961. Das auftreten der *Verticillium* tracheomykosen in Hamburg gartenbaukulturen. Die Gartenbaukulturen. Die Gartenbauwissenschaft 26:493–528.

STEPHAN, B. R. 1967. Untersuchungen über die variabilität bei *Colletotrichum gloeosporioides* Penzig in verbindung mit heterokaryose. III. Versuche zum nachweis der heterokaryose. Zentrabl. Bakteriol., Parasitenk., Infekionskrankh.-u Hyg., III Abt. 121:73–83.

TINLINE, R. D. 1962. *Cochliobolus sativum*. V. Heterokaryosis and parasexuality. Can. J. Botany 40:425–437.

TUVESON, R. W., and D. O. COY. 1961. Heterokaryosis and somatic recombination in *Cephalosporium mycophilum*. Mycologia 53:244–253.

VAN DEN ENDE, G. 1958. Untersuchungen über der phlanzenparasiten *Verticillium albo-atrum*. Acta Botan. Neerl. 7:665–740.

Mechanisms of Variation in Culture and Soil

J. R. PARMETER, JR.—*Department of Plant Pathology, University of California, Berkeley.*

Variation is the rule in most root-infecting fungi. In order to understand the biology and pathology of these fungi and to anticipate new variants that might arise following changes in crop culture, genetic modification of hosts, physical or chemical modification of the soil environment, or accidental introduction of new genetic material into a regional or local gene pool, it is necessary for us to understand how variants arise and how they are propagated and disseminated.

The mechanisms of genetic variation, excluding adaptation and cytoplasmic inheritance, can be conveniently discussed under four broad headings: (1) mutations, (2) sexual and parasexual mechanisms of genetic recombination, (3) mechanisms of plasmogamy and nuclear reassortment that permit (or preclude) sexual or parasexual recombination, and (4) mechanisms by which genetic diversity is conserved and expressed. Many aspects of these mechanisms have recently been reviewed and discussed in detail (Caten and Jinks, 1966; Davis, 1966; Emerson, 1966; Esser, 1966; Esser and Kuenen, 1967; Fincham and Day, 1965; Jinks, 1966; Raper, 1966a, b; Roper, 1966), and further review at this time is superfluous. I would like here to range briefly over the entire field, emphasizing aspects that have received little previous comment or that have special implications in the study of root-infecting fungi.

MUTATION AND GENETIC RECOMBINATION.—We can accept a priori that mutations are common to all fungi and ultimately underlie most of the observed variation. It is not known whether the "protective" soil environment reduces the likelihood of mutation or whether chemical and biological soil factors might increase mutation. Regardless of the rate at which it has arisen, extensive genetic diversity is characteristic of most, if not all, root-infecting fungi. Our main concern as plant pathologists is the "strategy" adopted by these fungi to "utilize" genetic diversity and how knowledge of the pathogen's strategy can aid in the study and control of plant disease.

Hypothetically, the simplest strategy would involve accumulation of mutations in a regularly haploid, imperfect fungus capable of prolific asexual sporulation. Until recently, this appeared to be the only possible strategy open to many fungi without perfect states. Single mutations affecting such characteristics as host range (Yang and Hagedorn, 1968) and chemical tolerance (Georgopoulos and Panopoulos, 1966) have been demonstrated; however, as pointed out by Flentje and Hastie (this volume), most mutations adversely affect basic processes. Undoubtedly, new pathogenic or ecologic capabilities arise occasionally from single mutations, but geneticists generally agree that mutation without additional mechanisms for genetic recombination is inadequate to account for most of the observed variations in pathogenic or ecologic behavior (Roper, 1966).

SEXUAL AND PARASEXUAL MECHANISMS.—Sexual and parasexual mechanisms per se in fungi are discussed in detail by Emerson (1966), Esser and Kuenen (1967), Fincham and Day (1965), and Roper (1966), and require little comment. Increasing evidence indicates that parasexual phenomena may be common in fungi and that these phenomena and those associated with sexual recombination provide the basic processes of evolution. Opportunities for sexual or parasexual genetic recombination depend on a variety of mechanisms governing plasmogamy and nuclear reassortment.

PLASMOGAMY AND NUCLEAR REASSORTMENT.—The capacity for genetic recombination or nuclear reassortment in root-infecting fungi, so far as we know, rests entirely with anastomosis, a process about which little is known. Whether anastomosis involves special sexual structures or unspecialized hyphae, cell contact, cell-wall dissolution, and plasmogamy appear essential to genetic interaction, as indicated by Flentje (this volume). Unfortunately, anastomosis is seldom considered when compatibility is discussed, and it is therefore difficult to determine to what extent anastomosis behavior mediates genetic exchange or at what phase of anastomosis incompatibility might be expressed.

Tropic responses have been observed frequently between hyphae of various fungi in cultures. We do not know if such responses are essential to anastomosis or if they merely increase the frequency of potential anastomosis encounters. Nor do we know if tropic responses operate well in soil or in plant tissue. The phenomenon exists, and its likely function in specialized sexual structures has been demonstrated (see Esser and Kuenen, 1967).

Following hyphal encounters in *Thanatephorus cucumeris*, further growth may be arrested on contact, or hyphae may grow around each other and continue terminal growth. Presumably, arrested tip growth and perhaps some form of "attachment" are necessary preludes to cell-wall dissolution, but mechanisms that might arrest growth or promote attachment have not been described.

Cell-wall dissolution must involve special physiological mechanisms that are operative only under special circumstances. Otherwise, cell-wall integrity would be difficult to maintain. The factors triggering mutual cell-wall dissolution at points of contact are unknown, but they undoubtedly affect the capacity to anastomose and, therefore, to exchange nuclei. Failure of cell-wall dissolution certainly constitutes a potential incompatibility mechanism. Work with *T. cucumeris* (see Flentje, this volume) indicates also that the cytoplasm of fused cells may fail to fuse, presumably because the cytoplasmic membranes remain intact. This would constitute an additional incompatibility mechanism, but no detailed investigations have been made.

Present concepts of compatibility (Esser, 1966; Esser and Kuenen, 1967; Raper, 1966a,b) are based mainly on the assumption that within a species hyphae anastomose freely and that incompatibility results from cytoplasmic or genetic phenomena that follow anastomosis. Failure to anastomose is usually considered an indication that mycelia represent different species (see Parmeter, Snyder, and Reichle, 1963). Work with *T. cucumeris* (Flentje, this volume; Kernkamp, et al., 1953; Richter and Schneider, 1953; Schultz, 1936) and *Ceratobasidium* sp. (Parmeter, Whitney, and Platt, 1967) indicates that intraspecific anastomosis failure is common in these species. Present evidence (Parmeter and Sherwood, unpublished) suggests that *T. cucumeris* includes several gene pools genetically isolated by anastomosis failure, and Flentje (this volume) has shown that some sibling isolates from the same parent may fail to anastomose.

Burnett (1965), in discussing intersterility among populations within a species, cites examples of restricted anastomosis between members of different populations, and Tinline (1962) indicates that anastomosis between hyphae of opposite mating types is rare in *Cochliobolus sativus*. The study of anastomosis failure as an isolating mechanism and perhaps as an indication of evolutionary divergence deserves emphasis. Flentje (in press) has suggested that genetic isolation may involve a series of steps from compatible anastomosis, to anastomosis with cell killing, to anastomosis without cytoplasmic fusion, to failure of cell-wall fusion and absolute isolation. Presently we have little information on anastomosis failure in root-infecting fungi. Clarification of anastomosis capacity within and between species of these fungi would provide a needed basis for assessing potential genetic interaction.

The study of anastomosis in root-infecting fungi may be complicated by environmental influences. Earlier studies (see Parmeter, Snyder, and Reichle, 1963) and recent studies by Hoffmann (1964) and McKenzie (1966) indicate that such factors as temperature, pH, and nutrient status may markedly influence the success or failure of anastomosis. Schreiber and Green (1966) have shown that materials in soil may promote anastomosis in *Verticillium albo-atrum* at some levels, but apparently inhibit it at others. Until the genetic and environmental factors mediating anastomosis in root-infecting fungi have been adequately investigated, our knowledge of compatibility and the operation of compatibility factors in field populations can only be considered fragmentary.

Where anastomosis occurs among isolates, mechanisms affecting sexual or heterokaryon compatibility may restrict nuclear exchange or genetic recombination. These mechanisms have been thoroughly reviewed in recent papers (Davis, 1966; Esser, 1966; Esser and Kuenen, 1967; Raper, 1966a, b). Regrettably, few workers have considered compatibility in root-infecting fungi, and the situation is not clear in those that have been studied, especially since the mechanisms conditioning heterokaryon compatibility in the vegetative state may not be the same as those conditioning sexual compatibility.

Most of the studies involving root-infecting fungi have relied on formation of the perfect state as a criterion of compatibility. As discussed by Davis (1966), vegetative heterokaryons occur among like mating types in *Neurospora crassa*. Thus, crosses compatible for sexual recombination are incompatible for heterokaryon formation. If similar compatibility patterns are common to root-infecting Ascomycetes, then isolates considered incompatible on the basis of perfect-state formation may be compatible on the basis of heterokaryon formation and parasexuality. Until details of incompatibility in these fungi are clearly understood, the capacity for genetic recombination or nuclear reassortment is difficult to assess.

T. cucumeris, a basically homothallic fungus, provides a good example of compatibility complexities involving failure of cell-wall fusion, failure of cytoplasmic fusion, and death of fused cells. Flentje (this volume), Whitney and Parmeter (1963) suggested a bipolar heterokaryon compatibility among sibling isolates. Vest and Anderson (1968) indicated that multiple alleles were involved. Flentje, Stretton, and Parmeter (unpublished) found that heterokaryon compatibility among many isolates could not be explained by any simple bipolar or tetrapolar system.

Among other root-infecting fungi, the mechanisms moderating the capacity to produce sexual recombinants or heterokaryons and parasexual recombinants are even less clearly understood. Virtually nothing is known about compatibility factors in such widespread and damaging fungi as *Armillaria mellea*, *Fomes annosus*, *F. lignosus*, *Polyporus tomentosus*, or *Helicobasidium purpureum*, to name only a few important Basidiomycetes. Recent studies showing hyphal aversion among four strains of *Sclerotium rolfsii* (Weera-pat and Schroeder, 1966) suggest that anastomosis failure may occur in this species.

Few root-infecting ascomycetous or imperfect fungi have been studied. Available evidence (Hastie, this volume) suggests that anastomosis and parasexuality are common among isolates of *Verticillium dahliae* and *V. albo-atrum*. Details on possible incompatibility factors are yet wanting. Heterokaryosis and parasexuality have also been demonstrated in *Fusarium oxysporum*, *F. solani*, *F. fujikuroi*, *Helminthosporium sativum*, *Ascochyta imperfecta*, and *Cephalosporium mycophilum* (Hastie, this volume). Again, details on possible compatibility factors are needed. Nelson and Kline (1964) have shown that 11 among 79 cultures of *Helminthosporium* with similar conidial morphology were sexually isolated, while 68 could be divided into two compatibility types with varying degrees of fertility. Whether the 11 cultures were isolated by anastomosis failure or postanastomosis incompatibility was not indicated.

The situation among Phycomycetes is equally confused. As indicated by Gallegly (this volume), intraspecific matings between the two recognized compatibility types are unrestricted in heterothallic species. This suggests that anastomosis, at least between sexual organs of compatible types, is unrestricted. Mechanisms preventing interaction in incompatible mating types are unknown. Anastomosis of vegetative hyphae has not been studied, but the possibility that vegetative hyphae are diploid presents an interesting aspect for study. Can a hypha carry two distinct diploid genomes? If not, is such a form of heterokaryosis prevented by anastomosis failure or some postanastomosis mechanism? Evidence that more than two genomes from different sources can exist compatibly in a hypha is limited (see Parmeter, Snyder, and Reichle, 1963).

This brief discussion of mechanisms affecting nuclear exchange and genetic interaction serves to point out that we still know very little about the factors that govern them in root-infecting fungi. The need for such knowledge is more than academic. Our ability to understand the present occurrence and distribution of variants and to anticipate potentially damaging changes in pathogenic or ecologic behavior of root-infecting fungi depends on this knowledge.

Present limited information on discontinuities in the geographic distribution of compatibility factors emphasizes the need for more study. Burnett (1965) cites a number of examples, mostly among higher Basidiomycetes, of geographic discontinuity in the occurrence of intersterile groups within species; sympatric occurrence of intersterile groups is also described. Among root-infecting fungi, Schippers and Snyder (1967) have shown that sexual mating types of *F.* sp. *cucurbitae* race I include only ♀ + isolates in California, ♂ − isolates in the Netherlands, and ♂ + isolates in Australia. Perithecia are unknown in nature. It is not known whether there is a center of origin where both mating types occur or whether these three occurrences represent independent origins. Similarly,

Cook, Ford, and Snyder (1967) have shown that sex and mating types of *F. solani* f. sp. *pisi* are discontinuously distributed and that knowledge of discontinuities is useful in determining the likely pattern of spread by man from one geographic area to another. Mating types appear to be geographically separated in *F. rigidiusculum* f. sp. *theobromae* (Ford, Bourret, and Snyder, 1967). Mating types in *Phytophthora infestans* show a similar discontinuity (Emerson, 1966), supposedly resulting from introduction into Europe and North America of only one mating type from the presumed center of origin in Central or South America.

Future studies with root-infecting fungi will no doubt show discontinuities in the distributions of compatibility types of many species. These might have arisen either from evolutionary divergence or from introductions by man. In either case, delineation of discontinuities may provide valuable information about past spread of root-infecting fungi by man and may also point up dangers that might result from future introductions of exotic isolates with the potential to interact genetically with local isolates.

Compatibility studies may also provide a valuable basis for ecological investigations among various hymenomycetes, many of which infect roots. Adams and Roth (1967) discuss some previous attempts to use the interaction of paired isolates as a method of identifying clones and their distribution in nature. Unfortunately, the nature of the "barrage" or "line of demarcation" between dikaryons or between di-mon pairings is not clearly understood. Clarification of this phenomenon might provide a better understanding of the interrelationships among diverse isolates on the local or intercontinental level.

Care in evaluating the significance of discontinuities in the occurrence of compatibility types is important. If, as appears to be the case with some Ascomycetes, sexual compatibility reaction is the opposite of heterokaryon compatibility reaction, then introduction of isolates of the same mating type, while leading to no danger of sexual recombinations, may lead to parasexual recombinations. Clearly, all of the details of compatibility must be understood before we can say that we know the biology of a root-infecting fungus.

CONSERVATION AND EXPRESSION OF VARIATIONS.—The previous section has dealt with origins of variation and the mechanisms promoting or precluding genetic recombination or nuclear reassortment. Knowledge of mechanisms by which variations are conserved and expressed is also important in understanding the biology of root-infecting fungi. These mechanisms generally must involve some form of mosaic, heterokaryotic, or diploid association, since mutations frequently are recessive and detrimental and would rapidly be eliminated from any population restricted to homokaryotic, haploid vegetative growth.

Little is known about conservation of genetic variation in root-infecting fungi. Basidiomycetes appear characteristically to grow vegetatively as heterokaryons, excepting monokaryotic phases in rust fungi, with either regularly binucleate or multinucleate hyphal cells. Nuclear balance is maintained through the mechanism of dikaryotic pairing (Raper, 1966a). Ascomycetous and imperfect fungi apparently lack a mechanism for maintaining nuclear balance in the vegetative state, and heterokaryons are therefore usually unstable. A mosaic condition arising from repeated anastomosis among homokaryotic branch hyphae appears to be more characteristic of these fungi (Davis, 1966). Limited evidence suggests that diploidy may also occur commonly in hyphae of root-infecting fungi (Brushaber, Wilson, and Aist, 1967) and that stable diploids may exist (Hastie, this volume).

Evidence suggests that vegetative diploidy in *Pythium* and *Phytophthora* (Gallegly, this volume) may conserve genetic variants in the thallus. None of the above associations has been investigated in sufficient depth, and nuclear behavior patterns are not well established.

Dissociation of heterokaryon components may be essential to survival. We have found recently in *T. cucumeris* that certain heterokaryons are tolerant to PCNB, but that such heterokaryons are slow growing and would not likely compete with faster growing isolates in the absence of PCNB. As long as PCNB is present, the nuclear association is stable, but in the absence of PCNB, faster growing homokaryotic components quickly sector out and dominate culture plates. No doubt there are many nuclear associations among root-infecting fungi that have survival value in one situation but not in another. The capacity to dissociate, therefore, may be a critical mechanism conferring flexibility in the expression of variation.

At least four possible mechanisms for nuclear dissociation are apparent: (a) formation of conidia with one or only a few of the possible variant nuclei present; (b) failure of clamp connections to fuse with the preceding cell, resulting in independent homokaryotic growth of the clamp process; (c) septal partitioning of hyphae into cells with fewer nuclei than occur in advancing tip cells; and (d) adjustment of nuclear ratios to produce a shift toward the predominance of one genome. Again, information on these mechanisms in root-infecting fungi is very limited.

Dissociation of nuclei by means of uninucleate conidia (or conidia derived from uninucleate conidiophores) has been demonstrated repeatedly among ascomycetous and imperfect fungi, including a number of root-infecting fungi (Davis, 1966). Conidia are rarely produced by root-infecting Basidiomycetes, and fungi in the Mycelia Sterilia lack this mechanism for nuclear dissociation. The situation in Oomycetes needs clarification (Caten and Jinks, 1968; Galindo and Zentmyer, 1967). The possibility of stable diploidy in the vegetative state suggests that vegetative dissociation of haploid genomes during sporangial or zoospore formation is unlikely.

Independent growth of clamp-connection processes, as described by Miles and Raper (1956), might serve as a mechanism of dissociation in such root-

infecting fungi as *Sclerotium rolfsii* or *Rhizoctonia carotae*, but this has not been demonstrated.

Evidence indicates that partitioning of multinucleate cells by secondary septa provides a mechanism of dissociation in the multinucleate *T. cucumeris*. Flentje and Stretton (1964) have shown that short cells from fragmented colonies yield variant cultures, whereas hyphal-tip isolates yield colonies like the parent. These short cells contain fewer nuclei than do cells without secondary septa, and apparently different combinations of the several nuclei that constitute the normal complement can be released by germination of the short cells. The extent to which this mechanism may operate among other species with normally multinucleate hyphal cells is unknown.

Among ascomycetous and imperfect fungi, sectoring is common. Since these fungi apparently lack a mechanism for insuring dikaryotic pairing, and since they are often quite irregular with regard to the nuclear condition of hyphal cells, accidental separation of nuclear components during septation must be common. In those forms, such as *Verticillium* spp. (MacGarvie and Isaac, 1966), possessing multinucleate hyphal-tip cells and predominately uninucleate cells behind the tip, separation of nuclei is certain. Branching and growth of such uninucleate cells must lead to nuclear dissociation. Hoffmann (1964) has shown that temperature, pH, and substrate may affect the numbers of nuclei in cells of *F. oxysporum* f. sp. *callistephi*. This suggests that the soil environment might influence the capacity of some fungi to sector out nuclear components.

Differential adjustment of nuclear ratios, as discussed by Davis (1966), may also provide a mechanism for dissociation, given selection pressure for a change in ratio. Information about this phenomenon is less than adequate to determine how ratios might be shifted in single cells such as multinucleate hyphal tips. Presumably, nuclei might divide at different rates. However, nuclear ratios measured by conidial isolation or by isolating from interior cells of a colony might reflect differential growth of homokaryotic branches. This phenomenon certainly warrants additional study in root-infecting fungi.

The above information suggests that the conservation and expression of genetic variation in many fungi involves some sort of balance between factors leading to the formation of heterokaryons and factors that promote dissociation of the nuclei in heterokaryons. Whether vegetative heterokaryosis serves mainly to provide for nuclear reassortment and conservation of genetic diversity or for parasexual recombination, a mechanism for the liberation of nuclei is essential to the flexible expression of variants carried in heterokaryotic hyphae. Aspects of conservation and expression of variants in root-infecting fungi in nature are poorly outlined at present, and we know virtually nothing about the moderating influence of the soil environment on these aspects.

It has been suggested that heterokaryotic thalli of many fungi, especially ascomycetous and imperfect fungi, may consist mainly of mosaics of cells containing different nuclei but connected by anastomosis into a continuous cytoplasm in which the different nuclei contribute to the whole through cytoplasmic streaming (Davis, 1966). Such a system requires the maintenance of thallus integrity, which may be more difficult in soil than in the laboratory. In soil, mites, nematodes, insects, parasitic fungi, and other agents destroy cells, and it is likely that thallus integrity is jeopardized. Unfortunately, we have no clear picture of colony character or integrity in soil.

Apparent dissociation of nuclei in cells partitioned by secondary septa has been demonstrated in the laboratory by fragmenting and replating thalli. Replating is necessary presumably because "staling" products accumulate in the medium and inhibit growth of interior cells. Whether such "staling" products accumulate in soil substrates is questionable. If growth from old cells is less inhibited in soil, dissociations of this kind may be much more common than those which occur on laboratory medium. One type of apparent dissociation deserves special mention. The dual phenomenon, first described by Hansen (1938), is a common laboratory phenomenon that has never received adequate explanation. Hansen and Snyder (1943) and Bistis (1959, 1961) suggest that M types arise by mutation. Jinks (1960) indicates that extranuclear determinants may underlie the dual phenomenon. Regardless of whether the dual phenomenon has a genetic or cytoplasmic origin (or both), it apparently confers on internal colony cells the capacity to produce dissociated nuclear types which can grow through the staled substrate. Two possible ways in which such a mechanism might be significant are evident. It might provide for the expression of variants arising through partitioning of interior cells by secondary septa. It might also provide a mechanism whereby nuclear types left behind by various mechanisms of dissociation could grow through the "staled" substrate to the margin of a thallus where they could contribute to sexual reproduction or to the re-formation of heterokaryosis in marginal cells. Certainly the possible function of the dual phenomenon in the biology of fungi deserves attention. Again, knowledge of "staling" in natural soil and plant substrates needs investigation.

One of the functions of sclerotia may also involve conservation of variation. In *T. cucumeris*, where isolations from cells partitioned by secondary septa yield variant colonies, isolations from sclerotial cells generally yield colonies similar to the parent thallus. Thus, available evidence indicates that sclerotial cells tend to preserve all of the genomes originally present in the thallus. Since hyphal cells of *T. cucumeris* are long lived and quite resistant to adverse conditions, sclerotial cells may function as much to conserve genetic diversity that might be lost in partitioned hyphal cells as they do to withstand adverse conditions.

CONCLUSIONS.—This discussion has included comments on only a few of the many ramifications associated with mechanisms of variation. Because of the excellent reviews on mechanisms of variation available now, I have elected to treat the subject in general terms and to point up some of the problems involved in relating our limited knowledge to what might be involved in the variation of root-infecting fungi in nature.

The occurrences in nature of varieties, formae, races, and variant clones of root-infecting fungi differing in host range, mode of pathogenesis, and reaction to physical and chemical factors of the environment are too numerous and too well known to receive detailed citation. In spite of this diversity, the introduction of resistant varieties, new cultural practices, or new chemical controls has not led generally to the rapid development and spread of new strains of root-infecting fungi. Whether this involves limits in the capacity to vary or in the spread and expression of variance is a question in need of an answer.

Where new strains have appeared, we rarely know whether these have arisen from genetic mechanisms within existing populations or whether they represent the spread of existing strains into new areas. We need only the example of *Phytophthora infestans*, however, to indicate that, given appropriate mating types, new strains can arise rapidly.

Information on spread of root-infecting fungi, especially those that are soil inhabitants, is sketchy. Most examples where massive invasions of new areas by root-disease fungi have occurred can be traced to the activities of man. It is likely, as is apparently the case with *P. infestans*, that entire gene pools have not been included in man's introductions. It is also likely that geographic isolation has led to differences in local gene pools of many root-infecting fungi. This may have some bearing on the comparatively infrequent rate at which new strains of root-infecting fungi appear, at least as contrasted to such fungi as rusts and smuts.

Thorough understanding of the genetic diversity of our major root-infecting fungi, the mechanisms by which this diversity is developed and expressed, and the present geographic distribution of gene pools and their potentials to interact is essential in assessing the problems in variation and distribution we observe today and the future problems that may arise in the control of root-disease fungi.

LITERATURE CITED

ADAMS, D. H., and L. F. ROTH. 1967. Demarcation lines in paired cultures of *Fomes cajanderi* as a basis for detecting genetically distinct mycelia. Can. J. Botany 45:1583–1589.

BISTIS, G. N. 1959. Pleomorphism in the dermatophytes. Mycologia 51:440–452.

BISTIS, G. N. 1961. Pleomorphism and growth cycles in *Trichophyton mentagrophytes*. Mycologia 52:394–409.

BRUSHABER, J. A., C. L. WILSON, and J. R. AIST. 1967. Asexual nuclear behavior of some plant pathogenic fungi. Phytopathology 57:43–46.

BURNETT, J. H. 1965. The natural history of recombination systems, p. 98–113. *In* K. Esser and J. R. Raper [ed.], Incompatibility in fungi. Springer Verlag New York, Inc. 124 p.

CATEN, C. E., and J. L. JINKS. 1966. Heterokaryosis:

its significance in wild homothallic Ascomycetes and Fungi imperfecti. Trans. Brit. Mycol. Soc. 49:81–93.

CATEN, C. E. and J. L. JINKS. 1968. Spontaneous variability of single isolates of *Phytophthora infestans*. I. Cultural variation. Can. J. Botany 46:329–348.

COOK, R. J., E. J. FORD, and W. C. SNYDER. 1968. Mating types, sex, dissemination, and possible sources of clones of *Hypomces (Fusarium) solani* f. *pisi* in South Australia. Australian J. Agr. Res. 19:253–259.

DAVIS, R. H. 1966. Mechanisms of inheritance. 2. Heterokaryosis, Vol. II, p. 567–588. *In* G. C. Ainsworth and A. S. Sussman [ed.], The fungi. Academic Press, New York and London.

EMERSON, S. 1966. Mechanisms of inheritance. 1. Mendelian, Vol. II, p. 513–566. *In* G. C. Ainsworth and A. S. Sussman [ed.], The fungi. Academic Press, New York and London.

ESSER, K. 1966. Incompatibility, Vol. II, p. 661–676. *In* G. C. Ainsworth and A. S. Sussman [ed.], The fungi. Academic Press, New York and London.

ESSER, K., and R. KUENEN. 1967. Genetics of fungi (Transl. E. Steiner). Springer Verlag New York, Inc. 500 p.

FINCHAM, J. R. S., and P. R. DAY. 1965. Fungal genetics, 2nd ed. Blackwell, Oxford. 326 p.

FLENTJE, N. T., and HELENA M. STRETTON. 1964. Mechanism of variation in *Thanatephorus cucumeris* and *T. praticolus*. Australian J. Biol. Sci. 17:686–704.

FORD, E. J., J. A. BOURRET, and W. C. SNYDER. 1967. Biological specialization in *Calonectria (Fusarium) rigidiuscula* in relation to green point gall of cocoa. Phytopathology 57:710–712.

GALINDO, J., and G. A. ZENTMYER. 1967. Genetical and cytological studies of *Phytophthora* strains pathogenic to pepper plants. Phytopathology 57:1300–1304.

GEORGOPOULOS, S. G., and N. J. PANOPOULOS. 1966. The relative mutability of the cnb loci in *Hypomyces*. Can. J. Genet. Cytol. 8:347–349.

HANSEN, H. N. 1938. The dual phenomenon in imperfect fungi. Mycologia 30:442–455.

HANSEN, H. N., and W. C. SNYDER. 1943. The dual phenomenon and sex in *Hypomyces solani* f. *cucurbitae*. Am. J. Botany 30:419–422.

HOFFMANN, G. M. 1964. Untersuchungen über die kernverhaltnisse bei *Fusarium oxysporum* f. *callistephi*. Arch. Mikrobiol. 49:51–63.

JINKS, J. L. 1960. The genetic basis of "duality" in imperfect fungi. Heredity 13:525–528.

JINKS, J. L. 1966. Mechanisms of inheritance. 4. Extranuclear inheritance, Vol. II, p. 619–660. *In* G. C. Ainsworth and A. S. Sussman [ed.], The fungi. Academic Press, New York and London.

KERNKAMP, M. F., D. J. DE ZEEUW, S. M. CHEN, B. C. ORTEGA, C. T. TSIANG, and A. M. KHAN. 1952. Investigations on physiologic specialization and parasitism of *Rhizoctonia solani*. Minnesota Agr. Exp. Sta. Tech. Bull. 200. 36 p.

MACGARVIE, Q., and I. ISAAC. 1966. Structure and behavior of the nuclei of *Verticillium* spp. Brit. Mycol. Soc. Trans. 49:687–963.

McKENZIE, A. R. 1966. Studies on genetically controlled variation in *T. Cucumeris*. Ph.D. thesis, University of Adelaide.

MILES, P. G., and J. R. RAPER. 1956. Recovery of the component strains from dikaryotic mycelia. Mycologia 48:484–494.

NELSON, R. R., and D. M. KLINE. 1964. Evolution of sexuality and pathogenicity. III. Effects of geographic origin and host association on cross-fertility between isolates of *Helminthosporium* with similar conidial morphology. Phytopathology 54:963–967.

PARMETER, J. R., JR., W. C. SNYDER, and R. E. REICHLE. 1963. Heterokaryosis and variability in plant-pathogenic fungi. Ann. Rev. Phytopathol. 1:51–76.

PARMETER, J. R., JR., H. S. WHITNEY, and W. D. PLATT. 1967. Affinities of some *Rhizoctonia* species that resemble mycelium of *Thanatephorus cucumeris*. Phytopathology 57:218–223.

RAPER, J. R. 1966a. Life cycles, basic patterns of sexuality, and sexual mechanisms, Vol. II, p. 473–511. *In* G. C. Ainsworth and A. S. Sussman [ed.], The fungi. Academic Press, New York and London.

RAPER, J. R. 1966b. Genetics of sexuality in higher fungi. The Ronald Press Co., New York. 283 p.

RICHTER, H., and R. SCHNEIDER. 1953. Untersuchungen zur morphologischen und biologischen Differenzierung von *Rhizoctonia solani* K. Phytopath. Z. 20:167–226.

ROPER, J. A. 1966. Mechanisms of inheritance. 3. The parasexual cycle, Vol. II, p. 589–617. *In* G. C. Ainsworth and A. S. Sussman [ed.], The fungi. Academic Press, New York and London.

SCHIPPERS, B., and W. C. SNYDER. 1967. Mating type and sex of *Fusarium solani* f. *cucurbitae* race 1 in the Netherlands. Phytopathology 57:328.

SCHREIBER, L. R., and R. H. GREEN, JR. 1966. Anastomosis in *Verticillium albo-atrum* in soil. (Nuclear exchange, genetic recombination, microsclerotia, relation to soil fungistatis.) Phytopathology 56:1110–1111.

SCHULTZ, H. 1936. Vergleichende Untersuchungen zur Ökologie, Morphologie, und Systematik des "Vermehrungpilzes." Arb. Biol. Reichanst. Land-u. Forstwirtsch., Berlin 22:1–41.

TINLINE, R. D. 1962. *Cochliobolus sativus*. V. Heterokaryosis and parasexuality. Can. J. Botany 40:425–437.

VEST, G., and N. A. ANDERSON. 1968. Studies on heterokaryosis and virulence of *Rhizoctonia solani* isolates from flax. Phytopathology 58:802–807.

WEERAPAT, P., and H. W. SCHROEDER. 1966. Effect of soil temperature on resistance of rice (*Oryza sativa*) to seedling blight caused by *Sclerotium rolfsii*. Phytopathology 56:640–644.

WHITNEY, H. S., and J. R. PARMETER, JR. 1963. Synthesis of heterokaryons in *Rhizoctonia solani* Kühn. Can. J. Botany 41:879–886.

YANG, S. M., and D. J. HAGEDORN. 1968. Cultural and pathogenicity studies of induced variants of bean and pea root rot *Fusarium* species. Phytopathology 58:639–643.

The Significance of Genetic Mechanisms in Soil Fungi

P. R. DAY—*The Connecticut Agricultural Experiment Station, New Haven.*

INTRODUCTION.—Few, if any, biologists today need to be reminded that the genetic mechanism of any organism has great significance. Twenty years ago few of us appreciated the importance of genetics and even then some of our ideas were naïve. I recall a comment at a meeting to discuss the genetic hazards of X rays in medicine and public health held in London in 1949. A physician suggested that since evolution was the result of mutation and selection, exposure of a man's gonads in an X-ray shoe-fitting machine might increase his chances of having a child who was a genius. Knowledge of genetics among laymen, like shoe-fitting methods, has evolved a good deal since then. A freshman undergraduate of today is familiar with the details of DNA replication in a test tube, but we may ask him and ourselves whether we are aware of what genetic mechanisms do in the field and what advantages this knowledge gives us to devise better control methods.

In thinking about this paper it seemed to me that a frankly speculative approach would at least hold your attention. The preceding papers have gone over the factual ground work. I must needs flounder in an area where few angels have had courage to tread. First I shall make some generalizations about soil fungi, then describe some of the controls which govern the chief genetic mechanisms, illustrating my remarks as far as possible with soil fungi, and then I will discuss approaches to the problem of controlling soil pathogens.

SOIL FUNGI.—The single most impressive feature of soil fungi is the bewildering complexity of their environment. Park (1963) diagramed the main interactions involving soil pathogens where H stands for host plant, S for soil microflora and P for pathogen (Fig. 1). All three components are living and their genetic systems interact along the 6 paths shown in the diagram. No stable equilibrium is reached at the level of the microenvironment, in fact there is a tendency for the microbial succession to reduce to zero as the substrates become exhausted (Garrett, 1956). On a larger scale, dynamic equilibria are reached, for example, in the development of natural climax vegetation characteristic of the soil type and climate. Agricultural practice, however, introduces important and systematic changes. For example, the host plants are often a single-crop variety which may be clonally propagated and hence made up of genetically identical individuals. Soil treatments and cultivation will disturb the interactions in the diagram. Many authors have pointed out the pitfalls of trying to study these interactions under laboratory conditions. Genetic experiments with soil fungi are particularly subject to these criticisms, for we often have little or no way of assessing whether or not behavior under laboratory or greenhouse conditions is representative of what takes place in the soil.

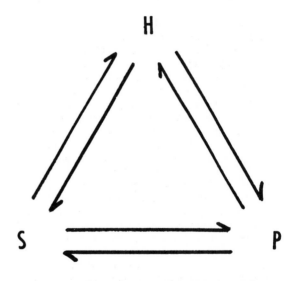

Fig. 1. The interactions involving soil pathogens (*P*), soil microflora (*S*) and host plant (*H*). From Park (1963).

In reading about the kinds of fungi isolated from soil, I find little that suggests they are peculiar and not representative of the fungi as a whole apart from the fact that they are well adapted for competition and survival in what appears to be a very inhospitable environment. Soil organisms are a source of antibiotics and other mycostatic materials (Jackson, 1965). Many soil fungi are resistant to many antibiotics but not all have been shown to produce such materials themselves. I cannot recall seeing an account of antibiotic production by a Phycomycete, for example, and yet these are an important group in the soil. The few pathogenic soil fungi that have been studied in some detail offer a chance to compare genetic mechanisms with similar forms which are not found in the soil and which are well known in the laboratory. But an important difficulty arises. Many geneticists restrict their studies to one wild type strain in order to minimize variation in genetic background. All the laboratory strains of *Aspergillus nidulans* developed in Glasgow are derived from 3 wild type strains (Pontecorvo et al., 1953). Only 5 wild type strains have contributed to our knowledge of *Neurospora crassa* (Frost, 1961). For our purpose we need to know the genetic properties of populations through study of many strains of fungi. However, detailed knowledge of a few strains must come first so that methods can be developed.

THE GENETIC MECHANISMS.—Several categories of genetic mechanism can be defined in the fungi. These are sexual mechanisms based on meiosis, heterokaryosis which recombines whole nuclei, parasexual mechanisms which recombine whole chromosomes or parts of chromosomes, cytoplasmic systems, and the mitotic cycle. We might also include adaptational responses which do not appear to involve genetic change but which may be important in contributing to what we recognize as phenotypic variation (Buxton, 1960).

The mitotic cycle.—Of all these mechanisms the mitotic cycle is undoubtedly the most important. It ensures that information coded in DNA is passed on from one cell generation to the next. The exactness of the replication mechanism guarantees identity among daughter cells and continuity of the genotype. An example of the inherent stability in the mitotic cycle comes from the work of Burnett and Evans (1966), who examined the number of mating-type factors present among fruit bodies of 8 fairy rings formed by the soil fungus *Marasmius oreades*. All fruit bodies from the same ring possessed the same two mating-type factors. *Marasmius* is bipolar. The two nuclei in every cell of a dikaryon each carry a different allele of the mating-type locus. There are a number of such alleles in the population at large which are currently presumed to have arisen by mutation. The largest ring, which had a radius of 3.87 meters, was estimated to be between 77 and 500 years old. All 12 fruit bodies from this ring segregated the same two mating-type alleles, so that the ring probably represents a single colony at least 100 years old. A crude estimate of the number of cell divisions to obtain these fruit bodies from the original inoculum can be made by dividing the average cell length into the colony radius and multiplying this by 12 on the assumption that the 12 radii defined by each fruit body are independent. If we assume the average cell length is 50μ, then close to 10^6 cells or that many mitotic divisions are involved. During this time there was no mutation or recombination involving the mating-type factor which was detected at the time of sampling. Spontaneous mutation rates in *Neurospora* have been estimated at less than 10^{-6} per nucleus per gene for forward mutations to auxotrophy. Reverse mutations, which are much easier to measure, may be as infrequent as 1 per 100 million conidia (Fincham and Day, 1965). However, these estimates were obtained over a short time span of a few days rather than 100 years or more. Auerbach (1959) has shown that spontaneous mutation occurs in dry *Neurospora* macroconidia kept at 32° C. At this temperature the proportion of conidia carrying recessive lethals increased at the rate of 0.3% per week. The rate at 4° C was less than half. One might expect mutations to accumulate during the growth of a dikaryon since only dominant lethals would be eliminated by selection. Indeed an example of this comes from Miles's (1964) comparison of old and newly isolated dikaryons of *Schizophyllum commune*, one of which was deposited with the culture collection at Baarn by Kniep in 1920. Newly isolated dikaryons gave sexual progenies with few aberrant forms, while progenies from the older dikaryons frequently had a high percentage of aberrant mycelia resulting from the presence of mutant genes. Mutations of the mating-type locus allowing homokaryotic or haploid fruiting would tend to be eliminated by the uncovering of recessive lethals. Mating-type mutations of this general kind are among the most common in tetrapolar agarics. The stability of the mating-type locus in *Marasmius* can thus be accounted for by selection against mutations permitting haploid fruiting. The problem of how the different alleles arose remains. Shantz and Piemeisel (1917) recorded rings of *Calvatia cyathiformis* as much as 200 meters in diameter in Colorado, likely to have been more than 420 years old. Examination of rings of this size could be very instructive.

A variety of chemical and physical agents are known which upset or derange the mitotic cycle. Many of these, including ultraviolet light, X-rays, heat shocks, and many chemicals, are mutagenic. It has become clear in recent years that sensitivity to these agents is under genetic control. For example, radiation-sensitive strains of *Aspergillus nidulans* (Lanier, Tuveson, and Lennox, 1968), *A. rugulosus* (Lennox and Tuveson, 1967), and *Ustilago maydis* (Holliday, 1967) are killed by relatively small doses of UV because they lack the enzymic repair mechanisms which are normally present in wild strains of these species and which correct errors in the DNA. However, most of these mutants are reported to be almost completely sexually sterile in the homozygous condition. The reverse situation, radiation resistance, is not well known among the fungi.

The difficulty of obtaining auxotrophic mutants in cultures of *Phytophthora infestans* following treatment with mutagens has been used as evidence to support the idea that this organism is predominantly diploid in nature (Clarke and Robertson, 1966). It seems likely that fungi which are diploid also carry a burden of recessive deleterious mutations in the intervals between haploidization or recombination.

The relative extent of haploid, dikaryotic, and diploid mycelia or resting structures in the soil is a matter for conjecture. Only those Basidiomycetes which form clamp-bearing mycelia can be readily identified. The ingenious method used by Ingram (1968) (see Hastie, this volume) to establish that an isolate of *Verticillium dahliae* was diploid is by no means adaptable to surveys. On the other hand, diploid solo-pathogenic lines of *Ustilago maydis* have a characteristic phenotype on certain media which would make them readily detectable (Puhalla, 1968).

In *Schizophyllum* Koltin and Raper (1968) reported a gene *dik* which when present in two mated compatible homokaryons resulted in the formation of a diploid mycelium instead of a dikaryon. This diploid is particularly interesting since it is phenotypically like the haploid parents except for its universal compatibility and, when the parent strains are auxotrophic, its prototrophy. The mycelium is made up of uninucleate cells and has no clamp connections. While *dik* has so far only been encountered once by chance, it is difficult to be sure that it is uncommon, since it may well have been observed before under circumstances where the phenotype of the diploid would have been interpreted as an incompatible reaction. The dominant allele *dik+* evidently defers nuclear fusion until the basidia are formed, at which time the diploid nucleus does not divide mitotically but immediately undergoes meiosis.

The readiness with which derived haploid products of the diploid form of *Verticillium* from nature reform the diploid suggests that a control analagous to *dik* may be found here. There is certainly reason to expect that our increasing familiarity with what these organisms do in the laboratory on defined media will enable us to find out far more about their life in the soil.

The replication and separation of the genetic information in the nucleus is a main function of the mitotic cycle, but the formation of cross walls after mitosis also results in a partitioning of cytoplasm between the daughter cells. In many fungi this partitioning process is never complete because a cytoplasmic continuity exists from cell to cell in the vegetative hyphae through a pore in each cross wall. The chief cytoplasmic discontinuity occurs at spore formation when a packet of cytoplasm containing one or more nuclei is constricted off from the main body of the mycelium. Unless the information contained in this cytoplasm is complete, a mutant phenotype will result on spore germination if such a spore is viable. In practice, from Jink's (1956) work on *Aspergillus glaucus* it seems that if the cytoplasm is defective, spore formation does not even take place. The ability to produce perithecia is thus lost rather easily during vegetative transfer of mycelium, suggesting that only a complete cytoplasm will produce the sexual apparatus. We will return to cytoplasmic mechanisms later.

The mitotic cycle ensures that those genotypes which give each species its peculiar competitive advantage in a particular ecological niche are reproduced precisely and in large numbers. However, as an environment changes, so the fitness of an es-

tablished genotype will usually decline. Evolution therefore calls for variation, so that a range of niches can be occupied. Most genetic mechanisms reach a compromise between the need for variation and the stable reproduction of genetic combinations with high survival value. Sometimes the compromise seems at first sight surprising. The smuts, for example, do not reproduce clonally as dikaryons. Dissemination appears to occur largely through the brand spores which germinate and undergo meiosis. Since 50% of meiotic products of most smuts are compatible (Whitehouse, 1951), and probably are not efficiently dispersed, inbreeding may well be common. Although the haploid meiotic products form discrete cells, the sporidia, it is not known if they exist independently in the soil or what their role is if they do. Puhalla (1968) finds that haploid sporidial lines of *U. maydis* of certain types produce an inhibitor which prevents the growth of nearby sensitive strains. Such an antibiotic material may have a role in the soil or on the host surface. It is much more difficult to be sure that an organism shows no recombination, sexual or parasexual, between different clones and thus relies solely on mutation as a source of variation. I must admit to being sorely tempted to think that *Cladosporium fulvum*, the tomato pathogen, might be an example.

Sexual mechanisms.—In the fungi self-fertilization is minimized or entirely prevented in a variety of ways. The resultant outbreeding pattern ensures genetic reassortment and, in diploid or dikaryotic forms, delays elimination of recessive genes which may reduce fitness in certain environments when homozygous. However, there are inbreeders or forms which are homothallic. Here the need for genetic diversity is either met by chance outcrossing or by other recombination mechanisms, or else sexual reproduction offers other advantages such as more efficient spore-dispersal mechanisms, or dormant or resistant structures which facilitate overwintering. In the Phycomycetes, as Burnett (1965) has pointed out, zygospores are very rarely observed in the soil in spite of the equal distribution of the two mating types. They are generally only observed on homothallic forms and, even when formed, do not appear to germinate readily. In the main potato-growing regions of the world only one mating type (A2) of *Phytophthora infestans* is found, which partly explains why these fungus populations are sexually sterile. Common sense suggests that it is advantageous to man for them to remain so, for the host population is also sexually sterile in the sense that it consists of a fairly limited number of clonally propagated potato varieties. Even so, the degree of variation available to *P. infestans* by other means than sexual reproduction has been sufficient to cause a trend by breeders to abandon use of major-gene, or hypersensitivity, resistance in favor of minor-gene or field resistance. It should be pointed out of course that such a trend was considerably accelerated by the discovery of a prolific source of physiologic races in central Mexico, attributed by Niederhauser (1956) to the occurrence there of the sexual stage.

In many heterothallic Basidiomycetes the predominant form of the organism is the dikaryon. In these species the two component nuclear types carry different mating-type alleles at one or two loci. The question arises as to whether the mating-type loci play a part in determining either pathogenicity or decay-producing ability. There is now good evidence from the work of Amburgey (1967, and unpublished) on *Lenzites trabea* and Bell and Burnett (1966) on *Polyporus betulinus* that decay activity or cellulase activity are under the control of several genes which are independent of mating type. Indeed, the activity of certain homokaryotic strains is only barely exceeded by the best dikaryons. The question then arises, are there any soil Basidiomycetes in which a differentiation of function between haploid homokaryon and dikaryon occurs which involves more than the production of fruit bodies? Apart from the smuts where homokaryons are nonpathogenic, I cannot think of such an example. Indeed, the economically important pathogenic soil Basidiomycetes, such as *Rhizoctonia* and *Armillaria,* appear to be homothallic and to lack a clear distinction between the two phases.

In heterothallic fungi mating-type genes regulate the occurrence of meiosis. There is now evidence that the degree of recombination which takes place is controlled by genes which determine how much crossing over occurs in a given chromosome interval. In crosses between stocks of *Schizophyllum* carrying different wild-type factors, Simchen (1967) and Stamberg (1968) have shown that the frequency of recombination between the subunits of the A and B mating-type loci is controlled by several genes. For example, recombination between the two subunits of the A factor ranges from less than 1% to more than 20% depending on which controlling alleles are present. In *Neurospora crassa* there are similar controls (Catcheside, 1966; Smith, 1966). Their adaptive significance may be that they permit the buildup and stabilization of chromosome segments or gene clusters.

Heterokaryosis and parasexual mechanisms.—It is almost an axiom in fungal genetics that for different strains of the same fungus to effect any kind of genetic transfer or recombination a cytoplasmic continuity has to be established in which the exchange can take place. Attempts to carry out transformation in fungi have generally been inconclusive (Shockley and Tatum, 1962). However, Sen, Mishra, and Nandi (1968) working with *Aspergillus niger* recently claimed to have obtained transformation of 6 different auxotrophic strains with DNA preparations from wild type in frequencies of the order of one or two per 100,000 cells.

In sexual mechanisms fusion sometimes occurs as a result of hormonal interaction bringing about directed growth of the two organs. In *U. maydis* fusion appears to be determined by a single locus with only two alleles. In still other forms, e.g. *Coprinus,*

Schizophyllum, hyphal fusion is apparently uncontrolled and takes place in compatible and incompatible matings. On the other hand, many examples of nonsexual fusion which lead to the establishment of heterokaryons have been shown to be under strict genetic control. For the most part these controls work in such a way that only mycelia which carry common alleles are able to fuse, establish cytoplasmic continuity, and form a functional heterokaryon. As Caten and Jinks (1966) have pointed out, such a mechanism may be supposed to greatly restrict the extent of heterokaryon formation among natural isolates and therefore genetic exchange among them. These systems are mainly found in Ascomycetes and related forms of imperfect fungi. Since heterokaryon formation is controlled in this way, we might ask what advantage accrues. One possibility is protection against genetic infection and we will return to this presently. Another possibility is that parasexual exchange occurs in nature more frequently than we suspect. Pontecorvo (1959) has estimated that in *Aspergillus nidulans* genetic recombination occurring through the parasexual cycle is about 500 times less than that through the sexual cycle. In *A. niger* and *Penicillium chrysogenum,* which are both asexual, its frequency, in the laboratory, is much higher. Heterokaryon incompatibility systems may function like the isolating mechanisms found in higher plants. They are a feature of highly evolved forms which are particularly well adapted to their ecological niches.

In *Rhizoctonia solani* Garza-Chapa and Anderson (1966) reported that the establishment of heterokaryons was controlled by alleles at a single locus such that only stocks carrying different alleles would form heterokaryons. Three such alleles were described. The evidence for heterokaryosis, formation of a mycelial tuft at the contact zone, and a difference in pathogenicity, included the fact that the heterokaryon would not form mycelial tufts with tester stocks carrying the parental alleles or the third allele. If strictly correct, the incompatibility of a heterokaryon with either parental isolate is an unusual situation. In all heterothallic Basidiomycetes di-mon matings, that is, matings between a dikaryon and a homokaryon with or without a nucleus common to the dikaryon, occur freely. Although now known to occur in a number of fungi (Roper, 1966), the significance of the parasexual cycle in nature cannot yet be assessed because, as others have pointed out, we have little or no idea of its frequency.

Even while we know so little about what happens in the soil there are still many questions to be answered from laboratory studies. In the Phycomycetes, hyphal anastomosis has been reported but rarely (Griffin and Perrin, 1960). Could this be a consequence of diploidy? Storage of genetic variation in a heterokaryon is of little value in a diploid since heterozygosity achieves the same end. In forced heterokaryons, do fusion nuclei occur as readily as they do in some haploid forms? Although zygospore germination is difficult in the laboratory, it may be sufficiently frequent in the soil for the fusion product to be useful both as a resistant spore and as a source of variation. On the other hand, sexual fusion and zygospore formation seem to occur readily even in certain interspecific matings in the Peronosporales (Gallegly, this volume).

Is the low rate of germination an isolating mechanism and zygospores an evolutionary dead end? If Phycomycetes are predominantly diploid, can we isolate haploid forms? We might try using para fluorophenylalanine (Lhoas, 1961).

The cytoplasm.—The studies of cytoplasmic inheritance in fungi at Birmingham have clearly established its major role in variation (Jinks, 1966). The distinction between somatic segregation due to the cytoplasm and that due to heterokaryosis is not always easy to make (Caten and Jinks, 1968). Like heterokaryons, heteroplasmons may arise by mutation or by fusion of forms with different cytoplasmic determinants. Frequently the result of such heteroplasmic interactions is suppression of the normal by the mutant, a state of affairs which rarely occurs in heterokaryons. This leads me to suggest that hyphal-incompatibility systems which restrict anastomosis to closely related strains may have value in protecting mycelia from invasion by suppressive cytoplasmic determinants. We need to know whether transmission of suppressive elements only occurs between strains which are heterokaryon compatible.

THE FUTURE.—The complexity of the soil as an environment and the sophistication of the genetic mechanisms available to soil fungi could make the pathologist throw up his hands in despair over the problem of selective control of soil pathogens. One method which has worked is to let these mechanisms take care of themselves. Flax grown on flax-sick soil is killed outright if susceptible to Fusarium wilt. Continued testing and selection of wilt-resistant varieties has produced a succession which can control the disease in spite of new races (Allard, 1960). The present tendency for breeders to use what van der Plank (1963) has called horizontal resistance, or field resistance, should be encouraged, since physiologic specialization seems not to occur.

Our knowledge of the genetics of soil fungi suggests the possibility of treating the soil, or the host, with materials designed to have specific effects on the pathogens. Agents which either completely block, or greatly promote, hyphal anastomosis or which have similar extreme effects on sexual development might be sought. These materials would upset the balance of the organism in such a way that it would no longer compete in the soil environment. The search for fungicides which suppress sporulation (Horsfall and Lukens, 1968) is based on a similar rationale. If the pathogen cannot reproduce, it will no longer menace crops. Materials which induce suppressive cytoplasmic variants with high frequency could be effective eradicants.

But such speculations stray too far from the real world of disease-control problems. The answer is clear. Few can tell what applications of basic knowledge of these genetic mechanisms may be made in the future, but we are certain that our increasing familiarity with these adversaries will one day pay dividends.

LITERATURE CITED

ALLARD, R. W. 1960. Principles of plant breeding, p. 363. Wiley, New York.

AMBURGEY, T. L. 1967. Decay capacities of monokaryotic and dikaryotic isolates of *Lenzites trabea*. Phytopathology 57:486–491.

AUERBACH, C. 1959. Spontaneous mutations in dry *Neurospora* conidia. Heredity 13:414. (Abstr.)

BELL, M. K., and J. H. BURNETT. 1966. Cellulase activity of *Polyporus betulinus* Basidiomycetes. Ann. Appl. Biol. 58:123–130.

BURNETT, J. H. 1965. The natural history of recombination systems, p. 98–113. *In* K. Esser and J. R. Raper [ed.], Incompatibility in fungi. Springer-Verlag New York Inc., N.Y.

BURNETT, J. H., and E. J. EVANS. 1966. Genetical homogeneity and the stability of the mating-type factors of 'Fairy Rings' of *Marasmius oreades*. Nature *(London)* 210:1368–1369.

BUXTON, E. W. 1960. Heterokaryosis, saltation and adaptation, Vol. 2, p. 359–405. *In* J. G. Horsfall and A. E. Dimond [ed.], Plant pathology. Academic Press, London and New York.

CATCHESIDE, D. G. 1966. A second gene controlling allelic recombination in *Neurospora crassa*. Australian J. Biol. Sci. 19:1039–1046.

CATEN, C. E., and J. L. JINKS. 1966. Heterokaryosis: its significance in wild homothallic Ascomycetes and Fungi imperfecti. Trans. Brit. Mycol. Soc. 49:81–93.

CATEN, C. E., and J. L. JINKS. 1968. Spontaneous variability of single isolates of *Phytophthora infestans*. 1. Cultural variation. Can. J. Botany 46:329–348.

CLARKE, D. D., and N. F. ROBERTSON. 1966. Mutational studies on *Phytophthora infestans*. Eur. Potato J. 9:208–215.

FINCHAM, J. R. S., and P. R. DAY. 1965. Fungal genetics. 2nd ed. 326 p. Blackwell, Oxford.

FROST, L. C. 1961. Heterogeneity in recombination frequencies in *Neurospora crassa*. Genetical Res. 2:43–62.

GARRETT, S. D. 1956. Biology of root-infecting fungi. Cambridge Univ. Press, 293 p.

GARZA-CHAPA, R., and N. A. ANDERSON. 1966. Behaviour of single basidiospore isolates and heterokaryons of *Rhizoctonia solani* from flax. Phytopathology 56:1260–1268.

GRIFFIN, D. M., and H. N. PERRIN. 1960. Anastomosis in the Phycomycetes. Nature (London) 187:1039.

HOLLIDAY, R. 1967. Altered recombination frequencies in radiation sensitive strains of *Ustilago*. Mutation Res. 4:275–288.

HORSFALL, J. G., and R. J. LUKENS. 1968. Aldehyde traps as antisporulants for fungi. Conn. Agr. Exp. Sta. Bull. 694. 23 p.

INGRAM, R. 1968. *Verticillium dahliae* Kleb. var. *longisporum* Stark: A stable diploid. Trans. Brit. Mycol. Soc. 51:339–341.

JACKSON, R. M. 1965. Antibiosis and fungistasis of soil microorganisms, p. 363–369. *In* K. F. Baker and W. C. Snyder [ed.], Ecology of soil-borne plant pathogens. Univ. of California Press, Berkeley and Los Angeles.

JINKS, J. L. 1956. Naturally occurring cytoplasmic changes in fungi. Compt. Rend. Trav. Lab. Carlsberg, Ser. Physiol. 26:183–203.

JINKS, J. L. 1966. Mechanisms of inheritance: Extranuclear inheritance, Vol. II, p. 619–660. *In* G. C. Ainsworth and A. S. Sussman [ed.], The fungi. Academic Press, London and New York.

KOLTIN, Y., and J. R. RAPER. 1968. Dikaryosis: Genetic determination in *Schizophyllum*. Science 160:85–86.

LANIER, W. B., R. W. TUVESON, and J. E. LENNOX. 1968. A radiation-sensitive mutant of *Aspergillus nidulans*. Mutation Res. 5:23–31.

LENNOX, J. E., and R. W. TUVESON. 1967. The isolation of ultraviolet-sensitive mutants from *Aspergillus regulosus*. Radiation Res. 31:382–388.

LHOAS, P. 1961. Mitotic haploidization by treatment of *Aspergillus niger* diploids with para-fluorophenylalanine. Nature (London) 190:744.

MILES, P. G. 1964. Possible role of dikaryons in selection of spontaneous mutants in *Schizophyllum commune*. Botan. Gaz. 125:301–306.

NIEDERHAUSER, J. S. 1956. The blight, the blighter and the blighted. Trans. N.Y. Acad. Sci. Ser. II. 19:55–63.

PARK, D. 1963. The ecology of soil-borne fungal disease. Ann. Rev. Phytopathol. 1:241–258.

PLANK, J. E. VAN DER. 1963. Plant diseases: epidemics and control. Academic Press, New York and London. 349 p.

PONTECORVO, G. 1959. Trends in genetic analysis. Columbia Univ. Press, New York. 145 p.

PONTECORVO, G., J. A. ROPER, L. M. HEMMONS, K. D. MacDONALD, and A. W. J. BUFTON. 1953. The genetics of *Aspergillus nidulans*. Adv. in Genetics 5:141–238.

PUHALLA, J. E. 1968. Compatibility reactions on solid medium and interstrain inhibition in *Ustilago maydis*. Genetics 60:461–474.

ROPER, J. A. 1966. Mechanisms of inheritance: The parasexual cycle, Vol. 2, p. 589–617. *In* G. C. Ainsworth and A. S. Sussman [ed.], The fungi. Academic Press, New York and London.

SEN, K., A. K. MISHRA, and P. NANDI. 1968. DNA induced transformation of nutritional markers in *Aspergillus niger*. Microb. Gen. Bull. 28:8–9.

SHANTZ, H. L., and R. L. PIEMEISEL. 1917. Fungus fairy rings in eastern Colorado and their effect on vegetation. J. Agr. Res. 11:191–245.

SHOCKLEY, T. E., and E. L. TATUM. 1962. A search for genetic transformation in *Neurospora crassa*. Biochem. Biophys. Acta 61:567–572.

SIMCHEN, G. 1967. Genetic control of recombination and the incompatibility system in *Schizophyllum commune*. Genet Res., Cambridge 9:195–210.

SMITH, B. R. 1966. Genetic controls of recombination. 1. The recombination-2 gene of *Neurospora crassa*. Heredity 21:481–498.

STAMBERG, J. 1968. Ph.D. thesis, Harvard University, Cambridge, Mass.

WHITEHOUSE, H. L. K. 1951. A survey of heterothallism in the Ustilaginales. Trans. Brit. Mycol. Soc. 34: 340–355.

◄

EFFECT OF SOIL MOISTURE AND AERATION ON FUNGAL ACTIVITY WITH ROOT DISEASES

Effect of Soil Moisture and Aeration on Fungal Activity: an Introduction

D. M. GRIFFIN—*Department of Agricultural Botany, University of Sydney, New South Wales, Australia.*

Change in the water content of a soil causes change in many physical and chemical properties of the soil and these changes are of biological significance. Fig. 1 presents with some simplification the relationships between some of the factors in the "soil water" complex. For the plant pathologist the situation is further complicated by the necessity to consider simultaneously two diverse organisms—the higher plant and the pathogenic microorganism. I have recently noted pertinent concepts in soil physics and have reviewed much of the relevant literature of mycology and plant pathology (Griffin, 1963a, 1969). The role of water as a factor influencing the incidence of root diseases is considered in detail by Cook and Papendick (this volume). I shall, therefore, now confine myself to a short introduction to the effect of soil water and aeration on fungal and bacterial activity and root diseases.

SOIL WATER—*Matric suction.*—The activity of a fungus in soil at a given matric suction depends not only upon the direct response of the fungus to the suction but also on the ability of the fungus to compete with the rest of the microflora in the colonization of the substrate under the specific environmental conditions. This effect of the environment on competitive saprophytic colonization is now well demonstrated for temperature (Persson-Huppel, 1963; Taylor, 1964; Harris, Allen, and Chesters, 1966; Burgess and Griffin, 1967) and for matric suction (Griffin, 1963b, 1966; Kouyeas, 1964; Chen and Griffin, 1966a, b). Fig. 2 illustrates the effect of matric suction on some fungi colonizing hair at 25°C (data from Chen, 1965). It is known that the details of the pattern of colonization change with substrate (Chen and Griffin, 1966a; Griffin, 1966) and temperature (Chen and Griffin, 1966b) (Fig. 3), but its general validity

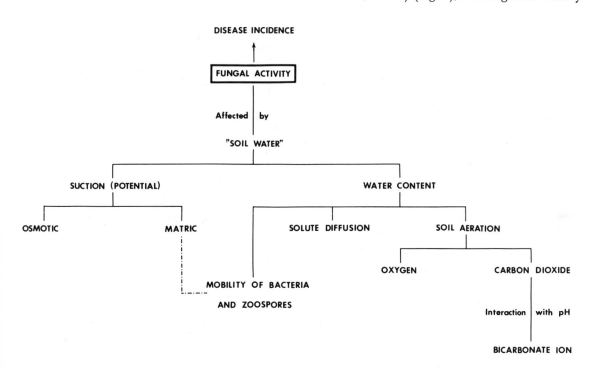

Fig. 1. Relationship between various factors, associated with the soil water regime, that affect the activity of fungi and the incidence of plant disease.

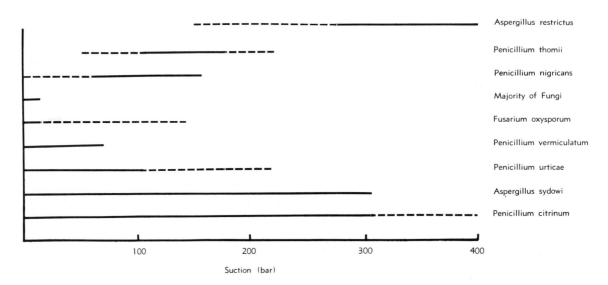

Aspergillus restrictus
Penicillium thomii
Penicillium nigricans
Majority of Fungi
Fusarium oxysporum
Penicillium vermiculatum
Penicillium urticae
Aspergillus sydowi
Penicillium citrinum

Suction (bar)

Fig. 2. Relationship between matric suction and frequency of isolation of fungi from hair which had been colonized by soil-borne fungi at 25°C. (Full line—frequent; broken line—less frequent.)

is supported by other data (Dommergues, 1962; Griffin, 1963*b*, 1966; Kouyeas, 1964). Thus, at suctions greater than pF 4.2, there is a gradual impoverishment of the active mycoflora and an increasing dominance of *Penicillium* and especially *Aspergillus.*

Osmotic suction.—The influence of the osmotic suction exerted by the soil solution on fungi has been little investigated but in moist soils it is probably negligible. Chen (1964) showed that many soil fungi were able to grow on media of high osmotic suction. Members of the *Aspergillus glaucus* group were particularly noteworthy in this respect, in accord with their known activity under high moisture stress in many other situations. There is good evidence that soil salinity is important in determining the distribution in soil of the human pathogen *Coccidioides immitis* (Egeberg, Elconin and Egeberg, 1964; Elconin, Egeberg and Egeberg, 1964) but no similar dependence has been demonstrated for plant pathogens.

BACTERIAL MOBILITY.—The interactions between the various components of the soil microflora are still poorly understood. One factor of likely importance is the difference between the means by which bacteria (and other motile microorganisms) and fungi spread in soil. Fungi, by virtue of their filamentous structure, can cross air-filled pores whereas bacteria are restricted to movement in water-filled pores or water films. The limitations imposed on bacterial mobility by the physical nature of a wet soil have been analysed for a small number of systems (Griffin and Quail, 1968). If these results have general validity, there can be little movement of bacteria by Brownian or flagellar activity in soils much drier than field capacity. This is probably important in the relationship between fungi and bacteria.

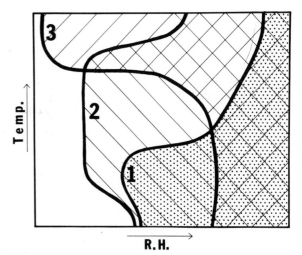

R.H.

Fig. 3. Diagram of the interaction between temperature (15–35°C) and relative humidity (75–100% R.H.) in determining the pattern of activity of fungi in colonizing hair. Group 1 includes *Penicillium chrysogenum, P. frequentans;* group 2, *P. citrinum, Aspergillus versicolor;* group 3, *A. flavus, A. niger, A. terreus.* (A. W.-C. Chen and D. M. Griffin. 1966*b*.)

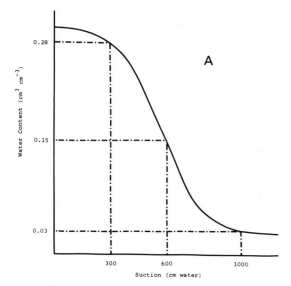

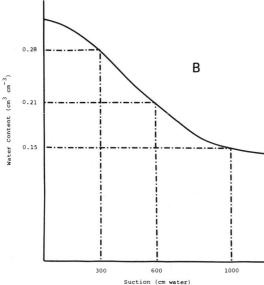

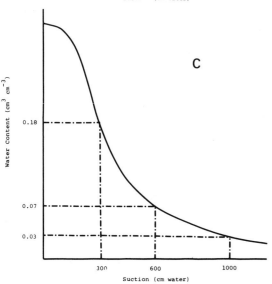

Fig. 4. Partial moisture characteristics (drying) for three hypothetical soils (*A*, *B*, and *C*). Bacteria are unlikely to move rapidly and over appreciable distances in a soil at *h* cm water suction unless at least 0.11 cm³ water per cm³ soil are held in pores which would drain between *h* and 1,000 cm suction (Griffin and Quail, 1968). The moisture characteristics reveal that such movement of bacteria will occur in the three soils at 300 cm suction but only in soil *A* at 600 cm suction, even though the water content of soil *B* is then much greater than that of soil *A*.

The analysis of bacterial movement in particulate systems cannot be performed without at least a partial moisture characteristic, that is, a relationship between water contents and the corresponding suctions (Griffin, 1963*a*). The utility of these characteristics is demonstrated in Fig. 4: they are essential for the precise evaluation of many other problems in soil ecology.

Aeration.—The importance of oxygen in soil ecology has been widely recognised. I have recently discussed many of the issues and have referred to much of the literature relevant to plant pathology (Griffin, 1968).

The interrelated effects of soil water and aeration are well illustrated by reference to the rhizomorph-producing basidiomycete, *Armillariella elegans*. (I am indebted to Mr. A. M. Smith, of the University of Sydney, for permission to refer to his as-yet unpublished work.) Whereas the undifferentiated mycelium of this fungus is not physiologically remarkable in the present context, the rhizomorphs exhibit many unusual features. For effective growth, a high rate of oxygen diffusion to the meristematic centre of the rhizomorph tip is necessary and this is maintained by diffusion in the gas phase along the hollow medulla. Factors which reduce the rate of diffusion, for example, the presence of a water film over the origin of the rhizomorph, retard or prevent elongation. An apparent second requirement for rhizomorph growth is that the tip must be covered by a film of water. Currently, this requirement is interpreted in terms of the reduced rate of melanin formation at slightly reduced oxygen concentrations. Such an explanation is supported by the high value of the Michaelis constant (*ca.* 5×10^{-5} M) for the reaction between oxygen and the phenol oxidase, a reaction which yields the precursors of melanin. In turn, it is thought that melanins are mycostatic (Farkas and Ledingham, 1959), protecting the mature rhizomorph from lysis (Potgeiter and Alexander, 1966; Bloomfield and Alexander, 1967; Kuo and Alexander, 1967) but inhibiting elongation of the tip if they are formed too rapidly. Clearly, the distribution of water films within the soil must play a key role in the ecology of this fungus, and presumably of the closely related *Armillaria mellea*.

Interactions Between Physical Factors.—It has been noted above that temperature and suction interact to affect fungal activity. This is but one example of many interactions between various components of the physical regime of the soil which need elucidation. Other examples are the interactions between temperature and aeration (Bunt and Rovira, 1955) and between concentration of CO_2 and pH. This latter interaction determines the concentration of the biochemically-active bicarbonate ion, which is probably more important than the concentration of CO_2 itself in affecting the growth of some fungi (Macauley, 1968).

LITERATURE CITED

BLOOMFIELD, B. J., and M. ALEXANDER. 1967. Melanins and resistance of fungi to lysis. J. Bacteriol. 93:1295–1280.

BUNT, J. S., and A. D. ROVIRA. 1955. The effect of temperature and heat treatment on soil metabolism. J. Soil Sci. 6:129–136.

BURGESS, L. W., and D. M. GRIFFIN. 1967. Competitive saprophytic colonization of wheat straw. Ann. Appl. Biol. 60:137–142.

CHEN, A. W.-C. 1964. Soil fungi with high salt tolerance. Trans. Kansas Acad. Sci. 67:36–40.

CHEN, A. W.-C. 1965. Studies in the ecology of soil fungi. Ph.D. thesis, University of Sydney.

CHEN, A. W.-C., and D. M. GRIFFIN. 1966a. Soil physical factors and the ecology of fungi. V. Further studies in relatively dry soils. Trans. Brit. Mycol. Soc. 49:419–426.

CHEN, A. W.-C., and D. M. GRIFFIN. 1966b. Soil physical factors and the ecology of fungi. VI. Interaction between temperature and soil moisture. Trans. Brit. Mycol. Soc. 49:551–561.

DOMMERGUES, Y. 1962. Contribution à l'étude de la dynamique microbienne des sols en zone semi-aride et en zone tropicale sèche. Ann. Agron. 13:265–324, 379–469.

EGEBERG, R. O., A. F. ELCONIN, and M. C. EGEBERG. 1964. Effect of salinity and temperature on *Coccidioides immitis* and three antagonistic soil saprophytes. J. Bacteriol. 88:473–476.

ELCONIN, A. F., R. O. EGEBERG, and M. C. EGEBERG. 1964. Significance of soil salinity on the ecology of *Coccidioides immitis*. J. Bacteriol. 87:500–503.

FARKAS, G. L., and G. A. LEDINGHAM. 1959. Studies on the polyphenol-polyphenoloxidase system of wheat stem rust uredospores. Can. J. Microbiol. 5:37–46.

GRIFFIN, D. M. 1963a. Soil moisture and the ecology of soil fungi. Biol. Rev. Cambridge Phil. Soc. 38:141–166.

GRIFFIN, D. M. 1963b. Soil physical factors and the ecology of fungi. III. Activity of fungi in relatively dry soil. Trans. Brit. Mycol. Soc. 46:373–377.

GRIFFIN, D. M. 1966. Fungi attacking seeds in dry seedbeds. Proc. Linnean Soc. N.S. Wales 91:84–89.

GRIFFIN, D. M. 1968. A theoretical study relating the concentration and diffusion of oxygen to the biology of organisms in soil. New Phytol. 67:561–577.

GRIFFIN, D. M. 1969. Soil water in the ecology of fungi. Ann. Rev. Phytopathol. 7: (In press.)

GRIFFIN, D. M., and E. G. QUAIL. 1968. Movement of bacteria in moist particulate systems. Australian J. Biol. Sci. 21:579–582.

HARRIS, R. F., O. N. ALLEN, and G. CHESTERS. 1966. Microbial colonization of soil aggregates as affected by temperature. Plant Soil 25:361–371.

KOUYEAS, V. 1964. An approach to the study of moisture relations of soil fungi. Plant Soil 20:351–363.

KUO, M.-J., and M. ALEXANDER. 1967. Inhibition of the lysis of fungi by melanins. J. Bacteriol. 94:624–629.

MACAULEY, B. J. 1968. A study of the effects of carbon dioxide and the bicarbonate ion on soil fungi. Ph.D. thesis, University of Sydney.

PERSSON-HUPPEL, A. 1963. The influence of temperature on the antagonistic effect of *Trichoderma viride* Fr. on *Fomes annosus* (Fr.) Cke. Studia For. Suec. 4, 13 p.

POTGEITER, J. H., and M. ALEXANDER. 1964. Susceptibility and resistance of several fungi to microbial lysis. J. Bacteriol. 91:1526–1533.

TAYLOR, G. S. 1964. *Fusarium oxysporum* and *Cylindocarpon radicicola* in relation to their association with plant roots. Trans. Brit. Mycol. Soc. 47:381–392.

Effect of Soil Water on Microbial Growth, Antagonism, and Nutrient Availability in Relation to Soil-borne Fungal Diseases of Plants

R. J. COOK and R. I. PAPENDICK—*Respectively, Pathologist, Crops Research Division, and Soil Scientist, Soil and Water Conservation Research Division, Agricultural Research Service, U. S. Department of Agriculture, Washington State University, Pullman.*

Soil water is one of the most important and probably the most poorly understood aspects of the soil environment influencing the growth and survival of soil-borne plant pathogens. A knowledge of soil water in relation to microbial growth, antagonism, host exudation, and other factors which affect pathogens in the soil is basic to plant disease control. It is essential, therefore, that investigations of soil water–soil-borne plant-disease relationships be pursued.

Understanding the effects of soil water on the host-pathogen-soil microfloral system is becoming possible only because of an increased understanding and application of physical concepts of soil water. Griffin (1963a), Raney (1965), and Couch, Purdy, and Henderson (1967) have discussed the importance of concepts of soil water physics in plant disease investigations, and Griffin (1963a) in particular has stressed soil water energy. Soil water energy, better known as the soil water potential, is determined by the forces binding water in soil. Unless an organism (either a pathogen or an antagonist to a pathogen) can overcome the forces that bind water in soil, it will not obtain water for its life processes. Organisms differ in ability to remove bound water; thus, water potentials limiting to one species may not limit another species (Griffin, 1963a, 1963d, Chen and Griffin, 1966; and Clark and Kemper, 1967). Knowledge of growth and reproduction of organisms in soil in relation to the soil water potential is basic, therefore, to an understanding of how soil water affects belowground fungal diseases of plants.

THE SOIL WATER POTENTIAL AND ITS MEASURE-MENT.—Consider an organism in contact with pure, free water and assume that the amount of work that must be done by that organism to obtain a given quantity of water under this condition is zero. Now consider the same organism having to obtain the same quantity of water from unsaturated soil. Such water is bound by certain forces associated with the presence of soil. The difference in work that must be done on the soil water relative to pure free water in order for the organism to obtain water is the soil water potential. By definition, the soil water potential decreases with decreased soil water content.

Total soil water potential may be divided into components based on origin of forces acting on the water (International Society of Soil Science, 1963). These are: (1) *osmotic potential* due to solutes in the soil water; (2) *matric* or *capillary potential* which includes both adsorption and capillary effects due to the solid phase; (3) *gravitational potential* caused by elevation differences from the reference; and (4) *pressure potential* caused by external atmospheric pressure on the soil water. The components are additive, the sum being equal to the total water potential.

Water potential is commonly expressed in units of pressure, i.e., bars, atmospheres (1 atm.=1.013 bars), or cm of water or mercury rather than in units of free energy with which it is more properly identified. Free energy units used are joules/cm^3 or ergs/cm^3 (joules/gm or ergs/gm if expressed on a unit mass basis). To convert from free energy units to pressure, multiply the former by the specific volume of water which is unity.

Total pressure and *total suction* have the same meaning as total soil water potential. Total pressure is "the pressure (positive or negative) relative to external gas pressure on the soil water to which a pool of pure water must be subjected in order to be in equilibrium through a semipermeable membrane with the water in the soil" (Soil Science Society of America, 1965). For suction, as the name implies, the "pressure" is explicitly stated to be "negative gauge pressure" (International Society of Soil Science, 1963), in order to avoid the use of negative quantities for expression of soil water energy. As with water potential, pressure and suction terminologies are expressed in pressure units such as cm of water or mercury, atmospheres, or bars. Moreover, total pressure and total suction both may be divided into parts, i.e., *osmotic pressure* or *suction* [identified with the osmotic potential] and *soil water pressure*, or matric suction (International Society of Soil Science, 1963; Soil Science Society of America, 1965) [identified with the matric potential]. Again the sum of the components equals the total.

Soil water tension is "the negative gauge pressure relative to the external gas pressure to which a solution identical in composition to the soil water must

PART IV. *SOIL MOISTURE AND AERATION*

be subjected in order to be in equilibrium through a porous permeable wall with the soil water." By definition, soil water tension is identical to the matric potential and not the total soil water potential. The term "total soil water stress" has no formal definition but is sometimes used to convey the concept of total soil water potential.

A common misconception among plant pathologists and microbiologists is that liquid soil water is more available to organisms than the associated vapor phase. However, the liquid and vapor phases within any given microsite are in equilibrium and thus both phases are at the same potential.

Thermodynamically, total soil water potential is identical to the partial molar free energy of the soil water relative to free water at the same temperature. In this case, if water vapor can be treated as an ideal gas (a valid assumption over the range of vapor pressures normally encountered), the difference in the partial molar free energy of the soil water and of free water at a given temperature is a function of the relative humidity, i.e.,

$$\Psi' = \frac{RT}{V} \log \text{ relative humidity}$$

where Ψ' is the water potential, R is the ideal gas constant, T is the absolute temperature, V is the volume of a mole of water, and log is the natural or Naperian logarithm. In essence, the equation states that in a system containing liquid water in equilibrium with its vapor, measurements of the relative humidity can be used to estimate the water potential. Thus, knowing the relative humidity limiting to growth of an organism, one can infer the water potential at which growth may be limited in soil. Fig. 1 shows the relationship between water potential and the equilibrium relative humidity of the soil atmosphere.

Soil water potential is related to soil water content; however, one property cannot be inferred from the other, since the relations are multiple-valued functions that depend on the wetting history of the system, and on certain soil properties. Thus, coarse-textured soils (sands) retain less water at a given water potential than do finer-textured soils. Also, the amount of water retained at a given energy depends on whether the point was reached by wetting or by drying the soil, a phenomenon known as hysteresis. The hysteresis effect and water retention, as influenced by soil type, have been discussed in relation to soil water-soil biology studies by Griffin (1963a).

Much of the work in the past on the measurement of soil water energy in relation to growth or activity of higher plants and microorganisms has been concentrated on matric potential rather than the total soil water potential. Apparatuses such as the tensiometer, gypsum or fiberglass electrical resistance unit, the pressure plate and pressure membrane, and the sintered glass plate-Büchner funnel are used to measure matric potential (Bouyoucous and Mick, 1940; Richards, 1947; Taylor, Evans, and Kemper, 1961; and U.S. Salinity Laboratory Staff, 1954) and, in many cases, the results approximate the total soil water

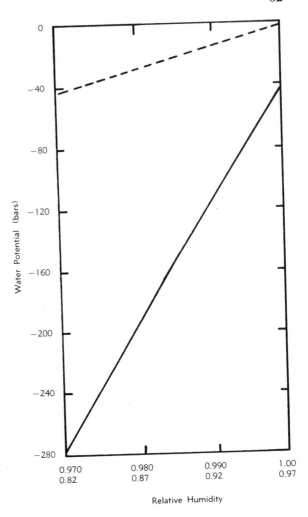

Fig. 1. Water potential-equilibrium relative humidity relationship at 25°C. The dashed curve shows water potential for relative humidities between 1.0 and 0.970. A reduced scale is used for the heavy curve, which shows water potential for relative humidities between 0.97 and 0.82.

potential. With these methods, however, the salt effect (osmotic potential) is not measured. The osmotic potential may be important in highly fertilized soils or in soils amended with nutrients for experimental stimulation of microbial growth. Moreover, the tensiometer measures matric potential to only about -0.75 bar. The gypsum and fiberglass resistance units will measure somewhat lower potentials but are slow to equilibrate in other than fairly wet soil. The pressure-plate and pressure-membrane apparatuses measure the desorption arm only of the water content-matric potential curve and are limited to laboratory use. None of these above methods cover the entire range of potentials important to the soil microflora.

More recently, thermocouple psychrometry has been developed for obtaining highly accurate estimates of total water potential in soils and plants

(Montieth and Owen, 1958; Spanner, 1951; and Richards and Ogata, 1958). This method involves a measurement of the equilibrium relative humidity and is highly reliable over a wide range of water potentials, including those important to most plants and to many microorganisms. Determination of soil relative humidity by thermocouple psychrometry is accomplished by effectively surrounding a thermocouple junction pair with the soil sample. In practice, this is best accomplished using a multiple-sample changer described by Campbell, Zollinger, and Taylor (1966). During the measurement one junction is kept wet and the other junction, which serves as a reference junction, is kept dry. If the soil water content is less than saturation (relative humidity <1.0), water will evaporate from the wet junction into the surrounding atmosphere. Evaporation causes a temperature depression of the wet junction, the amount of depression being dependent on the rate of evaporation which in turn is dependent on the relative humidity of the soil. An electromotive force (EMF) develops between the junction pair which is proportional to the relative humidity. The relationship between EMF and relative humidity can be determined by calibrating the thermocouple psychrometer against salt solutions of known equilibrium vapor pressures.

Two types of thermocouple psychrometers are employed for determining water potential of soils and plants. In one (Richards and Ogata, 1958), one junction is kept wet by the addition of a droplet of water in a small ring soldered to the junction. In the other method, Peltier cooling (Montieth and Owen, 1958; and Spanner, 1951) is used to condense water on the thermocouple prior to measurement of the temperature depression. In both instruments, the wet junction is made from Chromel-P-constantan wire usually 0.025 mm or less in diameter. Where Peltier cooling is used the reference junction is considerably more massive, so its temperature does not change significantly during the cooling process. The sensing element of the Peltier psychrometer can be enclosed within a ceramic bulb and buried in the soil. The element is ideal for use under field conditions and is referred to as a field psychrometer (Rawlins and Dalton, 1967).

One difficulty in thermocouple psychrometry originates from the fact that a very slight change in relative humidity is accompanied by a considerable change in water potential. For example, at 25°C a change in relative humidity from 1.0 to 0.99 causes the water potential to decrease from 0 to about −14 bars (Fig. 1). This high degree of sensitivity of water potential to humidity indicates the importance of temperature control during such measurements. A small temperature change in the measurement system can markedly affect the relative humidity and hence water potential, if equilibrium is upset. For example, if no vapor entered or left the air at a point under consideration, a 0.01°C change in atmosphere temperature would produce roughly a 1 bar change in the water potential. Small short-time temperature

decreases in systems at high humidity also favor condensation, which greatly increases the time for equilibrium reestablishment.

For precise work in the determination of water potential by thermocouple psychrometry, short-time temperature fluctuations should not exceed ±0.001°C, a degree of control readily accomplished with a bath designed by Richards (1965). With such control, together with the use of a highly sensitive d.c. amplifier, it becomes possible to infer water potentials within ±0.1 bar. Where soil samples were handled in a multiple-sample changer (Campbell, Zollenger, and Taylor, 1966) temperature changes of about 0.1°C in 4 to 5 hours were tolerated with no difficulty (Cook and Papendick, 1969). The field psychrometer has been used successfully to determine soil water potential to within ±0.5 bar in the presence of temperature fluctuations greater than 5°C per day (Rawlins and Dalton, 1967).

DIRECT EFFECTS OF WATER POTENTIAL ON GROWTH OF PLANT PATHOGENIC FUNGI IN SOIL.— Griffin (1963a) was among the first to recognize the importance of soil water energy to plant pathogenic fungi in soil. Using the sintered glass plate-Büchner funnel technique, he showed that growth and reproduction of *Curvularia ramosa* is not limited over a range of suctions equivalent to 1 to 600 cm of water (Griffin, 1963b) and that growth and reproduction of *Pythium ultimum* is not restricted over a suction range from 1 to 400 cm of water (Griffin, 1963c). This technique is limited to low suctions (high potentials) only, however, and thus the water potentials limiting to his test fungi probably could not be determined. Moreover, this method measures matric potential and not total soil water potential, but since the system used by Griffin (aluminum oxide grits) was probably salt free, the measurements he obtained approximate the total water potential.

Schneider (1954) and Kouyeas (1964) both present data on the influence of relative humidity on growth of certain fungi. Assuming that the relative humidity in each of their respective systems was at equilibrium, one may estimate the water potentials limiting to their test fungi (Fig. 1). For example, Kouyeas (1964) found that linear hyphal growth of *Pythium ultimum* from an agar plug on a glass slide diminished greatly as the relative humidity was lowered from 1.0 to 0.995. Hyphal growth of other *Pythium* species was prevented at 0.990 relative humidity, and with still others, growth was not detected at relative humidities below 0.981 and 0.970. The minimum relative humidity for hyphal growth of *Phythophthora parasitica* was 0.981. One could conclude that few if any of the species of *Pythium* and *Phythophthora* studied by Kouyeas (1964) made significant growth at water potentials below about −25 to −30 bars. Indeed, with some species of *Pythium*, growth may have been limited at −15 bars, often considered the lower limit of available water for higher plants. However, the temperature of the system used by Kouyeas (1964) varied ±0.5°C. It is impossible to determine whether his system was at equilibrium (and therefore at constant water potential) since this would depend largely upon the geometry of the system and how rapidly the temperature fluctuations occurred.

The pathogens and the respective relative humidities at which hyphal growth was prevented in the work by Schneider (1954) were: *Rhizoctonia solani,* 0.965; *Ophiobolus graminis,* 0.96; *Thielaviopsis basicola,* 0.94; *Verticillium albo-atrum,* 0.93; *Fusarium culmorum,* 0.92; *F. avenaceum, F. solani,* and *F. lini,* 0.91; and *F. lateritium, F. bulbigenum* var. *lycopersici,* and *F. coeruleum,* 0.90 to 0.89. The temperature (20°C) used by Schneider apparently was accurate only to the nearest whole degree, and thus, as with the work of Kouyeas (1964), the actual water potentials to which the pathogens were subjected might not have been constant. The data of Schneider (1954) suggest, nevertheless, that the pathogens studied can grow at water potentials at and well below the lower limits of water available for plant growth and that with certain *Fusarium* species, growth apparently occurs at potentials approaching −100 bars.

Purdy and Kendrick (1957, 1963) showed that teliospores of *Tilletia caries* germinated and infected wheat (*Triticum aestivum*) in a Palouse silt loam containing about 10% water by weight. The same soil held 10.8% water at 15 atm of pressure in a pressure-membrane apparatus. This implies that *T. caries* is active in soil at matric potentials at or below −15 bars. In their work, however, soil water contents used for studies of fungus activity were adjusted by wetting the soil, whereas the matric potential-water content relationship was determined from a desorption curve (drying down). Owing to hysteresis, the water potential corresponding to a given water content may be higher on an adsorption curve (wetting up) compared with soil dried down to the same water content. For example, the water potential determined by thermocouple psychrometry of a Palouse silt loam wetted to 10% water content was about −4 bars. The same soil dried down in the pressure-membrane apparatus, held 10.4% water under a pressure of about 15 bars (Cook, unpublished data). *Tilletia caries* apparently is active in soil at water potentials approaching the lower limits of available water for growth of higher plants, but the actual optimal and minimal water potentials for germination of teliospores of *T. caries* cannot be determined with certainty from the work of Purdy and Kendrick because of a possible hysteresis effect.

Cook and Papendick (1969) studied the lower limits of water available to *F. roseum* f. sp. *cerealis* 'Culmorum' [=*F. culmorum*] in natural soils. A Peltier psychrometer was used for direct measurement of water potential, and soil smears (Nash, Christou, and Snyder, 1961) were used to observe chlamydospore germination and germling growth. Nutrients (2,500 ppm C as glucose and 250 ppm N as ammonium sulfate) as a water solution were atomized and mixed into a Ritzville (RSL) and Palouse silt loam (PSL)

(both of which were highly infested with chlamydospores of Culmorum) to stimulate chlamydospore germination. Aliquots of the nutrient-amended soils were adjusted to different water contents, and hence water potentials, by adding more water or by allowing evaporation down to predetermined levels. Percentage spores with germ tubes at different times after adding the nutrients and water, and the soil water potential for a given treatment, both were determined by using samples of soil from the same sealed container. Since the thermocouple psychrometer measures total soil water potential, the osmotic potential created by the nutrients was included in these measurements and hysteresis was not a factor.

Percentage germination was uniformly high (40 to 50% in 24 hours) in the wetter soils and nil in the dry soils. Although both soils are classed as silt loams, the RSL contained considerably more sand and less clay than the PSL. As expected, the water content at which germination did not occur was lower in the RSL. In both soils, however, the point of reduced chlamydospore germination corresponded to about −50 to −60 bars. Some germination occurred by 72 hours at −80 to −85 bars in both soils, but none occurred below this water potential. Apparently, at potentials below −80 to −85 bars water in sufficient quantity was not available for growth of this fungus. This water potential corresponds to an equilibrium relative humidity of about 0.92 to 0.93, the humidity at which growth of Culmorum was restricted in the work by Schneider (1954).

The fact that growth of so many soil-borne plant parasitic fungi is not limited at 0.99 relative humidity, the wilting point of most higher plants, suggests that over the range of availability to plants, water will seldom if ever be in short supply to fungi (Griffin, 1963a). On the other hand, further research may reveal examples of root parasites unable to grow at relative humidities below 0.99. Such examples apparently exist among fungi parasitic on above-ground parts of plants (Clayton, 1942; Gottlieb, 1950). Moreover, evaporation may cause very low potentials to develop in surface soils and thus limit fungal parasitism in infection courts near the soil surface. Dryland wheat in the Pacific Northwest of the U.S.A. remains turgid as the plants approach maturity in June and July, yet potentials of −50 to −100 bars or lower often develop in the surface 6 to 12 inches of soil containing the plant crowns. Deep-feeding fibrous roots supply such plants with water. Likewise, seeds planted in very dry soils in anticipation of a coming rainy season may offer infection courts to few if any fungi because of the extremely low water potentials (Griffin, 1966).

Water potentials that develop in specific infection courts because of water absorption should also be mentioned. Potentials approaching −15 bars may develop and persist for some time in infection courts such as the root-hair zone or imbibing seeds while higher potentials exist in nearby soil, sometimes as little as a few millimeters away. Water of high potential (wetter soils some distance removed from an absorbing plant organ) will flow to areas of low water potential (the absorbing surface-soil interface) due to the water potential gradient which develops outward from the absorbing surface-soil interface, but rate of flow may often be limiting, especially when the transpiration rate is high.

Observing infection of plants in soil presumed to be at or near $-1/3$ bar therefore does not necessarily mean that infection occurred under a stress as slight as $-1/3$ bar. Presumably steep water potential gradients are less common in soil near below-ground stems, bulbs, tubers, root tips, and mature sections of roots, since these plant parts are less active in the absorption of water.

WATER POTENTIAL AS A FACTOR INFLUENCING MICROBIAL ANTAGONISM OF PLANT PATHOGENS IN SOIL.—Knowing that fewer kinds of organisms will be capable of growth and reproduction as the soil water potential is lowered, one could predict that, in soil, the lower the soil water potential the less possibility of antagonism among different species. More particularly, a root parasite growing in soil at a high water potential may be subject to the competitive and other antagonistic effects of more organisms than one growing in soil at a low water potential. This concept may help explain why some root diseases are not serious in wet soils even when ample water is available for diffusion of exudates and for growth of the parasite.

Indications are that bacteria are relatively inactive in soil at water potentials approaching the wilting point of plants. Clark (1967) reports reduction in bacterial activity at matric potentials of -3 atm or less, and that at -15 atm, bacterial activity is markedly reduced. Nitrification apparently occurs only very slowly at -15 bars (Justice and Smith, 1962) and the activity of the sulfur oxidizers in some soils is virtually prevented at -10 atm (Moser and Olsen, 1953). Cook and Papendick (1968) made soil dilution plate counts of total bacteria (using a medium of Bunt and Rovira [1955]) in a Ritzville silt loam 24 and 72 hours after the soil was amended with nutrients (2,500 ppm C as glucose and 250 ppm N as ammonium sulfate). After both time periods, total bacteria per gm of soil was from two to three hundred times higher in soil at water potentials of -5 bars or more than at about -15 bars or less. The point of reduced activity appeared to coincide with a water potential of -9 to -10 bars. Dommerques (1962) and Kouyeas (1964) likewise noted greatest bacterial activity in wetter soils.

Recognizing the relatively high water requirements of bacteria in soil, it seems probable that antagonism by this entire group of soil organisms may be minimized at water potentials below -10 to -15 bars, which is still within the range of available water for plant growth and which may exist in infection courts provided by certain absorbing roots and seeds. Using Culmorum as a test fungus, Cook and Papendick (1969) obtained evidence to suggest that bacterial antagonism does indeed decrease with decreases in the soil water potential. As discussed above, chlamydospore germination of Culmorum was 40 to 50% in 24 hours at all water potentials tested above -50 to -60 bars. After 72 hours, however, fungus germlings had lost germ tubes in soil at -10 bars or wetter, whereas in drier soils they survived and appeared to make growth for at least 6 days before converting into new chlamydospores. The loss of germ tubes at -10 bars or above reflected a lysis of, and new chlamydospore formation within, the germ tubes. The amount of lysis and new chlamydospore formation increased proportionally as the soil water potential was increased above -10 bars. When streptomycin and neomycin were added to the soil (300 ppm each, in the soil) in addition to glucose + ammonium sulfate, little or no lysis was detected at potentials as high as -0.5 bars.

Under field conditions, the seedling blight and foot rot of wheat caused by Culmorum has been associated with dry soils (Cook, 1968). Since chlamydospore germination and continuing growth of the fungus germlings both are essential to infection, and since soil bacteria apparently limit germling growth of Culmorum in soils at high water potentials, it may be that bacteria protect the wheat from Culmorum in wetter soils.

The severity of pea root and foot rot caused by *F. solani* f. sp. *pisi* likewise is suppressed in wetter soils (Kraft and Roberts, 1968), possibly because the higher soil water contents favor lysis of germlings of the pathogen. Cook and Flentje (1967) observed that, as the initial soil water potential was adjusted towards $-1/3$ bar, the amount of germ-tube lysis in soil near germinating pea seeds increased markedly. The greater lysis in wetter soils corresponded with fewer thalli of the fungus ultimately reaching the foot region of the plants and also with a reduced number of lesions on the plants.

Examples of other *Fusarium*-incited diseases suppressed in wet soils are: seedling blight of wheat caused by *F. roseum* f. sp. *cerealis* 'Graminearum' (Dickson, 1923); damping-off of Ladino clover caused by *F. roseum* (Graham, Sprague, and Robinson, 1957); and stem rot of sweet potatoes caused by *F. solani* f. sp. *batatas* (Harter and Whitney, 1927). Stover (1953) found greatest increases in populations of *Fusarium* species at the lower soil water contents. Since reduced activity of *Fusarium* in wetter soils probably is not related to insufficient water for growth of this fungus group, the effect of wet soil must be indirect. In some cases, the indirect effect may relate to reduced aeration (Griffin, 1963a; Clark and Kemper, 1967). With Culmorum, however, germ-tube lysis occurs in soil at -0.25 bars under conditions of continuous aeration. Moreover, the amount of *Fusarium* activity may change markedly with but a slight change in the soil water content (Cook and Papendick, 1969). These slight changes in water content relate to marked changes in the water potential, which in turn may affect bacterial activity. The greater *Fusarium* activity in dry soil may occur because they escape antagonism by soil bacteria but can still extract water for their own growth and reproduction.

A differential effect of the soil water potential on a pathogen and its antagonists is also suggested in work by Bruehl and Lai (1968). *Cephalosporium gramineum*, cause of Cephalosporium stripe disease of wheat, produces an antibiotic which aids its persistence in straw refuse between crops. In infested straws in soil at relative humidities of 0.90 to 0.86, however, species of *Penicillium* apparently enjoy an advantage over *C. gramineum* and colonize *C. gramineum*-occupied straw. At R.H. of 0.95 and higher or 0.82 and lower, *C. gramineum* retains possession of the straw. Bruehl and Lai (1968) proposed that when adequate water is present for its metabolism (0.95 RH or higher) *C. gramineum* competes well with other soil microorganisms. Relative humidities between 0.90 and 0.86, on the other hand, limit metabolism (and hence antibiotic production) of *C. gramineum* but do not restrict growth of species of *Penicillium*; hence the latter colonize the straw in spite of the presence of *C. gramineum*. At 0.82 relative humidity, water in sufficient quantity is not available to the fungal competitors, and *C. gramineum* again persists in an inactive state as "sole" occupant of the straw.

Unlike soil bacteria, the actinomycetes apparently can grow in fairly dry soils (Sandford, 1923; Goss, 1937; Dommerques, 1962; Kouyeas, 1964). Bumbieris and Lloyd (1966), in a study on lysis of hyphae of *F. solani* f. sp. *phaseoli*, *Helminthosporium sativum*, and *Thielaviopsis basicola* in soil, observed greatest lysis in wet soil but also detected lysis in soil which they considered too dry for bacterial activity. They concluded that bacteria were responsible for the lysis in wet soil and actinomycetes in dry soil.

Kouyeas (1964) noted that bacteria were the primary colonists of wheat and tomato stems buried in wet soil (water potential assumed to be −1 bar or higher), but as the soil dried, actinomycetes and fungi colonized the stems. Upon rewetting the soil, bacteria again were the main colonists, presumably at the expense of the fungi and actinomycetes.

Under field conditions, the soil water potential varies with time, position in the soil profile, and as discussed earlier, proximity to different below-ground plant parts because of precipitation, irrigation, drainage, and evapo-transpiration. With a concept of fluctuating potentials in mind, and recognizing that different species and groups of species are favored or disfavored by different water potentials, one might ask whether man can be successful in controlling soil-borne pathogens in soil near plants through biological means. It seems unlikely that any one group of potential antagonists will operate on a "full time" basis in soil near plants. With evapo-transpiration, bacteria in general will probably become less effective. With rains, irrigations, and water movement by unsaturated flow toward the host, actinomycetes and possibly fungi may become less effective antagonists. Obviously soil water, and more particularly the soil water potential,

needs more careful consideration in future studies directed toward biological control of soil-borne fungal diseases of plants.

INFLUENCE OF SOIL WATER ON HOST EXUDATION.— The importance of host exudation to the nutrition of soil-borne plant pathogenic fungi and of the general soil microflora is well known (Schroth and Hildebrand, 1964; Rovira, 1965). There is now evidence that soil water affects this biologically significant process and in this way affects soil-borne fungal diseases of plants.

Kerr (1964) showed that loss of dry weight by pea seeds in natural soil increases with increases in soil water content and that loss of dry weight in turn is a reflection of the amounts of sucrose exuded. Infection of pea seeds by *Pythium ultimum* likewise increases with increased soil water content and is dependent on loss of sugars from pea seeds (Flentje and Saksena, 1964; Kerr, 1964).

The influence of soil water content on pea seed exudation and the biological significance of the exudation was also studied by Cook and Flentje (1967), who used *Fusarium solani* f. sp. *pisi* as a test organism. Dry weight lost per seed 20 hours after planting in a sandy loam was uniformly high (8 to 10 mg/seed) provided the soil water content at time of planting was 8.7% or more. Dry weight lost per seed was progressively less as the initial soil water content was lowered below 8.7%. Chlamydospore germination of *F. solani* f. sp. *pisi* also was uniformly high (40–50%) in 20 hours in soil contiguous to the seeds, provided the initial soil water content was 8.7% or more. As the initial soil water content was lowered below 8.7%, percentage chlamydospore germination near the pea seeds was progressively less. Nutrients and not water per se were apparently the limiting factor to germination in drier soils near seeds, for when a mixture of glucose and ammonium sulfate was added to the soil, percentage chlamydospore germination was about 53% in 20 hours at all water contents tested down to about 6%, the lowest water content tested.

To determine whether the reduced chlamydospore germination of *F. solani* f. sp. *pisi* in drier soil near seeds was important to infection, Cook and Flentje (1967) examined the foot region of 3-to-6-day-old pea seedlings growing in soil at 6 to 7% water content (where chlamydospore germination was reduced) and found the numbers of *Fusarium* thalli and of lesions caused by the fungus to be less than on seedlings growing in soil at 8 to 9% content water. As discussed in the preceding section, water contents greater than 8 to 9% favored lysis of the germlings in soil near seeds. Maximal infection of peas by this fungus occured at about 8% water content, presumably because this water content favored maximal chlamydospore germination yet did not promote significant amounts of lysis of the developing germlings.

To determine whether loss of sugar from pea seeds is influenced by the ability of sugar to diffuse through soil, Kerr (1964) measured loss of dry weight by pea seeds in soils of different bulk densities, the idea being that for a given water content per unit soil mass, the greater the bulk density, the greater the water content per unit volume of soil and hence the greater the amount of seed exudation. Loss of dry weight increased with increases in the soil bulk density, which supports the hypothesis that soil water is important to diffusion of sugar. Apparently, a reduction of water content in soil contiguous to an exuding host surface results in less movement of exudates away from the exuding surface. Such reduction in outward movement would result in fewer propagules being contacted by the exudates and apparently restricts continued diffusion of exudates from the host itself.

With pea seeds, however, exudation is accompanied by water absorption which means that nutrient availability through exudation may depend not only on the water content of soil near the seeds but also on the movement of water toward the seeds. The data of Kerr (1964) show that in 72 hours water absorption amounts to 300 to 400 mg/seed; this water must be replenished by unsaturated flow toward the seeds to maintain the same medium for diffusion of sugar. Unsaturated water flow is greater in soil with fine pores than in soil with coarse pores. This implies that where roots or seeds are actively absorbing water, the water potential gradient from the host surface outward would be less in heavier soils. This immediate vicinity of an absorbing plant surface therefore may be drier in sandy soil than in clay soil because the former, in addition to retaining less water at a given potential, also transmits water at a slower rate. This probably explains why, in the study by Kerr (1964), a given percentage of infection of pea seeds by *P. ultimum* occurred with progressively less available water as the soil used contained a greater proportion of fine particles, i.e. sand, light sandy loam, and loam. The increase in loss of dry weight and infection of pea seeds with increases in soil bulk density can also be explained in terms of water flow toward the seed; as the bulk density increased, water movement toward the seed would also increase, enhancing the opportunity for diffusion of sugar.

Raney (1965) suggested that diffusion of exudates in soil is dominated by the thickness of water films and their continuity. At roughly $-1/3$ bar water potential, the equivalent film thickness of water around solid particles is calculated to be only 18 to 24 monomolecular water layers thick (Richards, 1960). At -15 bars, the adsorbed films are considerably thinner, about 6 to 8 monomolecular layers. The soil pores, on the other hand, are actually tortuous channels of various sizes and geometries surrounding the solid particles and, when filled with water, can provide continuous paths of some magnitude for diffusion of exudates. As the water content is lowered, more and more water is present as thin films on the surface of the solid particles and as wedges at points of contact between the particles. It seems likely that diffusion of exudates will be dominated primarily by the degree of water-filled pores rather than by the thickness of water films.

With some root parasites, a greater release or availability of nutrients through host exudation may be inimical because of effects of the nutrients on microbial antagonism. For example, with *F. solani* f. sp. *phaseoli*, exudates rich in amino acids and sugars promote high percentages of chlamydospore germination in natural soil but continuing growth and survival of the germlings is poor. In contrast, exudates weak in amino acids and consisting primarily of sugars promote low percentages of chlamydospore germination, but germlings persist for days even in soil at $-1/3$ bar water potential and higher (Cook and Snyder, 1965). The increased lysis of germ tubes of *F. solani* f. sp. *pisi* in wetter soils near pea seeds also may relate to greater amounts of exudation and not just higher water potentials per se (Cook and Flentje, 1967). Bumbieris and Lloyd (1966) have also shown that nutrients influence the rapidity of hyphal lysis. Presumably a richer and more varied nutritional environment supports a more active and varied soil microflora, increasing the chances for antagonism among different species. Thus, the effect of an increasing water potential on antagonism in soil near exuding host parts may be direct because more organisms can obtain the water necessary for growth, and indirect because greater exudation supports a larger and more varied soil microflora.

LITERATURE CITED

BOUYOUCOS, G. J., and A. H. MICK. 1940. An electrical resistance method for the continuous measurement of soil moisture under field conditions. Mich. Agr. Exp. Sta. Tech. Bull. 172.

BRUEHL, G. W., and P. LAI. 1968. Influence of soil pH and humidity on survival of *Cephalosporium gramineum* in infested wheat straw. Can. J. Plant Sci. 48:245–252.

BUMBIERIS, M., and A. B. LLOYD. 1966. Influence of soil fertility and moisture on lysis of fungal hyphae. Australian J. Biol. Sci. 20:103–112.

BUNT, J. S., and A. D. ROVIRA. 1955. Microbiological studies of some subantarctic soils. J. Soil Sci. 6:119–128.

CAMPBELL, G. S., W. D. ZOLLINGER, and S. A. TAYLOR. 1966. Sample changer for thermocouple psychrometers: Construction and some applications. Agron. J. 58: 315–318.

CHEN, ALICE WHEI-CHU, and D. M. GRIFFIN. 1966. Soil physical factors and the ecology of fungi. V. Further studies in relatively dry soils. Trans. Brit. Mycol. Soc. 49:419–426.

CLARK, F. E. 1967. Bacteria in soil, p. 15–49. *In* A. Burges and F. Raw [ed.], Soil biology. Academic Press, London and New York.

CLARK, F. E., and W. D. KEMPER. 1967. Microbial activity in relation to soil water and soil aeration, p. 472–480. *In* R. M. Hagan et al. [ed.], Irrigation of Agricultural Lands, American Society of Agronomy, Madison, Wisconsin.

CLAYTON, C. N. 1942. The germination of fungous spores in relation to controlled humidity. Phytopathology 32:921–943.

COOK, R. J. 1968. *Fusarium* root and foot rot of cereals in the Pacific Northwest. Phytopathology 48:127–131.

COOK, R. J., and N. T. FLENTJE. 1967. Chlamydospore germination and germling survival of *Fusarium solani* f. *pisi* in soil as affected by soil water and pea seed exudation. Phytopathology 47:178–182.

COOK, R. J., and R. I. PAPENDICK. 1969. Soil water potential as a factor in the ecology of *Fusarium roseum* f. sp. *cerealis* 'Culmorum'. Plant Soil. (In Press.)

COOK, R. J., and W. C. SNYDER. 1965. Influence of host

exudates on growth and survival of germlings of *Fusarium solani* f. *phaseoli* in soil. Phytopathology 55:1021–1025.

COUCH, H. B., L. H. PURDY, and D. W. HENDERSON. 1967. Application of soil moisture principles to the study of plant disease. Dept. Plant Pathol., Research Division, Virginia Polytechnic Inst., Bull. 4. 23 p.

DICKSON, J. G. 1923. Influence of soil temperature and moisture on the development of the seedling-blight of wheat and corn caused by *Gibberella saubinetii*. J. Agr. Res. 23:837–869.

DOMMERQUES, Y. 1962. Contribution à l'étude de la dynamique microbienne des sols en zone semi-aride et en zone tropicale sèche. Ann. Agron. 13:265–324, 391–468.

FLENTJE, N. T., and H. K. SAKSENA. 1964. Pre-emergence rotting of peas in South Australia. III. Host-pathogen interaction. Australian J. Biol. Sci. 17:665.

GOSS, R. W. 1937. The influence of various soil factors upon potato scab caused by *Actinomyces scabies*. Nebraska Agr. Exp. Sta. Res. Bull. 93. 40 p.

GOTTLIEB, D. 1950. The physiology of spore germination. Botan. Rev. 16:229–257.

GRAHAM, J. H., V. G. SPRAGUE, and R. R. ROBINSON. 1957. Damping-off of Ladino clover and Lespedeza as affected by soil moisture and temperature. Phytopathology 57:182–185.

GRIFFIN, D. M. 1963a. Soil moisture and the ecology of soil fungi. Biol. Rev. 38:141–166.

GRIFFIN, D. M. 1963b. Soil physical factors and the ecology of fungi. I. Behavior of *Curvularia ramosa* at small soil water suctions. Trans. Brit. Mycol. Soc. 46:273–280.

GRIFFIN, D. M. 1963c. Soil physical factors and the ecology of fungi. II. Behavior of *Pythium ultimum* at small soil water suctions. Trans. Brit. Mycol. Soc. 46:368–372.

GRIFFIN, D. M. 1963d. Soil physical factors and the ecology of fungi. III. Activity of fungi in relatively dry soil. Trans. Brit. Mycol. Soc. 46:373–377.

GRIFFIN, D. M. 1966. Fungi attacking seeds in dry seedbeds. Proc. Linn. Soc. N. S. Wales. 91:84–89.

HARTER, L. B., and W. A. WHITNEY. 1927. The relation of soil temperature and soil moisture to the infection of sweet potatoes by the stem-rot organism. J. Agr. Res. 34:435–441.

INTERNATIONAL SOCIETY SOIL SCIENCE. 1963. Soil physics technology committee. Soil physics terminology. Intern. Soc. Soil Sci. Bull. 23. 4 p.

JUSTICE, K. J., and R. L. SMITH. 1962. Nitrification of ammonium sulfate in calcareous soil as influenced by combinations of moisture, temperature, and levels of added nitrogen. Soil Sci. Soc. Amer. Proc. 26:246–250.

KERR, A. 1964. The influence of soil moisture on infection of peas by *Pythium ultimum*. Australian J. Biol. Sci. 17:676–685.

KOUYEAS, V. 1964. An approach to the study of moisture relations of soil fungi. Plant Soil. 20:351:363.

KRAFT, J. M., and D. D. ROBERTS. 1968. Influence of soil water and temperature on the pea root rot complex caused by *Pythium ultimum* and *Fusarium solani* f. *pisi*. Phytopathology. (In Press.)

MONTIETH, J. L., and P. C. OWEN. 1958. A thermocouple method for measuring relative humidity in the range of 90–100%. J. Sci. Inst. 34:443–446.

MOSER, U. S., and R. V. OLSEN. 1953. Sulfur oxidation in four soils as influenced by soil moisture tension and sulfur bacteria. Soil Sci. 76:251–257.

NASH, SHIRLEY, T. CHRISTOU, and W. C. SNYDER. 1961. Existence of *Fusarium solani* f. *phaseoli* as chlamydospores in soil. Phytopathology 41:308–312.

PURDY, L. H., and E. L. KENDRICK. 1957. Influence of environmental factors on the development of wheat bunt in the Pacific Northwest. I. Phytopathology 47:591–594.

PURDY, L. H., and E. L. KENDRICK. 1963. Influence of environmental factors on the development of wheat bunt in the Pacific Northwest. IV. Phytopathology 53:416–418.

RANEY, W. A. 1965. Soil physical factors as they affect soil microorganisms, p. 115–119. *In* K. F. Baker and W. C. Snyder [ed.], Ecology of soil-borne plant pathogens. Univ. of California Press, Berkeley and Los Angeles.

RAWLINS, S. L., and F. N. DALTON. 1967. Psychrometric measurement of soil water potential without precise temperature control. Soil Sci. Amer. Proc. 31:297–301.

RICHARDS, L. A. 1947. Pressure-membrane apparatus, construction and use. Agr. Engr. 28:451–454.

RICHARDS, L. A. 1960. Advances in soil physics. 7th Intern. Congress of Soil Science 1:67–79.

RICHARDS, L. A. 1965. Metallic conduction for cooling a thermocouple psychrometer bath. Soil Sci. 100:20–24.

RICHARDS, L. A., and GEN OGATA. 1958. Thermocouple for vapor pressure measurement in biological and soil systems at high humidity. Sci. 128:1089–1090.

ROVIRA, A. D. 1965. Plant root exudates and their influence upon soil microorganisms, p. 170–186. *In* K. F. Baker and W. C. Snyder [ed.], Ecology of soil-borne plant pathogens. Univ. of California Press, Berkeley and Los Angeles.

SANDFORD, G. B. 1923. The relation of soil moisture to the development of common scab of potato. Phytopathology 13:231–236.

SCHNEIDER, R. 1954. Untersuchungen über feuchtigkeitsansprüche parasitischer Pilze. Phytopathol. Z. 21:61–78.

SCHROTH, M. N., and D. C. HILDEBRAND. 1964. Influence of plant exudates on root-infecting fungi. Ann. Rev. Phytopathol. 2:101–132.

SOIL SCIENCE SOCIETY OF AMFRICA. Committee on Terminology. 1965. Glossary of soil science terms. Soil Sci. Soc. Amer. Proc. 29:330–351.

SPANNER, D. C. 1951. The Peltier effect and its use in the measurement of suction pressure. J. Exp. Botany 2:145–168.

STOVER, R. H. 1953. The effect of soil moisture on *Fusarium* species. Can. J. Botany 31:693–697.

TAYLOR, S. A., D. D. EVANS, and W. D. KEMPER. 1961. Evaluating soil water. Utah Agr. Exp. Sta. Bull. 426.

U.S. SALINITY LABORATORY STAFF: 1954. Diagnosis and improvement of saline and alkali soils. Agr. Handbook No. 60. USDA.

Effect of Aeration and of Concentration of Carbon Dioxide on the Activity of Plant Pathogenic Fungi in the Soil

J. LOUVET—*Station de Recherches sur la Flore Pathogène dans le sol. Institut National de la Recherche Agronomique, Dijon, France.*

The action of carbon dioxide and oxygen on fungi, particularly on soil fungi, has received special attention in the last few years. Various authors have reviewed the subject: Cochrane (1958), Domsch (1962), Griffin (1963 and 1966), Louvet and Bulit (1964), Chapman (1965), Sewell (1965), Eckert and Sommer (1967). Certain pertinent papers were omitted from earlier reviews, and some additional papers on the subject have appeared recently. Bewley (1963) indicated that CO_2 increased infection of tomatoes by *Colletotrichum atramentarium*. The inhibitory effect of CO_2 on the sporulation of various fungi was shown by the work of Niederpruem (1963), Tschierpe and Sinden (1964), Ingold and Nawaz (1967), Medeiros and Alvim (1967). CO_2 inhibits the germination of conidia of *Erysiphe polygoni* (Jhooty 1967) but is essential for the germination of *Chaetomium globosum* (Buston, Moss, and Tyrrell, 1966). Jaffe (1966) studied the autotropism of *Botrytis cinerea* in atmospheres enriched with CO_2. The influence of CO_2 on mycelial growth and chlamydospores was demonstrated by Bourret, Gold, and Snyder (1965, 1968), Buston, Moss, and Tyrrell (1966), Lockhart (1967), Raabe and Gold (1967). The effect of oxygen on various fungi was investigated by Stolzy et al. (1965 *a*, *b*), Klotz et al. (1965), Caldwell (1965), Robb (1966), Jhooty (1967), and Lockhart (1967).

In earlier work Louvet and Bulit (1964, 1965) studied the effect of CO_2 on growth *in vitro* and parasitic activity of *Fusarium oxysporum* f. sp. *melonis* and of certain sclerotia-forming fungi. The present work deals with effect of CO_2 on population changes of *Fusarium oxysporum* f. sp. *melonis* and *Fusarium solani* in soil.

INFLUENCE OF AERATION AND OF CONCENTRATION OF CARBON DIOXIDE ON THE POPULATION OF FUSARIUM OXYSPORUM AND OF FUSARIUM SOLANI IN UNSTERILIZED SOIL.—Greenhouse soil was artificially infested with *F. oxysporum* f. sp. *melonis* without preliminary sterilization. Muskmelons planted in this soil developed severe wilt. The soil was then stored dry for 15 months to allow the inoculum to convert to chlamydospores.

The soil was moistened to 40% of its water-holding capacity at the start of the experiment and placed in glass cylinders 16 mm in diameter and 220 mm long. The cylinders were fitted at both ends with stoppers and connected in series by tubing so that air or a mixture of air with CO_2 in the desired concentrations could be passed through the respective series. The concentrations ranged from 0 to 22% of CO_2 by volume; they were regulated by means of a system of capillary tubes (Louvet and Bulit, 1964). Temperature was maintained at 25°C. Population counts were determined at the start of the experiment, one week later, and every two weeks thereafter, using the method of Bouhot, Bulit, and Louvet (1964). The medium used was modified from that described by Papavizas (1967): V-8 agar with the addition of 5% oxgall, 0.125% tetrachloronitrobenzene, 100 ppm sterptomycin, and 50 ppm chlortetracycline.

A marked increase in the number of propagules of *F. oxysporum* in the first three weeks of the experiment was noted at all concentrations of CO_2 (Fig. 1). This was attributed to the moistening of the soil which induced germination of the chlamydospores. The number of propagules remained almost constant from

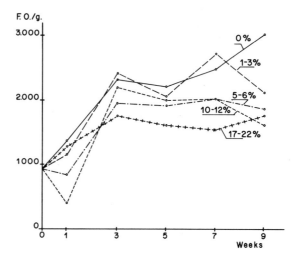

Fig. 1. Effect of CO_2 concentration on the population level of *Fusarium oxysporum* f. sp. *melonis* in soil.

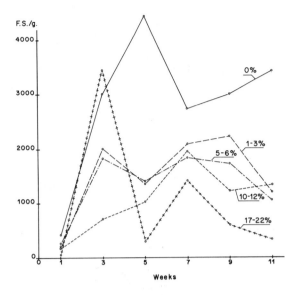

Fig. 2. Effect of CO_2 concentration on the population level of *Fusarium solani* in soil.

the third to the ninth week at CO_2 concentrations of 5 to 22%, but increased in soil exposed to air. The results for 1 to 3% CO_2 were similar to those for air, at least until the seventh week.

The soil used was naturally infested with *F. solani*. As with *F. oxysporum*, an increase in the number of propagules was noted during the first 3 weeks. The population level was higher in soil exposed to air than to a mixture of air and CO_2, and the number of propagules continued to increase until the end of the experiment (Fig. 2). At the end of the experiment there were only one tenth as many propagules in soil maintained under 17 to 22% CO_2 as in soil exposed to a current of air: 359/g compared to 3424/g.

DISCUSSION.—These results show clearly that aeration of soil favors, whereas CO_2 particularly at high concentrations inhibits, multiplication of both species of *Fusarium*. The effect of CO_2 is more pronounced on *F. solani* than on *F. oxysporum*.

Stover and Freiberg (1958), Newcombe (1960), and Bourret et al. (1965, 1968) have shown that CO_2 favors germination of chlamydospores and mycelial development of *F. oxysporum* f. sp. *cubense* and *F. solani* f. sp. *phaseoli* but inhibits the formation of new chlamydospores, thus leading to a decrease in the population level. Our results agree with this concept and confirm the importance of aeration and of CO_2 as ecological factors influencing the development of *Fusarium* in the soil.

LITERATURE CITED

BEWLEY, W. F. 1963. Commercial glasshouse crops, p. 75. Country Life Ltd., London. 523 p.

BOUHOT, D., J. BULIT, and J. LOUVET. 1964. Recherche d'une méthode sélective et quantitative d'analyse de *Fusarium oxysporum* f. *melonis* dans le sol. Ann. Epiphyties 15:57–72.

BOURRET, J. A., A. H. GOLD, and W. C. SNYDER. 1965. Inhibitory effect of CO_2 on chlamydospore formation in *Fusarium solani* f. *phaseoli*. Phytopathology 55:1052.

BOURRET, J. A., A. H. GOLD, and W. C. SNYDER. 1968. Effect of carbon dioxide on germination of chlamydospores of *Fusarium solani* f. sp. *phaseoli*. Phytopathology 58:710–711.

BUSTON, H. W., M. O. MOSS, and D. TYRRELL. 1966. The influence of carbon dioxide on growth and sporulation of *Chaetomium globosum*. Brit. Mycol. Soc. Trans. 49:387–396.

CALDWELL, J. 1965. Effects of high partial pressures of oxygen on fungi and bacteria. Nature (London) 206: 321–323.

CHAPMAN, H. D. 1965. Chemical factors of the soil as they affect microorganisms, p. 120–141. *In* K. F. Baker and W. C. Snyder [ed.], Ecology of soil-borne plant pathogens. Univ. of California Press, Berkeley and Los Angeles.

COCHRANE, V. W. 1958. Physiology of fungi. John Wiley and Sons, New York. 524 p.

DOMSCH, K. H. 1962. Bodenatmung sammelbericht über methoden und ergebnisse. Zentralbl. Bakt. Parasit. Infectionskr. u. Hyg. 116:33–78.

ECKERT, J. W., and N. F. SOMMER. 1967. Control of diseases of fruits and vegetables by postharvest treatment. Ann. Rev. Phytopathol. 5:391–432.

GRIFFIN, D. M. 1963. Soil moisture and the ecology of soil fungi. Biol. Rev. 38:141–166.

GRIFFIN, D. M. 1966. Soil physical factors and the ecology of fungi. IV. Influence of the soil atmosphere. Trans. Brit. Mycol. Soc. 49:115–119.

INGOLD, C. T., and M. NAWAZ. 1967. Carbon dioxide and fruiting in *Sphaerobolus*. Ann. Botany (London) 31:351–357.

JAFFE, L. F. 1966. On autotropism in *Botrytis*: measurement technique and control by CO_2. Plant Physiol. 41:303–306.

JHOOTY, J. S. 1967. Effect of CO_2 and O_2 on germination of *Erysiphe polygoni* conidia. Trans. Brit. Mycol. Soc. 50:299–304.

KLOTZ, L. J., L. H. STOLZY, T. A. DeWOLFE, and T. E. SZUSZKIEWICZ. 1965. Rate of oxygen supply and distribution of root-rotting fungi in soils. Soil Sci. 99:200–204.

LOCKHART, C. L. 1967. Influence of controlled atmospheres on the growth of *Gleosporium album* in vitro. Can. J. Plant Sci. 47:649–651.

LOUVET, J., and J. BULIT. 1964. Recherches sur l'écologie des champignons parasites dans le sol. I. Action du gaz carbonique sur la croissance et l'activité parasitaire de *Sclerotinia minor* et de *Fusarium oxysporum* f. *melonis*. Ann. Epiphyties 15:21–44.

LOUVET, J., and J. BULIT. 1965. Travaux sur l'écologie de certains champignons à sclérotes. Phytiatrie-Phytopharm. 14:179–182.

MEDEIROS, A. G., and P. DE T. ALVIM. 1967. Influência do gas carbonico e da umidade do ar na esporulação do *Phytophthora palmivora* (Butl.) Butl. Turrialba, 17:18–22 (in R.A.M. 49:9, 2672).

NEWCOMBE, MARGARET. 1960. Some effects of water and anaerobic conditions on *Fusarium oxysporum* f. *cubense* in soil. Brit. Mycol. Soc. Trans. 43:51–59.

NIEDERPRUEM, D. J. 1963. Role of carbon dioxide in the control of fruiting of *Schizophyllum commune*. J. Bacteriol. 85:1300–1308.

PAPAVIZAS, G. C. 1967. Evaluation of various media and antimicrobial agents for isolation of *Fusarium* from soil. Phytopathology 57:848–852.

RAABE, R. D., and A. H. GOLD. 1967. Effects of different levels of carbon dioxide on the growth of *Armillariella mellea* in culture. Phytopathology 57:101.

ROBB, SHEILA M. 1966. Reactions of fungi to exposure

to 10 atmospheres pressure of oxygen. J. Gen. Microbiol. 45:17–29.

SEWELL, G. W. F. 1965. The effect of altered physical condition of soil on biological control, p. 479–494. *In* K. F. Baker and W. C. Snyder [ed.], Ecology of soil-borne plant pathogens. Univ. of California Press, Berkeley and Los Angeles.

STOLZY, L. H., J. LETEY, L. J. KLOTZ, and C. K. LABANANKAS. 1965a. Water and aeration as factors in root decay of *Citrus sinensis*. Phytopathology 55:270–275.

STOLZY, L. H., J. LETEY, L. J. KLOTZ, and T. A. DE-WOLFE. 1965b. Soil aeration and root-rotting fungi as factors in decay of citrus feeder roots. Soil Science 99:403–406.

STOVER, R. H., and S. R. FREIBERG. 1958. Effect of carbon dioxide on multiplication of *Fusarium* in soil. Nature (London) 181:788–789.

TSCHIERPE, H. J., and J. W. SINDEN. 1964. Weitere Untersuchungen über die Bedeutung von Kohlendioxyd für die Fruktification des Kulturchampignons : *Agaricus campestris* var. *bisporus* (L.). Arch. Microbiol. 49:405–425.

PART V

◄

EFFECT OF ROOT EXUDATES ON ROOT INFECTION

Nutrition and Pathogenesis of Fusarium solani f. sp. phaseoli

T. A. TOUSSOUN—*Institute for Fungus Research, San Francisco, California.*

▶

Much has been written about nutrients and soil fungi, and the reader is referred to review articles by Rovira (1965), Patrick and Toussoun (1965), Schroth and Hildebrand (1964), Lockwood (1964), and Baker (1968) for an introduction to this literature. The purpose of the present short review is to present a cohesive picture of the effect of nutrients on the parasitism of a soil-borne fungus from spore germination to lesion development. I have chosen to do this with *Fusarium solani* f. sp. *phaseoli* simply because it has been the focus of investigations along these lines for the past 12 years by W. C. Snyder and his colleagues and consequently more is known about the biology of this pathogen in soil than possibly any other.

CHLAMYDOSPORE GERMINATION.—As a consequence of this work it has been firmly established that the pathogen normally exists in soil as quiescent chlamydospores (Nash, Christou, and Snyder, 1961) and that it is the host that is responsible for its own pathos. This condition is brought about by the diffusion into the soil of nutrients in plant exudates. These nutrients cause the chlamydospores to germinate, and the fungus to grow and produce a thallus on the surface of the host, to penetrate the host, and to establish secondary infection centers by means of runner hyphae which radiate out on the host surface from the primary infection center.

I would like to emphasize now that each of these steps is sustained in very large measure by nutrients present in the soil solution surrounding the host. It is only after parasitism is firmly established that the fungus apparently derives most of its nutrition from within the host.

The nutrient requirements of the fungus in soil appear to be simple, for it has been shown (Cochrane et al., 1963) that carbon and nitrogen sources are all that is required to sustain the activities of the organism. Cook and Schroth (1965) made a detailed study of these interactions and showed that chlamydospore germination occurred only when carbon and nitrogen were available to the fungus. Thus the application of amino acids such as asparagine, glutamine, glycine, etc., stimulated 40–50% of the chlamydospores to germinate. Interestingly, the application of higher doses of these compounds did not result in a concomitant increase in germination, and Cook and

Schroth concluded that soil factors such as competition by other soil microorganisms come strongly into play and limit the pathogen to this level of activity.

The complications that the soil environment thrusts into the situation are also evident in the results Cook and Schroth (1965) obtained with glucose and inorganic nitrogen sources. Glucose applied alone resulted in a germination maximum of 16–20%, quite below that obtained with amino acids. Inorganic nitrogen, in the form of potassium nitrate or ammonium sulfate, did not cause chlamydospores to germinate when applied alone. However, when glucose and inorganic nitrogen, particularly in the ammonium form, were added together, the level of germination could be as high as that achieved with amino acids. These results indicated that both carbon and nitrogen are required for chlamydospore germination and that in the soils that they used, a readily available source of carbon was the limiting factor. Powelson (1966) in his work on *Verticillium* also found that diffusible carbon substrates are the limiting factor. In such situations, therefore, the addition of inorganic nitrogen alone is of no avail. This also helped to explain something of a puzzle, for others of us had frequently obtained higher germination percentages upon the addition of glucose alone. It now appears that the agricultural soil we had used had a higher content of nitrogen than the nonagricultural soil that Cook and Schroth used. There are of course any number of such factors that can and do becloud the issue in these kinds of experiments, and it is therefore all the more remarkable that Papavizas, Adams, and Lewis (1968) reported results with glucose and inorganic nitrogen amendments that are almost identical to those obtained by Cook and Schroth (1965).

Carbon and nitrogen sources continue to have an effect on the growth of the fungus. Thus when Cook and Snyder (1965) compared the effects of asparagine, glucose, and glucose plus inorganic nitrogen, they found that the destruction of the hyphae through lysis was a function of nutrient concentration, being greatest at the higher concentrations. Lysis also depended on the kind of nitrogen added, since it was greater with asparagine than with glucose plus inorganic nitrogen. Other unknown factors appear also to be involved inasmuch as lysis was stimulated by the addition of yeast extract.

When glucose and asparagine are used separately but at equal concentrations, as Cook and Snyder (1965) have done, some interesting differences become evident. Thus, while asparagine caused the germination of about 30% of chlamydospores, most of the growth that ensued lysed in three days, so that less than 4% survived at this time. On the other hand, glucose (plus the background nitrogen of the soil) caused 15% of the chlamydospores to germinate, but after three days there was so little lysis that almost all survived. Raising the concentration of glucose does not raise the germination level, as I have mentioned previously. In fact, these higher concentrations are detrimental, since the hyphae became vacuolate and stopped growing under these conditions. From this it would appear, in the soil used by Cook and Snyder, that 2,000 ppm of glucose together with 10–20 ppm of background soil nitrogen are more than adequate for the fungus. In a separate study Schroth, Toussoun, and Snyder (1963) obtained 7% germination in the presence of 175 ppm of a mixture of three simple sugars and seven amino acids. These results also tend to show that the fungus may only require the added presence of a trace of sugar(s) or some similar energy source for germination and growth in soil.

PENETRATION.—In the ordinary course of events, the fungus upon reaching the hypocotyl forms a thallus on the host surface prior to penetration. Penetration was shown by Christou and Snyder (1962) to occur mostly through stomata and small wounds in the cuticle. In our earlier work (Toussoun, Nash, and Snyder, 1960), which gave our first direct evidence of the importance of fungus nutrition on pathogenesis, we observed that when conidia were placed in droplets of various nutrient concentrations on excised bean hypocotyls, lesions appeared earlier and developed most rapidly at the highest nutrient concentrations. The technique was refined in our subsequent experiments by using standardized optimum concentrations of washed spores and nutrient concentrations that were not toxic to the host. When glucose, potassium nitrate, L-arginine, and distilled water were compared under these conditions clear-cut differences were obtained. A fungal thallus developed in all solutions, but spore germination was most rapid in glucose. However, penetration of host tissues was observed first and was most abundant in the nitrogen solutions; it was sparse in the glucose solutions. Invasion of the host as determined by lesion size was greatest in the nitrogen solutions; it was least in glucose. These results were confirmed in glasshouse inoculations of seedlings grown in sand and watered with the various solutions. Disease was most severe with nitrogen, least severe with glucose.

Weinke ('962) made additional experiments along these lines using potassium nitrate and ammonium sulfate, and he refined the technique by adding these to soil infested with chlamydospores of the fungus. The concentrations of nitrogen he used were also much lower, in the order of 70–300 ppm. He showed that the development of the thallus on the hypocotyl was enhanced in the presence of nitrogen, particularly in the ammonium form, and that the lesions were produced earlier and were larger than in controls receiving only water. He concluded that ammonium nitrogen caused the pathogen to be more aggressive. By means of a double-pot technique Weinke also showed that the nitrogen enhancement of the disease occurred when the nitrogen was added to the soil in the upper pot, where the fungus and the hypocotyls were situated, and that it did not occur when the nitrogen was added to the lower pot, where the roots were situated, even though this nitrogen was taken up by the plant. Thus nitrogen added to a plant does not increase disease unless it is available *outside* the plant where the fungus can also get at it.

Cook (1964) also investigated the effects of nutrients on pathogenesis in soil on excised hypocotyls. He used nutrient concentrations of about 2,000 ppm and found that disease was most severe in the presence of ammonium nitrate or asparagine. Glucose, on the other hand, lessened the amount of disease. This effect was negated by the addition of nitrogen. He found that the addition of glucose, nitrogen, and yeast extract also resulted in lesser amount of disease, and he attributed this to increased competition of the microflora leading to lysis of the hyphae before penetration could occur.

The effect of nutrients placed outside the host does not stop after penetration. Weinke (1962) observed that runner hyphae grew out of the initial infection center and ramified over the surface of the host to cause the development of secondary infection centers. This phenomenon was found to take place abundantly in the presence of added nitrogen, particularly in the ammonium form. In fact, Weinke noted that runner hyphae and secondary-lesion development were frequently absent on plants where no nitrogen was added. Thus we see that nitrogen causes rapid invasion and the development of secondary lesions, all of which together with the growth of the fungus within the plant contribute to the coalescing of the discrete lesions to give the typical picture of Fusarium stem rot of bean as it is seen in the field.

CONCLUSIONS.—I hope I have advanced sufficient evidence to show that the events from spore germination to the establishment of the pathogen in its host are influenced by carbon and nitrogen sources available to the fungus in the soil solution. This effect of nutrition on pathogenesis has recently been shown to also occur with *Rhizoctonia* (Dodman, this volume).

In agricultural soils there appears to be sufficient nitrogen to support germination and growth. Ten–20 ppm is adequate and it could perhaps be as low as 5 ppm. The limiting factor in such soils would therefore be a readily available carbon source. Hence the addition of 100–1,000 ppm glucose suffices to stimulate the fungus. Low levels of nutrients may cause only 10% of the chlamydospores to germinate. This is not a low figure, however, especially if one recalls that Nash and Snyder (1962) found 1,500–3,000 propagules/g of soil in California bean fields where disease occurs. Indeed, the fungus appears to be fairly well off under these conditions and perhaps better off than when nutrients are abundant. For if glucose or nitrogen or both are increased very much, complications can arise. In the presence of excess glucose, other soil microorganisms compete vigorously for this energy substrate. This depletes the nitrogen supply, and the pathogen is nitrogen starved and stops growing. If nitrogen is in excess in the presence of glucose, and especially if it is in the organic form, the hyphae of the pathogen are destroyed by lysis. The same thing seems to occur if glucose together with inorganic or organic nitrogen are in excess.

The situation at the penetration and secondary infection stages appears to be the reverse of that obtaining during germination and growth, in that nitrogen appears to be the limiting factor. Increasing the nitrogen content of the soil at this time results in an increased aggressiveness of the pathogen.

In view of this it is clear that disease may be prevented by an insufficiency of readily utilizable carbohydrate or nitrogen. In practice (Baker, 1968), nitrogen withdrawal by the use of carbohydrate amendments seems to be the most promising method. That this should be the case may be clearer if one examines the locale for these activities. Germination, growth, etc., which leads directly to pathogenesis occurs in the immediate vicinity of the hypocotyl if not in actual contact with it (Toussoun and Snyder, 1961). Here, according to Schroth and Snyder (1961), the host exudes sugars almost exclusively. Cook (1964) is also of the opinion that the behavior of the fungus on the hypocotyl strongly resembles that obtained with a glucose amendment. Nitrogen, therefore, is on the threshold of unavailability to the pathogen, and any increase in carbohydrate level would render it unavailable through competition.

The whole situation is in delicate balance, for if the pathogen gains a foothold and a lesion is formed, exudation, especially of nitrogenous substances, increases, as was shown by Schroth and Teakle (1963). We can postulate therefore that more chlamydospores are caused to germinate, runner hyphae develop, new lesions are formed, and the process repeats itself, accelerating all the while. None of this may occur, however, if nitrogen is in too short a supply to support pathogen growth and development. There is the possibility, too, that higher levels of nitrogen are required for penetration than for growth. Recently, Patil and Dimond (1968) have shown that the production of extracellular polygalacturonase by *Fusarium oxysporum* f. sp. *lycopersici* is repressed in the presence of glucose. This glucose effect or catabolic effect may be the reason why glucose inhibits penetration of the pathogen.

It is evident that our knowledge of these matters must await a study of the enzymes involved in pathogenesis and of the effects of nutrition on these enzymes. This should be a rewarding area for fungus physiologists. We are also beginning to realize the importance of soil water on host-pathogen interactions (Cook and Papendick, this volume). This and other physical, chemical, and biological factors need to be understood and brought into the picture. Obviously no single discipline can accomplish such a huge task, and it should be evident that progress will come only when workers from related disciplines tackle the problem on a cooperative basis. When such becomes the case, what I have presented here may well seem rather rudimentary.

LITERATURE CITED

BAKER, R. 1968. Mechanisms of biological control of soil-borne pathogens. Ann. Rev. Phytopathol. 6:263–294.

CHRISTOU, T., and W. C. SNYDER. 1962. Penetration and host parasite relationships of *Fusarium solani* f. *phaseoli* in the bean plant. Phytopathology 52:219–226.

COCHRANE, F. C., V. W. COCHRANE, F. G. SIMON, and J. SPAETH. 1963. Spore germination and carbon metabolism in *Fusarium solani*. I. Requirements for spore germination. Phytopathology 53:1155–1160.

COOK, R. J. 1964. Influence of the nutritional and biotic environments on the bean root rot *Fusarium*. Ph.D. thesis, University of California, Berkeley.

COOK, R. J., and M. N. SCHROTH. 1965. Carbon and nitrogen compounds and germination of chlamydospores of *Fusarium solani* f. *phaseoli*. Phytopathology 55:254–256.

COOK, R. J., and W. C. SNYDER. 1965. Influence of host exudates on growth and survival of germlings of *Fusarium solani* f. *phaseoli* in soil. Phytopathology 55:1021–1025.

LOCKWOOD, J. L. 1964. Soil fungistasis. Ann. Rev. Phytopathol. 4:341–362.

NASH, SHIRLEY M., and W. C. SNYDER. 1962. Quantitative estimations by plate counts of propagules of the bean root rot *Fusarium* in field soils. Phytopathology 52:567–572.

NASH, SHIRLEY M., T. CHRISTOU, and W. C. SNYDER. 1961. Existence of *Fusarium solani* f. *phaseoli* as chlamydospores in soil. Phytopathology 51:308–312.

PAPAVIZAS, G. C., P. B. ADAMS, and J. A. LEWIS. 1968. Survival of root infecting fungi in soil. V. Saprophytic multiplication of *Fusarium solani* f. sp. *phaseoli* in soil. Phytopathology 58:414–420.

PATIL, SURESH S., and A. E. DIMOND. 1968. Repression of polygalacturonase synthesis in *Fusarium oxysporum* f. sp. *lycopersici* by sugars and its effect on symptom reduction in infected tomato plants. Phytopathology 58:676–682.

PATRICK, Z. A., and T. A. TOUSSOUN. 1965. Plant residues and organic amendments in relation to biological control, p. 440–459. *In* K. F. Baker and W. C. Snyder [ed.], Ecology of soil-borne plant pathogens. Univ. of California Press, Berkeley and Los Angeles.

POWELSON, R. L. 1966. Availability of diffusible nutrients for germination and growth of *Verticillium dahliae* in soils amended with oats and alfalfa residues. Phytopathology 56:895. (Abstr.)

ROVIRA, A. D. 1965. Plant root exudates and their influence upon soil micro-organisms, p. 170–186. *In* K. F. Baker and W. C. Snyder [ed.], Ecology of soil-borne plant pathogens. Univ. of California Press, Berkeley and Los Angeles.

SCHROTH, M. N., and D. C. HILDEBRAND. 1964. Influence of plant exudates on root infecting fungi. Ann. Rev. Phytopathol. 2:101–132.

SCHROTH, M. N., and W. C. SNYDER. 1961. Effect of host exudates on chlamydospore germination of the bean root rot fungus, *Fusarium solani* f. *phaseoli*. Phytopathology 51:389–393.

Schroth, M. N., and D. S. Teakle. 1963. Influence of virus and fungus lesions on plant exudations and chlamydospore germination of *Fusarium solani* f. *phaseoli*. Phytopathology 53:610–612.

Schroth, M. N., T. A. Toussoun, and W. C. Snyder. 1963. Effect of certain constituents of bean exudate on germination of chlamydospores of *Fusarium solani* f. *phaseoli* in soil. Phytopathology 53:809–812.

Toussoun, T. A., and W. C. Snyder. 1961. Germination of chlamydospores of *Fusarium solani* f. *phaseoli* in unsterilized soils. Phytopathology 51:620–623.

Toussoun, T. A., Shirley M. Nash, and W. C. Snyder. 1960. The effect of nitrogen sources and glucose on the pathogenesis of *Fusarium solani* f. *phaseoli*. Phytopathology 50:137–140.

Weinke, K. E. 1962. Influence of nitrogen on the root disease of bean caused by *Fusarium solani* f. *phaseoli*. Ph.D. thesis, University of California, Berkeley.

The Influence of Cottonseed Exudate on Seedling Infection by Rhizoctonia solani

D. S. HAYMAN—*Soil Microbiology Department, Rothamsted Experimental Station, Harpenden, Herts, England.*

INTRODUCTION.—Materials released from the underground parts of plants are an important source of nutrients for microorganisms in the soil, including plant pathogens. That such materials affect some fungus diseases has been shown with root exudates (Schroth and Hildebrand, 1964), but there have been few corresponding studies with seed exudates. The release and availability in soil of these materials are influenced by environmental conditions. This can indirectly affect pathogenesis, as when low temperatures increase root exudation and rotting of strawberry roots by *Rhizoctonia fragariae* (Husain and McKeen, 1963).

In this paper the results of work at the University of California, Berkeley (Hayman, 1967), on the effect of cottonseed exudate on preemergence damping-off of cotton seedlings by *Rhizoctonia solani* are discussed in relation to the general role of seed exudates in seedling infection.

MATERIALS RELEASED BY GERMINATING COTTON SEEDS.—Two acid-delinted seed lots of cotton (*Gossypium hirsutum* "Acala 4–42"), designated G and W, were compared. Their germination and growth rates were similar, but many more W seedlings damped off in field soil, especially at low temperatures.

Exudate was collected aseptically from seeds in moist sand, or in loose cheesecloth bags in a shallow layer of water, during the first 48 hours of germination. Total carbohydrates and ninhydrin-positive materials were determined photometrically, using anthrone and ninhydrin reagents, respectively. Individual sugars were identified by paper chromatography and individual amino acids recorded in an amino-acid analyser. CO_2 production was determined titrimetrically.

Carbohydrate was exuded faster and in greater amounts at 12° and 18°C than at 24°, 30°, and 36°C (Fig. 1A), whereas amino-acid exudation increased greatly at 36°C (Fig. 1B). Total carbohydrate and ninhydrin-positive materials were 4 to 10 times greater from W seeds than from G seeds. At 12°C, when exudation was greatest, the seeds did not germinate. At 30°C, when germination and growth were fastest, exudation was not correspondingly increased. Five sugars and sixteen amino acids were present in exudates from seeds of both lots at the five temperatures in about the same relative proportions. The sugars were glucose, fructose, sucrose, galactose, and probably lactose, and the amino acids were lysine, histidine, arginine, aspartic acid, threonine, serine, asparagine, glutamine, glutamic acid, glycine, alanine, valine, isoleucine, leucine, tyrosine, and phenylalanine, with traces of proline and methionine. CO_2 production was similar for both seed lots (Fig. 1C). This seed exudation seems to be a general leakage of materials, unrelated to respiration or germination and growth rates.

EFFECT OF COTTON SEED EXUDATE ON THE VIRULENCE OF RHIZOCTONIA SOLANI.—Dry weights of *R. solani* grown on exudate from W and G seeds corresponded to the quantity of sugars and amino acids present; on exudate collected at 24°C from 50 seeds of each kind during the first 48 hours of germination, dry weights were 66 ± 2.6 and 19 ± 1.7 mg respectively. Growth of *R. solani* on agar was greatly stimulated by seed exudate, and mycelial growth was much denser around W than G seeds, especially when low temperatures preceded high temperatures (Fig. 2). This directly affected pathogenesis. The accumulation of exudate at 12°C, before incubating at 30°C, caused the fungus to grow rapidly and kill the seeds before germination, whereas seeds kept at 30°C produced young seedlings before the fungus killed them. Corresponding results were obtained in field soil; when seeds were incubated for 3 days at 12°C to allow exudate to accumulate, and the temperature was then increased to 24°C, only half as many emerged as when seeds were kept at 24°C; by contrast, germination and emergence rates were the same with the two treatments in sterilized soil. Adding exudate to seeds in field soil decreased seedling emergence. Only half as many G seedlings emerged when G seeds were treated with W seed exudate as when treated with G seed exudate. Planting W seeds next to G seeds also decreased emergence of G seedlings. Leaching the seeds before sowing did not increase seedling emergence. Most cotton seedlings that failed to emerge were infected by *R. solani* close to the seed coat, showing lesions on the adjacent hypocotyl, cotyledons and emerging radicles.

Differences in hypocotyl susceptibility could not

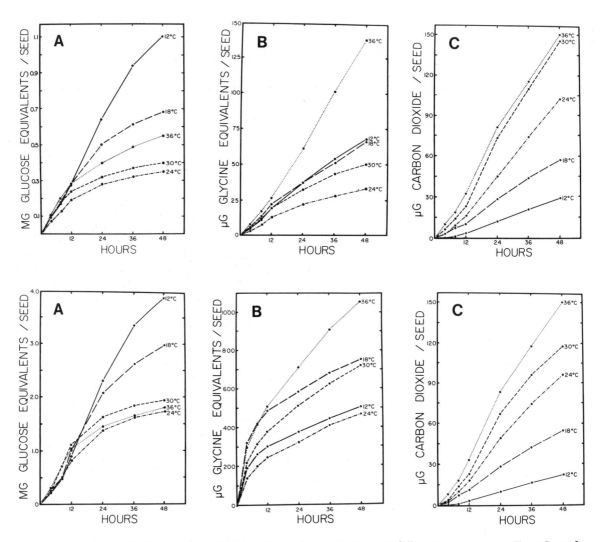

Fig. 1. Quantities of substances released by cotton seeds germinating at different temperatures. Top, G seeds; bottom, W seeds. *A*, total carbohydrates; *B*, ninhydrin-positive materials; *C*, carbon dioxide.

account for the differences in damping-off of G and W seedlings. When seedlings were inoculated with discs of the fungus growing on Ionagar, Ionagar plus G seed exudate, and Ionagar plus W seed exudate, percentage infection was 0, 40, and 89 for G seedlings and 0, 36, and 89 for W seedlings, respectively.

DISCUSSION.—The release of nutrient materials by germinating seeds stimulates a zone of increased microbial activity in the adjacent soil, termed the spermatosphere (Slykhuis, 1947). The materials released include large amounts of sugars from pea (Flentje and Saksena, 1964; Kerr, 1964), bean (Schroth, Toussoun, and Snyder, 1963), soybeans (Brown and Kennedy, 1966), and rice (Rajagopalan and Bhuvaneswari, 1964), also amino acids and vitamins (Brown and Kennedy, 1966; Rajagopalan and Bhuvaneswari, 1964; Verona, 1963; Woodcock,

1962). The nutrition of a soil-borne plant pathogen outside its host is an important factor in pathogenesis (Toussoun, Nash, and Snyder, 1960). Nutrients in seed exudates stimulate germination of chlamydospores of *Fusarium solani* f. sp. *phaseoli* (Schroth et al., 1963), and may stimulate propagules of *R. solani* to produce hyphae able to infect (Kamal and Weinhold, 1967). The virulence of *R. solani* on cotton hypocotyls is affected by the amount of nitrogen (asparagine) utilized (Weinhold and Bowman, 1967). Hexoses increase its virulence as a mycoparasite (Butler, 1957). In soil, infection cushions of *R. solani* can form on the surface of nylon bags containing sterile solutions of sugars such as glucose, fructose, and galactose, present in cottonseed exudate (Martinson and Holder, 1967). Seed exudation is greatest during the first day or two of germination, and is affected by age of seed (Verona, 1963), temperature (Hayman,

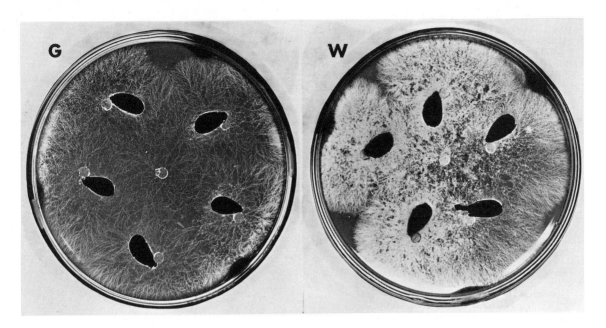

Fig. 2. Cotton seeds from seed lots G and W and *Rhizoctonia solani* on Ionagar for 4 days at 12°C followed by 1 day at 30°C.

1966; Schroth, Weinhold, and Hayman, 1966), aeration (Brown and Kennedy, 1966; Woodcock, 1962), soil moisture (Kerr, 1964) and soil type (Kerr, 1964). The results reported here show that exudation from cotton seeds is greatest during the first 24 hours of germination and is markedly increased at low temperatures. The increased damping-off of seedlings at low temperatures may be due primarily to accumulation of nutrients utilized by the pathogen while host growth is retarded. The exudation of nutrients by seeds and the formation of lesions near the seed coat suggest that the seed coat can provide a food base from which the pathogen can attack the susceptible juvenile tissues. The amount of nutrients exuded affected the inoculum potential of the fungus (*sensu* Garrett, 1956) at the infection site and hence the degree of infection.

With the two seed lots of cotton tested, the one that produced more exudate also damped off more in field soil. Similar correlations have been shown with bean (Schroth and Cook, 1964) and pea (Flentje and Saksena, 1964). Further correlations of this type involving several seed lots and varieties of a crop would indicate the extent of this phenomenon.

LITERATURE CITED

BROWN, G. E., and B. W. KENNEDY. 1966. Effect of oxygen concentration on *Pythium* seed rot of soybean. Phytopathology 56:407–411.

BUTLER, E. E. 1957. *Rhizoctonia solani* as a parasite of fungi. Mycologia 49:354–373.

FLENTJE, N. T., and H. K. SAKSENA. 1964. Pre-emergence rotting of peas in South Australia. III. Host-pathogen interaction. Australian J. Biol. Sci. 17:665–675.

GARRETT, S. D. 1956. Biology of Root-Infecting Fungi. Cambridge University Press, London and New York. 292 p.

HAYMAN, D. S. 1966. Seed exudate in relation to temperature and pre-emergence damping-off. Phytopathology 57:814. (Abstr.)

HAYMAN, D. S. 1967. The role of cotton and bean seed exudate in pre-emergence infection by *Rhizoctonia solani*. Ph.D. thesis, University of California, Berkeley.

HUSAIN, S. S., and W. E. McKEEN. 1963. Interactions between strawberry roots and *Rhizoctonia fragariae*. Phytopathology 53:541–545.

KAMAL, M., and A. R. WEINHOLD. 1967. Virulence of *Rhizoctonia solani* as influenced by age of inoculum in soil. Can. J. Botany 45:1761–1765.

KERR, A. 1964. The influence of soil moisture on infection of peas by *Pythium ultimum*. Australian J. Biol. Sci. 17:676–685.

MARTINSON, C. A., and ELIZABETH HOLDER. 1967. Stimulation of infection cushion formation by *Rhizoctonia solani* in soil. Phytopathology 57:342. (Abstr.)

RAJAGOPALAN, K., and K. BHUVANESWARI. 1964. Effect of germination of seeds and host exudations during germination on foot-rot disease of rice. Phytopathol. Z. 50:221–226.

SCHROTH, M. N., and R. J. COOK. 1964. Seed exudation and its influence on pre-emergence damping-off of bean. Phytopathology 54:670–673.

SCROTH, M. N., and D. C. HILDEBRAND. 1964. Influence of plant exudates on root-infecting fungi. Ann. Rev. Phytopathol. 2:101–132.

SCHROTH, M. N., T. A. TOUSSOUN, and W. C. SNYDER. 1963. Effect of certain constituents of bean exudate on germination of chlamydospores of *Fusarium solani* f. *phaseoli* in soil. Phytopathology 53:809–812.

SCHROTH, M. N., A. R. WEINHOLD, and D. S. HAYMAN. 1966. The effect of temperature on quantitative differences in exudates from germinating seeds of bean, pea, and cotton. Can. J. Botany 44:1429–1432.

SLYKHUIS, J. T. 1947. Studies on *Fusarium culmorum*

blight of crested wheat and brome grass seedlings. Can. J. Res. 25(C):155–180.

TOUSSOUN, T. A., SHIRLEY M. NASH, and W. C. SNYDER. 1960. The effect of nitrogen sources and glucose on the pathogenesis of *Fusarium solani* f. *phaseoli*. Phytopathology 50:137–140.

VERONA, O. 1963. Interaction entre la graine en germination et les microorganismes telluriques. Ann. Inst. Pasteur, Paris 105:75–98.

WEINHOLD, A. R., and TULLY BOWMAN. 1967. Virulence of *Rhizoctonia solani* as influenced by nutritional status of inoculum. Phytopathology 57:835–836. (Abstr.)

WOODCOCK, W. P. 1962. Influence of aeration on exudation from seeds and on growth of root-rot fungi. Phytopathology 52:927. (Abstr.)

Some Factors Involved in the Accumulation of Phycomycete Zoospores on Plant Roots

YUNG CHANG-HO and C. J. HICKMAN—*Department of Botany, University of Western Ontario, London, Ontario, Canada.*

Studies of the biology of root-infecting fungi have provided a multitude of examples of the influences of root exudates on the host-parasite interaction (Schroth and Hildebrand, 1964). The germination and subsequent development of spores and other propagules of pathogens lying on the root surface, or in the rhizosphere, may be directly stimulated or inhibited by root exudates; or affected indirectly through the effects of these materials on other members of the soil microflora (Park, 1963). The motility of zoospores provides opportunity for an additional response which, typically, leads to their accumulation culminating in encystment in large numbers immediately behind the root tip (Hickman and Ho, 1966). This response has been widely observed, it occurs in a number of fungi, and with few exceptions (Zentmyer, 1961) is nonspecific, accumulation occurring alike on roots of host and nonhost plants.

Accumulation of zoospores is the result of a sequence of responses (Royle and Hickman, 1964b). All-important in the initial stages of accumulation are attraction to the source of stimulation and trapping in a zone close to this source. These responses are followed in turn by rapid encystment and germination. A feature of cyst germination is that the germ tube emerges from a point on the cyst wall facing the stimulus source and grows toward it.

At the present time the precise nature of the stimulus is not entirely clear or whether, in view of the several responses involved in accumulation, not one but several factors may be involved (Ho and Hickman, 1967).

Of the explanations advanced to account for the phenomenon, electrotaxis, suggested by Troutman and Wills (1964), has not been confirmed with other fungi though it is recognized that electrical charges may play a role, possibly in adhesion of zoospores to root surfaces (Schroth and Hildebrand, 1964; Hickman and Ho, 1966; Katsura, Masago, and Miyata, 1966). Ho and Hickman (1967) found that zoospores of *Phytophthora megasperma* var. *sojae* displayed toward hydrogen resin particles all responses except that of attraction, while Katsura and Miyata (1966) described zoospore aggregation in response to a current of water emerging from a capillary tube at a velocity equal to the swimming speed of the zoospore and introduced the concept of rheotaxis.

Most widely held is the view that zoospore accumulation is brought about by exudation of one or more chemical compounds from roots. Amongst the responses, attraction is identified as chemotaxis, directed growth of germ tubes as chemotropism. The sites of maximum accumulation on roots, immediately behind the root tip, and over wounds, are those of maximum exudation (Pearson and Parkinson, 1961; Schroth and Snyder, 1961) and zoospores display their characteristic responses toward root exudate or extract diffusing from capillary root models. Interest has thus centred on the identification of the compounds involved. Two approaches have been used in these studies, (a) examination of zoospore responses to known compounds and relating these latter to root exudate, (b) fractionation of root exudate into progressively more delimited compounds, their identification and response tests, and subsequent tests with the pure compounds.

Royle and Hickman (1964b) reported an analysis of the first type using pea and *Pythium aphanidermatum*. In the present work the second approach was adopted using the same plant-fungus combination.

COMPOSITION OF PEA EXUDATE.—Sterile root exudate was collected from static sand cultures of seedling peas (3 weeks old), concentrated in vacuo at 40C to 1 ml for every 10 plants (10 P) and separated into cationic, anionic, and neutral fractions using ion-exchange resins. These fractions were further analysed by thin-layer chromatography. Detailed methods of analysis will be published in another paper.

Pea exudate contained at least three organic acids, three sugars, seventeen known and four unknown amino acids (Table 1). Lactic, tartaric, and ketoglutaric acids were found in the anionic fraction in minute quantities. The total acidity of the anionic fraction (amount exuded by 10 plants dissolved in 1 ml of water) was approximately 0.006N, determined by titration with NaOH. Quantitative determination of the sugars was made using the phenol-sulphuric acid method after elution. Fructose and glucose were dominant; sucrose was present in much smaller amount.

TABLE 1. Quantitative analysis of exudate

Compounds exuded		Amount/Plant
Organic acids		0.0006 N
Sugars		
	fructose	0.18 mg
	glucose	3.38 mg
	sucrose	0.07 mg
Amino-acids		
	a-alanine	0.0021 mg
	arginine	trace*
	asparagine	trace
	aspartic acid	0.0175 mg
	glutamine	0.0033 mg
	glutamic acid	0.0085 mg
	glycine	0.0005 mg
	histidine	0.0132 mg
	homoserine	0.0262 mg
	isoleucine	trace
	leucine	trace
	lysine	trace
	phenylalanine	trace
	proline	0.0084 mg
	serine	0.0080 mg
	tyrosine	trace
	valine	0.0025 mg

* Trace = less than 0.0002 mg/plant.

Aspartic acid and homoserine were found to be the dominant amino acids by analysis.

ACCUMULATION.—Accumulation tests were carried out using capillary root models (Royle and Hickman, 1964b), prepared from micro-pipettes of 0.44 mm diameter. Quantitative estimations of accumulation were made using the principle adopted by Rai and Strobel (1966). An accumulation ratio, AR, was calculated as: AR = number of zoospores adjacent to root model / number of zoospores beyond influence of root model. Counts were made from photographs of zoospores taken at fixed time intervals after introduction of root models into zoospore suspensions.

Crude exudate and major fractions (See Fig. 1 and Table 2.)—Crude exudate caused zoospore accumula-

TABLE 2. Zoospore accumulation in response to pea root exudate and the Major Fractions

Fraction		Accumulation ratio	Encyst-ment
control (blank agar)		1.12 ± 0.20	—
crude exudate	(1 min.)	10.09 ± 5.48***	+
	(3 min.)	19.23 ± 6.53***	
	(10 min.)	23.82 ± 0.71***	
anionic fraction	(1 min.)	1.68 ± 0.47*	±
	(3 min.)	2.40 ± 0.69**	
	(10 min.)	2.26 ± 0.78**	
neutral fraction	(1 min.)	5.14 ± 1.62***	+
	(3 min.)	14.30 ± 0.34***	
	(10 min.)	10.80 ± 0.14***	
cationic fraction	(1 min.)	4.57 ± 1.39***	+
	(3 min.)	15.22 ± 1.36***	
	(10 min.)	17.93 ± 0.13***	

*** significant difference between test and control: p = 0.001
** p = 0.01   * p = 0.05
\+ = positive, − = negative, ± = slight or sporadic.

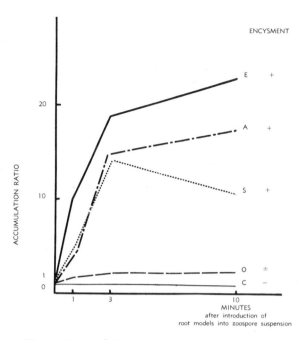

Fig. 1. Accumulation rations of zoospores towards crude exudate and major fractions. E = crude exudate, A = amino-acids (cationic fraction), S = sugars (neutral fraction), O = organic acids (anionic fraction), C = control.

tion within seconds. After 3 minutes a milling mass of zoospores was found at the mouth of the capillary tube. Accumulation continued after 3 minutes but at a slower rate. After 10 minutes most of the zoospores had encysted.

Compared with crude exudate, accumulation in response to the cationic fraction was slow in the first minute. The rate increased rapidly and the degree of encystment was nearly the same as with crude exudate. Accumulation in response to the neutral fraction at first followed that of the cationic fraction but fell off with time and encystment was slow. The organic acid fraction caused only very slight accumulation.

Eluted individual compounds.—(See Table 3.) The three major fractions were further separated by thin-layer chromatography into compounds or groups of compounds. These were eluted and their effects compared with those of substances eluted from a corresponding blank spot on the cellulose plate.

None of the organic acids significantly affected zoospore accumulation. All three sugars improved accumulation relative to the control, sucrose and glucose being most effective. The amino acids in the cationic fraction could not all be separated into individual compounds by one-way chromatography (which alone allowed location of spots on unsprayed plates). Of the groups of compounds tested the most active was a mixture of aspartic acid, arginine, glutamine, homoserine, and serine.

TABLE 3. Zoospore accumulation in response to eluted compounds

Compounds	Accumulation ratio
Control (blank cellulose)	2.13 ± 0.34
Anionic fraction	
ketoglutaric acid	2.44 ± 0.23
lactic acid	1.98 ± 0.13
tartaric acid	2.36 ± 0.45
Neutral fraction	
fructose	3.04 ± 0.39*
glucose	3.61 ± 0.30***
sucrose	4.51 ± 0.32***
Cationic fraction	
asparagine + histidine + lysine	3.16 ± 0.31**
arginine + aspartic acid + glutamine	
+ homoserine + serine	9.82 ± 0.71***
glycine + glutamic acid	3.15 ± 0.44*
alanine + proline	2.67 ± 0.35
valine	2.23 ± 0.13
isoleucine + leucine	2.13 ± 0.48

*** significant difference between test and control: p = 0.001
** p = 0.01   * p = 0.05

The results obtained with major fractions and with eluted compounds served to illustrate the effects of these compounds as found in the natural material. However, since the eluted substances also contained the soluble matter from the cellulose plates, no direct quantitative comparison could be made with the effects of crude exudate and major fractions.

Chemicals.—The analysis of effects of compounds present in natural material having proceeded as far as possible, attention was now turned to the effects of pure chemicals. Solutions of chemicals were made up at concentrations corresponding to the amount exuded by 10 plants dissolved in 1 ml of water (10 P). The organic acids were assumed to exist in equal amounts and each was made up in a 0.002 N solution. The trace amino acids were made up in a mixture containing 0.002 mg of each in 1 ml.

Zoospores were indifferent to ketoglutaric and lactic acid (Table 4). They accumulated in response to tartaric acid during the first minute but soon dispersed. At 3 minutes no accumulation was observed.

TABLE 4. Zoospore accumulation in response to organic acids and sugars

Compound	Accumulation ratio	Encystment
Control (blank agar)	1.05 ± 0.22	—
Organic acids		
ketoglutaric acid	1.27 ± 0.59	—
lactic acid	1.50 ± 0.31	—
tartaric acid (1 min.)	3.64 ± 0.83***	—
(3 min.)	1.22 ± 0.43	
Sugars		
fructose	1.19 ± 0.09	—
glucose	2.34 ± 0.50**	±
sucrose	3.04 ± 0.12**	±
complete mixture of sugars	4.45 ± 0.26***	+

*** significant difference between test and control: p = 0.001
** p = 0.01   * p = 0.05
+ = positive, — = negative, ± = slight or sporadic.

Of the sugars sucrose was most effective. Glucose also caused accumulation but fructose was without effect. The combination of all three sugars caused the greatest accumulation, but this effect was much less than that of the neutral fraction (Fig. 1; Table 2).

Of the eleven amino acids tested (Table 5), serine and glutamine were the most effective. The amino acids were also tested at a higher concentration (100P). At this level, glutamic acid, proline, and histidine became effective. The complete mixture of all the amino acids was stronger in effect than any single acid but much less effective than the cationic fraction (Fig. 1, Table 2). Of the chemicals tested, glucose, sucrose, glutamine, and serine were most effective in causing zoospore accumulation.

Combinations of compounds.—In addition to testing compounds singly, experiments were made with various combinations of compounds. With respect to amino acids, mixtures were built up, starting with aspartic acid, by adding one acid at a time and testing the effect (Fig. 2). These results confirmed the effectiveness of serine and glutamine in producing accumulation.

Royle and Hickman (1964b) showed that 1% casein hydrolysate alone caused mild accumulation by trapping of zoospores but a mixture of casein hydrolysate

TABLE 5. Zoospore accumulation in response to amino acids

Compound		Accumulation ratio	Encystment
Control (blank agar)		1.08 ± 0.21	—
a-alanine	(10P)	1.38 ± 0.40	—
	(100P)	1.35 ± 0.18	—
arginine	(10P)	1.30 ± 0.05	—
aspartic acid	(10P)	1.16 ± 0.11	—
	(100P)	1.18 ± 0.09	—
glutamine	(10P)	5.10 ± 1.90***	+
	(100P)	4.59 ± 1.43***	+
glutamic acid	(10P)	1.08 ± 0.26	—
	(100P)	2.34 ± 0.50**	+
glycine	(10P)	0.98 ± 0.13	—
	(100P)	1.19 ± 0.14	—
histidine	(10P)	1.14 ± 0.14	—
	(100P)	1.57 ± 0.46*	±
homoserine	(10P)	1.09 ± 0.18	—
	(100P)	1.14 ± 0.23	—
phenylalanine	(10P)	1.06 ± 0.06	—
	(100P)	1.11 ± 0.36	—
proline	(10P)	1.33 ± 0.39	—
	(100P)	2.25 ± 0.83**	±
serine	(10P)	2.31 ± 0.79**	+
	(100P)	3.30 ± 0.14***	+
traces	(arginine	1.02 ± 0.11	—
	+ asparagine		
	+ isoleucine		
	+ leucine + lysine		
	+ phenylalanine		
	+ tyrosine) (10P)		
valine	(10P)	1.10 ± 0.14	—
	(100P)	1.23 ± 0.13	—
complete mixture of amino-acids		5.69 ± 0.64***	+

*** significant difference between test and control: p = 0.001
** p = 0.01   * p = 0.05
+ = positive, — = negative, ± = slight or sporadic.

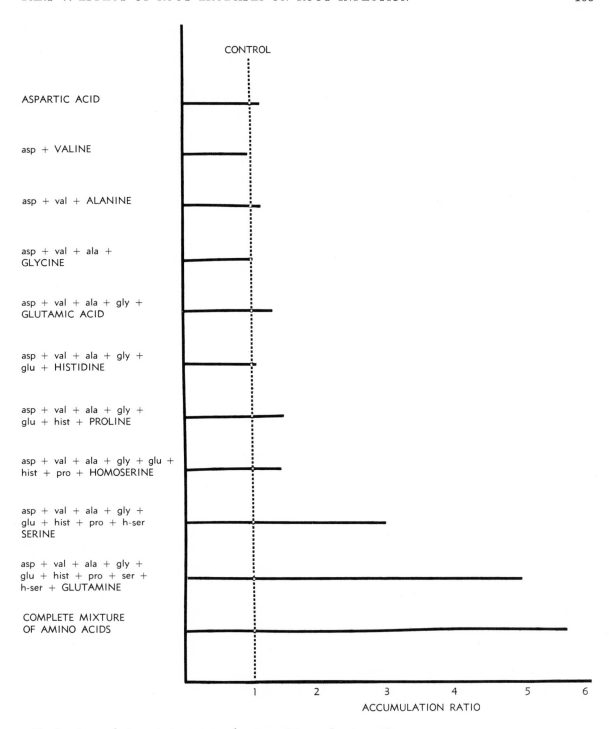

Fig. 2. Accumulation ratio in presence of various mixtures of amino-acids.

and sugars was very attractive. They also found 0.75% glutamic acid adjusted with ammonia to pH 5.4 and 7.0, respectively, to be very attractive. Casein hydrolysate does not contain glutamine, a compound found to be quite important in the present study. The effects of 1% casein hydrolysate, 1% casein hydrolysate +

sugars (10P), 1% casein hydrolysate + glutamine (10P) and 1% casein hydrolysate + glutamine (10P) + sugars (10P) are shown in Table 6. The effects of ammonium glutamate (0.75% glutamic acid) at pH 7.0, a mixture of amino acids and sugars based on analysis (10P), are also shown for comparison.

TABLE 6. Zoospore accumulation in response to various combinations of compounds

Compounds	Accumulation ratio	Encystment
Control (blank agar)	1.06 ± 0.19	—
1% Casein hydrolysate	3.35 ± 0.62***	+
1% Casein hydrolysate + glutamine (10P)	6.12 ± 0.48***	+
1% Casein hydrolysate + sugars (10P)	6.36 ± 0.71***	+
1% Casein hydrolysate + glutamine (10P) + sugars (10P)	7.40 ± 0.73***	+
ammonium glutamate, pH 7.0 (0.75 % glutamic acid)	4.28 ± 0.87***	
complete sugars and amino-acids (10P)	11.01 ± 0.24***	+
crude exudate (10P)	18.41 ± 3.26***	+

*** significant difference between test and control: p = 0.001
** p = 0.01   * p = 0.05
+ = positive, − = negative, ± = slight or sporadic.

DISCUSSION.—Among the many chemicals tested by Royle and Hickman (1964b) with *Pythium aphanidermatum*, very few approached the capacity of root material in inducing zoospore accumulation. Only ammonium glutamate and a mixture of sugars and amino acids possessed the activity of root exudates or extracts, but in terms of their carbon and nitrogen content the two could not be equated with root material. It seemed, therefore, that root material either must contain some other factor(s), or must consist of a more appropriately balanced mixture, as suggested by Carlile (1966) and Zentmyer (1966). This led to the present study in which chemicals were tested at concentrations as exuded by the plants.

It should be pointed out that the culture of *Pythium aphanidermatum* used for most of the work was not the same one as used by Royle and Hickman (1964 a, b) because it was no longer possible to obtain active zoospore suspension from that culture (ATCC 12192). However, this second culture (IMI 104926) behaved toward pea root, crude exudate, and the major fractions in all respects like the original one.

Royle (1963) described the response of zoospores to the cationic fraction as exhibiting all the characteristics shown by the root exudate and was deficient only in the degree of encystment. This observation was confirmed in the present study (Table 2). He reported that the neutral + anionic fraction stimulated good initial accumulation but this did not persist. This particular result is believed in the present study to be due to the action of the neutral rather than the anionic fraction (Table 2). The effect of the anions, not observed by Royle (1963), was much weaker.

Using chemicals, the results obtained from the different experimental approaches raised some points of interest. In general, Royle and Hickman (1964b) tested individual chemical compounds at concentrations of 1%, much higher than those used in the experiments reported here. At this concentration organic acids killed zoospores. In response to fructose, glucose,

and sucrose, some temporary aggregation of zoospores took place, the effect being much stronger with the latter two sugars. There was no encystment, however, nor did this occur in response to a mixture of these three sugars. In the present study, although organic acids, at the level of concentrations present in pea exudate, were virtually without effect, low levels of glucose and sucrose caused significant accumulation leading to some encystment, and a similar but stronger response was shown to a mixture of the three sugars at these concentrations (Table 4).

In both studies most amino acids were virtually ineffective when tested individually. Glutamic acid at 0.75% concentration was reported by Royle and Hickman (1964b) to be the only amino acid effective in inducing zoospore accumulation. This concentration, according to present analysis, approaches the amount exuded by 1,000 plants (Table 1). Zoospores accumulated in response to 0.085% glutamic acid (Table 5, 100P) but not at 0.0085% (10P). In the present study, glutamine and serine were found to be effective at the low concentrations used. The simulatory effect of glutamine was also obvious when it was added to a 1% casein hydrolysate (Table 6) and to a mixture of eight other amino acids (Fig. 2). With ammonium glutamate (0.75% glutamic acid) Royle and Hickman (1964b) obtained a response equivalent to that of root material. Though this compound had a significant effect in promoting zoospore accumulation in our experiments (Table 6), the level of response was well below that of crude exudate.

In studying the effects of various compounds on zoospore accumulation, several workers have observed the phenomenon of synergism. For example, Fisher and Werner (1958) demonstrated the phenomenon with mixtures of amino acids and inorganic salts on chemotaxis of *Saprolegnia mixta*. Royle and Hickman (1964b) showed that amino acids and sugars separately would only weakly attract P. *aphanidermatum* zoospores, but when the two groups of compounds were combined, they produced effects resembling that of pea-root exudate. We have confirmed this effect (Table 6).

In an investigation of the accumulation of zoospores of *Aphanomyces cochlioides* on beet roots, Rai and Strobel (1966) found that the amino-acid fraction of beet-root exudate promoted only germination whereas the organic-acid and sugar fractions attracted zoospores but did not affect germination and subsequent growth. In our own study, no specific function could be attributed to any one fraction alone. Rai and Strobel found that the natural material always exceeded the recombined fractions in effect and attributed this to the possibility that other biological factor(s) might be destroyed, or not isolated, during analysis. Lack of other active factor(s) could also be the explanation for the fact that mixtures of chemicals simulating exudate or fractions were always less effective than the original crude exudate (Tables 2, 4, 5, 6).

Interesting differences in zoospore response to different plants are sometimes obtained. Chemotaxis of zoospores of *Phytophthora cinnamomi* occurred to glutamic acid and aspartic acid, but these zoospores were rarely attracted to roots of nonhosts, e. g. tomato and pea; glutamic acid and aspartic acids, however, could be found in these nonhosts (Zentmyer, 1966). Chemotaxis of *A. cochlioides* zoospores occurred to gluconic acid in beet roots and to roots of other plants including pea (Rai and Strobel, 1966). In our own investigations, however, gluconic acid has not been found in Alaska pea-root exudate, which also attracted *A. cochlioides* zoospores.

The behaviour of zoospores toward plant roots is exceedingly complex and far from being fully understood. It is felt that unless accumulation tests are based on the actual composition of root material, the effect of certain compounds, though positive, cannot be taken as conclusive proof of root-zoospore interaction. Moreover, more detailed analysis of root material might contribute much in locating other, possibly important, factors. Also, it seems likely that more than one compound is responsible for the accumulation of the zoospores of a fungus, and that synergism may play an important part in the phenomenon. Finally, in the light of the examples mentioned above, comparative studies of several fungi in relation to different plants might also shed some light on the factors underlying zoospore response.

LITERATURE CITED

CARLILE, M. J. 1966. The orientation of zoospores and germ-tubes, p. 175–187. *In* M. F. Madelin [ed.], The fungus spore. Butterworths, London.

FISCHER, F. G., and G. WERNER. 1958. Die Chemotaxis der Schwärmsporen von Wasserpilzen (Saprolegniaceen). Hoppe-Seyler's Z. Physiol. Chem. 310:65–91.

HICKMAN, C. J., and H. H. HO. 1966. Behaviour of zoospores in plant-pathogenic Phycomycetes. Ann. Rev. Phytopathol., 4:195–220.

HO, H. H., and C. J. HICKMAN. 1967. Factors governing zoospore responses of *Phytophthora megasperma* var. *sojae* to plant roots. Can. J. Botany 45:1983–1994.

KATSURA, K., H. MASAGO, and Y. MIYATA. 1966. Movement of zoospores of *Phytophthora capsici*. I. Electrotaxis in some organic solutions. Ann. Phytopathol. Soc. Japan 32:215–220.

KATSURA, K., and Y. MIYATA. 1966. Movement of zoospores of *Phytophthora capsici*. III. Rheotaxis. Sci. Rep. Kyoto Prefect. Univ. Agr. 18:51–56.

PARK, D. 1963. The ecology of soil-borne fungal disease. Ann. Rev. Phytopathol. 1:241–258.

PEARSON, R., and D. PARKINSON. 1961. The sites of excretion of ninhydrin-positive substances by broad bean seedlings. Plant Soil 13:391–396.

RAI, P. V., and G. A. STROBEL. 1966. Chemotaxis of zoospores of *Aphanomyces cochlioides* to sugar beet seedlings. Phytopathology 56:1365–1369.

ROYLE, D. J. 1963. The behaviour of zoospores of *Pythium aphanidermatum* in response to roots, root substances and chemical compounds. Ph.D. thesis, Univ. of Western Ontario, London, Canada. 194 p.

ROYLE, D. J., and C. J. HICKMAN. 1964a. Analysis of factors governing *in vitro* accumulation of zoospores of *Pythium aphanidermatum* on roots. I. Behaviour of zoospores. Can. J. Microbiol. 10:151–162.

ROYLE, D. J., and C. J. HICKMAN. 1964b. Analysis of factors governing *in vitro* accumulation of zoospores of *Pythium aphanidermatum* on roots. II. Substances causing response. Can. J. Microbiol. 10:202–219.

SCHROTH, M. N., and W. C. SNYDER. 1961. Effect of host exudates on chlamydospore germination of the bean root rot fungus *Fusarium solani* f. *phaseoli*. Phytopathology 51:389–393.

SCHROTH, M. N., and D. C. HILDEBRAND. 1964. Influence of plant exudates on root-infecting fungi. Ann. Rev. Phytopathol. 2:101–132.

TROUTMAN, J. L., and W. H. WILLS. 1964. Electrotaxis of *Phytophthora parasitica* zoospores and its possible role in infection of tobacco by the fungus. Phytopathology 54:225–228.

ZENTMYER, G. A. 1961. Chemotaxis of zoospores for root exudates. Science 133:1595–1596.

ZENTMYER, G. A. 1966. Role of amino-acids in chemotaxis of zoospores of three species of *Phytophthora*. Phytopathology, 56:907. (Abstr.)

Tactic Responses of Zoospores of Phytophthora

G. A. ZENTMYER—*Department of Plant Pathology, University of California, Riverside.*

Zoospores of the genus *Phytophthora* react in a variety of ways to various types of stimuli. The primary purpose of this symposium is to develop information on the relation of root exudates to pathogenesis of root-invading fungi. Therefore, the object of this paper will be to indicate any significant relationships between exudation attraction and pathogenesis of zoospores of *Phytophthora* and related genera of Phycomycetes. Hickman and Ho (1966) and Carlile (1966) have recently summarized very well many of the aspects of zoospore taxis and tropic responses.

Root exudates have been reported by Rovira (1959) and others to contain a wide variety of chemicals, ranging from amino acids (23 identified in various exudates), sugars (arabinose, fructose, glucose, maltose, raffinose, sucrose, xylose), organic acids (acetic, butyric, citric, malic, oxalic, propionic, tartaric, valeric), vitamins (biotin, niacin, pantothenate, thiamine), enzymes (amylase, invertase, protease), and many miscellaneous substances, such as adenine, guanine, flavonone, glycosides, auxins. Our research on several of the hosts of *Phytophthora cinnamomi* has shown the presence of 12 amino acids in sterile root exudate (amino-acid tests with the cooperation of Dr. G. Bowen, CSIRO, Adelaide, Australia, and Dr. Brian Mudd, University of California, Riverside).

Root exudation varies with environmental conditions. Therefore, if there is a relationship between root exudation and infection, this relationship should be affected by varying environmental conditions. Environmental influences on exudation include the following effects. Amino acid exudation was increased by alternate wetting and drying of the soil as Katznelson, Rouatt, and Payne (1955) reported. This may bear some relation to reported increased incidence of some *Phytophthora*-induced diseases by alternate wetting and drying of the soil. Rovira (1959) showed that amino-acid exudation varies with light intensity, that exudation of some amino acids is decreased with decreasing light intensity while exudations of others is increased with decreased light intensity. Temperature effects on exudation are striking, with exudation generally increasing with high and low temperatures, compared to moderate temperatures of 20° to 27° C. Age of the plant also influences the amount of exudation.

ATTRACTION TO AMINO ACIDS.—Research, primarily with species of *Phytophthora* and *Pythium*, by several investigators over the past 7 or 8 years, has resulted in reports of attraction of zoospores to a variety of chemicals. The most consistent and numerous reports relate to attraction to amino acids. Royle and Hickman (1964) first noted the strong response of zoospores to glutamic acid and concluded that the response was similar in many ways to the attraction to roots but that additional factors involved in chemotaxis probably exist in roots.

Our research has confirmed some of the observations of Royle and Hickman, as will be noted below. Reports by Katsura and Hosomi (1963) and others have also dealt with attraction to glutamic acid as well as to other substances including sugars and various salts.

In our laboratory tests of chemotaxis we have used the method of Royle and Hickman (1964), incorporating chemicals in agar in capillary tubes.

Using this method, attraction to a number of amino acids, sugars, and organic acids has been studied, using five species of *Phytophthora*, and several different isolates of some of the species: *P. cinnamomi*, *P. cactorum*, *P. capsici*, *P. citrophthora*, and *P. palmivora*.

The most consistent and the strongest response has been to the dicarboxylic amino acids, glutamic, aspartic, and 4-aminobutyric acid. Response, particularly with *P. cinnamomi* was pH dependent, and was generally strongest when these amino acids were tested at low pH (pH3), near their isoelectric point. At pH6, the response was much less. Some of the species tested, including *P. citrophthora*, *P. cactorum*, and *P. capsici*, were usually attracted to a wide range of substances, amino acids and sugars, including asparagine, glutamine, sucrose, glucose, and organic acids such as citric acid, as well as the amino acids noted above. Chemoreceptors may be involved here, as Carlile (1966) postulates.

ATTRACTION TO ROOT EXUDATES.—Attraction has similarly been reported to roots, to root extracts and to root exudates. My first report (Zentmyer, 1961) emphasized the striking attraction of zoospores of *Phytophthora cinnamomi* to the region of elongation

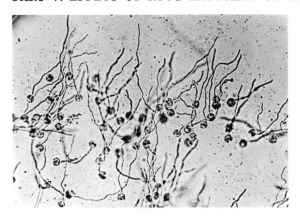

Fig. 1. Accumulation of zoospores of *Phytophthora cinnamomi* in region of elongation on avocado root, as result of chemotaxis.

on a root of a susceptible host (Fig. 1), and also reported attraction to aqueous extracts of roots. Goode (1956) and Bywater and Hickman (1959) had earlier reported encystment of zoospores of *P. fragariae* and of *P. erythroseptica* v. *pisi* in the root-tip zone. Many reports since 1961 have noted attraction to the region of elongation including those by Katsura and Hosomi (1963) with *Phytophthora capsici*, Troutman and Wills (1964) with *Phytophthora parasitica* v. *nicotianae*, Cunningham and Hagedorn (1962) with *Aphanomyces euteiches*. Dukes and Apple (1961) reported that *P. parasitica* v. *nicotianae* did not respond to intact roots, only to wounds on roots. A number of other investigators have also noted attraction of zoospores to wounds on roots.

There have been numerous reports of attraction of zoospores to host as well as nonhost roots, and to resistant as well as susceptible hosts. Most reports indicate lack of specificity in zoospore attraction and accumulation. The recent paper by Broadbent (1968) provides substantial evidence in this regard. Zoospores of *P. citrophthora* were equally attracted to roots of resistant and susceptible citrus rootstocks, and to tomato and black cowpea. Both resistant and susceptible citrus hosts were invaded by the germinating zoospores.

Roots of avocado sterilized with propylene oxide, or boiled, were not attractive to zoospores of *P. cinnamomi* (Zentmyer, 1961), whereas Goode (1956) found that zoospores of *P. fragariae* encysted on roots killed with boiling water. Our research has shown further some differences in hosts and nonhosts; these differences are not always consistent, however, and results have varied considerably from test to test. There are many factors affecting response of zoospores to roots and to chemicals, as has been evident in the hundreds of tests we have run with chemotaxis. Factors involved in these variable results include the vigor of the host, the type of root tested, the medium in

which the roots have been grown, the concentration and rate of motility of the zoospore suspension.

In other tests with *P. cinnamomi*, roots that were partially girdled were less attractive to zoospores than intact, vigorous roots. In our routine screening tests for resistance, roots that are damaged or partially severed do not show the lesions that are apparent within 4 or 5 days after inoculation of healthy, vigorous roots; these lesions always originate in the region of elongation. Removal of all or part of the foliage of an avocado seedling also significantly reduced the attraction of roots for zoospores, and reduced the subsequent infection of the roots. This response must reflect the change in quantity and possibly in quality of root exudate.

In further regard to possible specificity of response of zoospores to different plants or hosts with differing susceptibility, Turner's (1963) observations with *Phytophthora palmivora* are significant. He noted that zoospores responded differently to exudates of susceptible cacao hosts versus resistant selections.

It is apparent that at least some of the chemotactic response is the result of attraction to amino acids and sugars exuding from roots; this has not been demonstrated in soil, however. Another aspect of tactic responses which is not clearly defined is that of electrotaxis, first proposed for zoospores of *Phytophthora* by Troutman and Wills (1964) who reported attraction of zoospores of *P. parasitica* v. *nicotianae* to a weak electric current with attraction to the negative pole. Several other reports have indicated significant differences between this electrotaxis response and the reaction of zoospores to roots. The general conclusion has been that electrotaxis probably plays little role in attraction to roots in general. The relations of electrotaxis to chemotaxis are still somewhat obscure.

Tropic responses of germ tubes of germinating zoospores are also of significance. Orientation of germ tubes of zoospores of *P. cinnamomi* toward the region

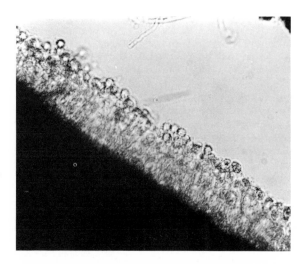

Fig. 2. Chemotropism of germ tubes of zoospores of *Phytophthora cinnamomi* toward avocado root.

of elongation of susceptible roots is striking (Zentmyer, 1961) (Fig. 2). Similar orientation has been observed to amino acids and to root exudates. This response must also play a role in root infection.

Following chemotaxis and chemotropism with susceptible hosts, penetration and infection takes place rapidly, as with invasion of avocado roots by *P. cinnamomi*. In some cases, as Broadbent (1968) has demonstrated, penetration may occur with resistant roots, but subsequent infection does not progress. Chemotaxis and chemotropism obviously are important phenomena in pathogenesis of root-invading fungi in the genus *Phytophthora,* and additional significant information in this regard is sure to develop in the next few years.

LITERATURE CITED

BROADBENT, PATRICIA. 1968. Observations on the mode of infection of *Phytophthora citrophthora* in resistant and susceptible citrus roots. Proc. Citrus Symp. (In press.)

BYWATER, J., and C. J. HICKMAN. 1959. A new variety of *Phytophthora erythroseptica* which causes a soft rot of pea roots. Trans. Brit. Mycol. Soc. 42:513–524.

CARLILE, M. J. 1966. The orientation of zoospores and germ-tubes, p. 175–187. *In* M. F. Madelin [ed.], The fungus spore, Colston Papers No. 19. Butterworths, London.

CUNNINGHAM, J. L., and D. J. HAGEDORN. 1962. Attraction of *Aphanomyces euteiches* zoospores to pea and other plant roots. Phytopathology 52:616–618.

DUKES, P. D., and J. L. APPLE. 1961. Chemotaxis of zoospores of *Phytophthora parasitica* v. *nicotianae* by plant roots and certain chemical solutions. Phytopathology 51:195–197.

GOODE, P. M. 1956. Infection of strawberry roots by zoospores of *Phytophthora fragariae*. Trans. Brit. Mycol. Soc. 39:367–377.

HICKMAN, C. J., and H. H. HO. 1966. Behaviour of zoospores in plant pathogenic Phycomycetes. Ann. Rev. Phytopathol. 4:195–220.

KATSURA, K., and T. HOSOMI. 1963. Chemotaxis of zoospores for plant roots in relation to infection by *Phytophthora capsici* Leonian. Sci. Rept. Kyoto Prefect. Univ. Agr. 15:27–32.

KATZNELSON, H., J. W. ROUATT, and T. M. B. PAYNE. 1955. The liberation of amino acids and reducing compounds by plant roots. Plant Soil 7:35.

ROVIRA, A. 1959. Root excretions in relation to the rhizosphere effect. IV. Influence of plant species, age of plant, light, temperature, and calcium nutrition on exudation. Plant Soil 11:53–64.

ROYLE, D. J., and C. J. HICKMAN. 1964. Analysis of factors governing *in vitro* accumulation of zoospores of *Pythium aphanidermatum* on roots. I. Behaviour of zoospores. Can. J. Microbiol. 10:151–162.

ROYLE, D. J., and C. J. HICKMAN. 1964. Analysis of factors governing *in vitro* accumulation of zoospores of *Pythium aphanidermatum* on roots. II. Substances causing response. Can. J. Microbiol. 10:201–219.

TROUTMAN, J. L., and W. H. WILLS. 1964. Electrotaxis of *Phytophthora parasitica* zoospores and its possible role in infection of tobacco by the fungus. Phytopathology 54:225–228.

TURNER, P. D. 1963. Influence of root exudates of cacao and other plants on spore development of *Phytophthora palmivora*. Phytopathology 53:1337–1339.

ZENTMYER, G. A. 1961. Chemotaxis of zoospores for root exudates. Science 133:1595–1596.

Germination of Chlamydospores of Phytophthora

SRECKO M. MIRCETICH and G. A. ZENTMYER—*Research Plant Pathologist, Crops Research Division, Agricultural Research Service, U. S. Department of Agriculture, Beltsville, Maryland, and Professor of Plant Pathology, University of California, Riverside.*

▶

Several *Phytophthora* species readily form chlamydospores (Waterhouse, 1956). Although germination of these spores has been observed occasionally (Blackwell and Waterhouse, 1931; Zentmyer and Paulus, 1957; Hendrix and Kuhlman, 1965; Ocana and Tsao, 1965; Mircetich and Zentmyer, 1966), the literature contains little data on the factors influencing their germination.

Elucidation of factors influencing chlamydospore germination should contribute to better understanding of the role of chlamydospores in survival and pathogenesis, as well as aid in designing more effective control measures.

GERMINATION OF CHLAMYDOSPORES UNDER ASEPTIC CONDITIONS.—Butler and Kulkarni (1913) and Dastur (1913) reported the germination of *P. colocasiae* and *P. parasitica*, respectively, upon addition of water to old cultures. Ashby (1920) observed germination of chlamydospores of *P. palmivora* within 24 hours. However, Rosenbaum (1917) failed to obtain germination of chlamydospores of *P. faberi (P. palmivora)*, *P. jatrophae*, and *P. parasitica* from young or old cultures. The amount and type of germination of *P. cactorum* chlamydospores were attributed to their maturity (Blackwell and Waterhouse, 1931). Mircetich, Zentmyer, and Kendrick (1968) found that chlamydospores of *P. cinnamomi* require an adequate supply of exogenous nutrients for germination. Availability of exogenous nutrient influenced both the amount of germination and type of germination. Under an optimum nutritional condition (V-8 juice: 1% nonsterile soil extract, 1:10 v/v) germination was 90% compared to 0% in the soil extract alone. Germinated spores developed 8-16 germ tubes/spore, and no germ tube bearing a sporangium was observed. However, when the same medium was diluted 1:100, germination was 41%; germinated spores developed 1-4 germ tubes/spore; and 13% of the chlamydospores had a short germ tube bearing a sporangium. The percentage of chlamydospores with sporangia remained approximately the same regardless of the maturity of the chlamydospores.

While vegetative growth of *P. cinnamomi* is significant only between pH 4.5–5.5 (Cameron and Milbrath, 1965), germination of chlamydospores was good to excellent at pH 3–9, but no germination occurred at pH 2.5 (Mircetich, Zentmyer, and Kendrick, 1968). Chlamydospores germinated over a wide range of temperatures: minimum 9°–12° C, optimum 18°–30° C and maximum 33°–36° C, but the isolate used in this study did not grow vegetatively at 33° C, although 54% of the spores germinated at this temperature (Mircetich, Zentmyer, and Kendrick, 1968).

Carbon and nitrogen requirements for germination.—Less is known about the physiology of chlamydospore germination than about the physiology of vegetative growth of *Phytophthora.*

Mircetich, Zentmyer, and Kendrick, (1968) reported that glucose, fructose, sucrose, $(NH_4)_2SO_4$, and KNO_3 failed to induce chlamydospore germination of *P. cinnamomi*, whereas citric acid (0.025 M) induced 53% germination. Asparagine (10^{-3} and $3.3 \times 10^{-4}M$) induced approximately 75% and 23% germination, respectively, but was ineffective at $1.6 \times 10^{-4}M$. Eleven amino acids (L-alanine, L-asparagine, L-aspartic acid, L-glutamic acid, L-glycine, L-leucine, L-lysine, L-methionine, L-phenylalanine, L-serine, and L-threonine) and casein hydrolysate, at a concentration of 350 mg total N/liter, were equally effective in stimulating germination (95–98%); while L-cysteine and L-arginine induced 82% and 72% germination, respectively. The percent germination in distilled water ranged from 0 to 6%. Similar requirements for chlamydospore germination of *P. parasitica* was reported by Tsao and Bricker (1968).

Various combinations of glucose, inorganic nitrogen, and mineral compounds used as the components of complete basal synthetic media (Erwin and Katznelson, 1961), with and without thiamine, failed to induce more germination of chlamydospores of *P. cinnamomi* than that obtained with asparagine and citric acid alone (Mircetich, Zentmyer, and Kendrick, 1968), indicating specificity in nutrient requirement for chlamydospore germination.

Roncadori (1965) demonstrated that glucose, fructose, $(NH_4)_2SO_4$ buffered with fumaric acid, and KNO_3 induce excellent vegetative growth of *P. cinnamomi* and *P. parasitica*. Vegetative growth of *P.*

112

cinnamomi was equally good in the presence of inorganic and amino nitrogen (Cameron and Milbrath, 1965).

Comparisons of the pH, temperature, and nutritional requirements for chlamydospore germination in *P. cinnamomi* and *P. parasitica* under aseptic conditions and those reported for vegetative growth suggest a very sharp physiological difference between these two stages of the same fungus.

GERMINATION OF CHLAMYDOSPORES IN NATURAL SOIL.—In natural soil, germination may be influenced by the surrounding environment and by the exchange between living and nonliving components in the soil. The nutritional requirement for germination may depend not only on the inherited character of the fungal spore, but also on the biotic and abiotic properties of the soil. Thus, the nutritional requirement for spore germination in natural soil may be different from that under aseptic conditions.

Carbon and nitrogen requirement.—Sugars and amino acids are commonly found in the root exudates. Inorganic nitrogen in the form of ammonium or nitrate may be present in soil, either as the product of mineralization of organic nitrogenous compounds or as introduced fertilizers. These compounds are often recognized as nutrients for spore germination of various fungi.

Several *Phytophthora* species exist in soil as chlamydospores (Hendrix and Kuhlman, 1965; Ocana and Tsao, 1965; Mircetich and Zentmyer, 1966). The effect of various factors on germination of chlamydospores may determine the role of this spore in survival and pathogenesis. Germination of chlamydospores of *P. cinnamomi* is enhanced in natural soil amended with certain nutrients, but varies in different soils amended with the same amount of nutrients (Mircetich, 1966). In contrast to results obtained under aseptic conditions, glucose (0.05 M) induced good germination of chlamydospores (60% in the Bonsall and 46% in the Vista soil series, compared to 9% in nonamended soils). Asparagine (0.0125 M) was less effective than glucose. Ammonium sulfate slightly enhanced chlamydospore germination in the Bonsall soil, but was ineffective in the Vista soil series, while NaNO$_3$ (0.012M) failed to enhance germination in either soil series. Germination of chlamydospores in natural soil amended with glucose (0.02 M) and asparagine (0.006 M) or glucose and (NH$_4$)$_2$SO$_4$ (0.006 M) was equal to that occurring on cornmeal agar control. Tsao and Bricker (1968) reported that germination of chlamydospores of *P. parasitica* in 5 citrus soils ranged from 43–78%, and was about the same or higher in sterile or glucose-amended, natural soils. Germination was reduced, however, when natural or sterile soils were amended with asparagine or NH$_4$NO$_3$.

Soil microflora and chlamydospore germination.—The microflora could presumably affect chlamydospore germination by competing for limited nutrients (Wen-Hsiung and Lockwood, 1967) and/or by elaborating materials toxic to chlamydospores (Dobbs, Hinson, and Bywater, 1960; Cook and Schroth, 1965). Glucose failed to stimulate germination of chlamydospores of *P. cinnamomi* under an axenic condition, but this compound was stimulatory in natural soil (Mircetich, 1966).

Germination of chlamydospores of *P. cinnamomi* in soil amended with various amounts of asparagine (0.0062–0.05 M) increased proportionally to asparagine concentration, as was noted also under aseptic condition. Apparently asparagine is directly utilized by chlamydospores, and the percentage of germination depends on the amount of available asparagine in the soil. However, in glucose-amended natural soil, the percent of germination remained the same regardless of glucose concentration (0.05–0.006 M). These results suggest that glucose was indirectly involved in stimulating germination of chlamydospores in the natural soil. Possibly the soil microflora played an important part in this phenomenon, by elaborating product(s) stimulatory to chlamydospore germination.

Mircetich (1966) studied the role of the soil microflora in germination of chlamydospores of *P. cinnamomi* in natural, asparagine- or glucose-amended and nonamended soils. In natural, nonamended soil supplemented with antibiotics Vancomycin (40μg/g soil) and Nystatin (8μg/g soil) to reduce activity of the microflora, germination of chlamydospores was approximately 55%, compared to 12% in nonamended, natural soil. Germination in natural soil amended with asparagine (0.006 M), with and without antibiotics, was approximately 76% and 45%, respectively, compared to 70% in cornmeal agar controls. Apparently, competition for available nutrient plays a significant role in germination of chlamydospores of *P. cinnamomi* in natural soil. Furthermore, the amount of fungitoxic materials present in the soils, either prior to or after the addition of antibiotics and nutrients, did not significantly inhibit germination of chlamydospores of *P. cinnamomi*. In natural, glucose-amended soil (0.006 M) germination was 77% compared to 72% on cornmeal agar controls. However, germination in natural soil amended with the antibiotics and glucose was 62% compared to 55% in the soils amended with the antibiotics alone. Thus, the full activity of soil microflora has a beneficial effect on chlamydospore germination of *P. cinnamomi* in natural soil when glucose is the source of exogenous energy.

Influence of root exudates on chlamydospore germination.—Influence of root exudates on root-infecting fungi and soil microorganisms was recently reviewed (Schroth and Hildebrand, 1964), although the literature is lacking in data on the effects of root exudates on the chlamydospore germination of *Phytophthora* species. Root exudates contain a number of compounds (Rovira, 1965) that may influence germination of chlamydospores of these fungi.

Zentmyer and Mircetich (1966) reported that *P. cin-*

namomi persisted in fallow soil over 6 years. This fungus was readily isolated each year for 11 years from soil under culture of avocado and macadamia nut. However, attempts failed to isolate the fungus from soil after 1 year under citrus culture, suggesting that the survival period of the pathogen was shorter in the presence of citrus roots than in fallow soil.

Chlamydospores of *P. cinnamomi* germinate readily in rhizosphere soil of host and of nonhost (Mircetich, 1966). Chlamydospore germination in the Bonsall soil series containing root exudates of avocado and rough lemon was equal to that observed in cornmeal agar controls, and about 10 times higher than in fallow soil. Although the germination of the spores in the Vista soil series containing the root exudates was somewhat less than that in the Bonsall soil, it was approximately 8 times higher than in the fallow soil.

Apparently, mass germination of chlamydospore of *P. cinnamomi* may occur in natural soils containing root exudates of host and nonhost plants. Thus, chlamydospore germination in the rhizosphere is relatively nonspecific. In the presence of a host plant, the germination of the resting structures may lead to infection and perpetuation of the pathogen. However, germination of chlamydospores in the presence of root exudates of a nonhost may lead to shortening of the survival period, providing that the nutrients are not sufficient to support saprophytic growth and formation of new resting spores.

Influence of exogenous nutrients on the type of germination of chlamydospores in natural soil.—The type of germination as well as the persistence of germ tubes in the natural habitat of chlamydospores may determine the importance of these propagules in survival and pathogenesis.

In natural, nonamended soils, most chlamydospores of *P. cinnamomi* (Mircetich, 1966) and *P. parasitica* (Tsao and Bricker, 1968) produced a short germ tube bearing a sporangium. In natural, glucose-amended soil most of the spores developed numerous germ tubes that continued to grow forming mycelia. Chlamydospores of *P. cinnamomi* in natural soil amended with excised avocado roots (1.5% w/w) germinated readily (over 80%, compared to approximately 10% in nonamended soil), and predominantly with numerous germ tubes. Microscopic examination revealed no lysis of the germ tubes within 8 days. Thus, is appears that germ tubes are relatively resistant. The relative persistence of germ tubes in soil, and the dependence of the spores on exogenous nutrients for their germination as well as the capability of chlamydospores to persist in the soil (Mircetich, 1966), suggest that these structures are capable of bridging hosts in time. Since a large number of these spores under certain nutritional conditions (e.g. fallow soil or in the soil at some distance from the roots of a host) may develop a functional sporangium bearing zoospores which are capable of moving actively toward the host roots, it is apparent that chlamydospores also are an effective device for bridging hosts in space.

CONCLUSION.—It is clear that the basic studies on the factors influencing germination of chlamydospores of *Phytophthora* species are long overdue. There is a wealth of data on the effect of root exudates on the spores of various root-infecting fungi (Schroth and Hildebrand, 1964), but there are few data concerning the factors influencing germination of resting structures of *Phytophthora*. Chlamydospores of *P. cinnamomi* and *P. parasitica* germinate readily only under specific nutritional conditions, and their germ tubes are relatively resistant to general antagonism in soil. Since the amount of germination and the type of germination are governed in a manner that is more likely to assure perpetuation of the fungus, it is obvious that chlamydospores of these two *Phytophthora* spp. are an effective evolutionary device in survival and pathogenesis.

LITERATURE CITED

ASHBY, S. F. 1920. Bud rot caused by *Phytophthora palmivora* Butler. West Ind. Bul. XVIII., p. 62.

BLACKWELL, E. M., and G. M. WATERHOUSE. 1931. Spores and spore germination in the genus *Phytophthora*. Trans. Brit. Mycol. Soc. 15:294–320.

BUTLER, E. J., and G. S. KULKARNI. 1913. Colocasia blight, caused by *Phytophthora colocasiae* Rac. Mem. Dep. Agric. India, Bot. Ser. V, p. 233.

CAMERON, H. R., and G. M. MILBRATH. 1965. Variability in the genus *Phytophthora*. I. Effect of nitrogen source and pH on growth. Phytopathology 55:653–657.

COOK, R. J., and M. N. SCHROTH. 1965. Carbon and nitrogen compounds and germination of chlamydospores of *Fusarium solani* f. *phaseoli*. Phytopathology 55:254–256.

DASTUR, J. F. 1913. On *Phytophthora parasitica* nov. spec. Mem. Dept. Agric. India, Bot. Ser. V, p. 177.

DOBBS, G. G., W. H. HINSON, and J. BYWATER. 1960. Inhibition of fungal growth in soils, p. 130–47. *In* D. Parkinson and J. S. Waid [ed.], The ecology of soil fungi. Liverpool Univ. Press, Liverpool, England.

ERWIN, D. C., and H. KATZNELSON. 1961. Studies on the nutrition of *Phytophthora cryptogea*. Can. J. Microbiol. 7:15–25.

HENDRIX, F. F., and E. G. KUHLMAN. 1965. Existence of *Phytophthora cinnamomi* as chlamydospores in soil. Phytopathology 55:499. (Abstr.)

MIRCETICH, S. M. 1966. Saprophitic behavior and survival of *Phytophthora cinnamomi* in soil. Ph.D. thesis, Univ. of Calif., Riverside, 159 p.

MIRCETICH, S. M., and G. A. ZENTMYER. 1966. Production of oospores and chlamydospores of *Phytophthora cinnamomi* in roots and soil. Phytopathology 56:1076–1078.

MIRCETICH, S. M., G. A. ZENTMYER, and J. B. KENDRICK, JR. 1968. Physiology of germination of chlamydospores of *Phytophthora cinnamomi*. Phytopathology 58:666–671.

OCANA, G., and P. H. TSAO. 1965. Origin of colonies of *Phytophthora parasitica* in selective pimaricin media in soil dilution plates. Phytopathology 55:1070. (Abstr.)

RONCADORI, R. W. 1965. A nutritional comparison of some species of *Phytophthora*. Phytopathology 55:595–599.

ROSENBAUM, J. 1917. Studies on the genus *Phytophthora*. J. Agr. Res. 8:233–276.

ROVIRA, A. D. 1965. Plant root exudates and their influence upon soil microorganisms, p. 170–184. *In* K. F. Baker and W. C. Snyder [ed.], Ecology of soil-borne pathogens. Univ. of California Press, Berkeley and Los Angeles.

SCHROTH, M. N., and D. C. HILDEBRAND. 1964. Influence of plant exudates on root-infecting fungi. Ann. Rev. Phytopathol. 2:101–132.

Tsao, P. H., and J. L. Bricker. 1968. Germination of chlamydospores of *Phytophthora parasitica* in soil. Phytopathology 58:1070. (Abstr.)

Waterhouse, G. M. 1956. The genus *Phytophthora*. Commonwealth Mycol. Inst. Miscellaneous Publication No. 12. p. 120.

Wen-Hsiung, Ko, and J. L. Lockwood. 1967. Soil fungistasis: Relation to fungal spore nutrition. Phytopathology 57:894–901.

Zentmyer, G. A., and A. O. Paulus. 1957. Phytophthora avocado root rot. Calif. Agr. Expt. Sta. Circular 465, 15 p.

Zentmyer, G. A., and S. M. Mircetich. 1966. Saprophitism and persistence in soil by *Phytophthora cinnamomi*. Phytopathology 56:710–712.

Factors Affecting the Prepenetration Phase of Infection by Rhizoctonia solani

R. L. DODMAN—*Research Fellow, Department of Plant Pathology, Waite Agricultural Research Institute, Adelaide, South Australia.*

INTRODUCTION.—In considering factors affecting cushion formation by *Rhizoctonia solani*, it is also necessary to examine growth of the fungus on and around the plant prior to cushion formation. Although there are distinct morphological and probably physiological differences between vegetative growth on the plant and the formation of cushions, it seems probable that both of these phases are important in the establishment of infection.

The following discussion will consider first the general factors affecting growth and cushion formation, followed by the more specific morphogenetic factors involved in the stimulation of cushion development. Finally the relation of these aspects to specificity is examined.

GENERAL FACTORS AFFECTING GROWTH AND CUSHION FORMATION.—It is now well known that plant materials influence the behavior of microorganisms in soil. There is considerable evidence that materials released from seeds and roots affect the saprophytic activity of *R. solani* and these effects may be reflected by changes in disease severity.

Observations of the fungus in soil have been made by Kerr (1956), who showed that exudates from plant roots were able to diffuse through cellophane and stimulate growth of various isolates of *R. solani*. Other studies by Wyllie (1962) and Martinson (1965) have provided similar evidence; Martinson also found that seed exudates induced a marked stimulation of hyphal growth.

Other workers have collected exudates and examined the effect of these materials on the fungus both *in vitro* and in soil. Kerr and Flentje (1957), Flentje, Dodman, and Kerr (1963), and de Silva and Wood (1964) showed that exudates from seedlings stimulated hyphal growth on cellophane *in vitro*. Nour El Dein and Sharkas (1964) found that root exudates from tomato seedlings supported vigorous growth of *R. solani* in liquid culture. The effect of exudates on the fungus in soil has been examined by Martinson and Baker (1962), who found that seedling exudates added to soil microbiological sampling tubes increased the frequency of isolation of *R. solani*. In other experiments, Martinson (1965) showed that seed exudates placed in cellophane bags caused a stimulation of hyphal development on the outside of the bags when they were placed in infested soil.

From the above evidence it can be concluded that plant exudates and other materials in soil have the potential to influence the activity of *R. solani* in soil. It may be suggested that the influences are mainly on saprophytic rather than parasitic activities, but these phases are often closely associated. Thus the support of growth on and near plant seeds and roots in soil will tend to increase the population of the fungus and provide better opportunities for penetration and infection.

There is now increasing evidence that the stimulation of growth of *R. solani* by plant materials plays an important role in determining fungal virulence and disease severity. Barker (1961) found that if bean seed was sown in soil infested with some isolates of *R. solani* there was almost 100% preemergence damping-off, whereas when seedlings were planted in the same soil there was little infection. Since it is known that bean seeds exude large amounts of materials in comparison with bean hypocotyls (Schroth and Snyder, 1961), it is possible that the differences in disease severity can be attributed to differences in growth induced by the seed and hypocotyl exudates. Nour El Dein and Sharkas (1964) reported that the level of disease caused by *R. solani* in three tomato varieties is directly correlated with the stimulation of growth induced by exudates from the different varieties. Similarly, Schroth and Cook (1964) found that bean varieties which are more susceptible to damping-off caused by *Pythium* and *Rhizoctonia* exude greater amounts of sugars and amino acids than do less susceptible varieties.

Finally, Hayman (this volume) has shown that differences in the susceptibility of two types of cotton are closely correlated with differences in the amounts of exudates released and with the effects of these exudates on stimulation of fungal growth. From these examples it is clear that the stimulation of growth induced by exudates is an important factor in infection by *R. solani*. However, the exact role that exudates play in growth and infection is not known. Recent studies by Weinhold and Bowman (1967) and Weinhold, Bowman, and Dodman (unpublished data) have helped to provide some understanding of this aspect of infection.

In studying damping-off of cotton caused by *R. solani*, it was found by Kamal and Weinhold (1967) that when cotton stem tissue colonized by the fungus was buried in soil for a short time and then removed and placed next to the stems of cotton seedlings, severe infection resulted. Similar tissue when left in soil for more than 8 weeks was unable to initiate infection on stems but did cause severe damping-off when placed alongside seed in soil at the time of planting. This observation led to the idea that the fungus present in the old tissue required materials from the seed to function effectively as a pathogen and that these materials were used as nutrients.

Further studies showed that the fungus grown on a medium rich in carbon and nitrogen could infect cotton stems, whereas when it was grown on water agar no infection occurred. The hypothesis was then developed that the nutritional status of the inoculum is an important factor in infection and that exudates and other materials in soil are able to alter this nutritional status, thereby influencing the ability of the fungus to infect.

In an attempt to clarify this complex interrelationship, detailed studies were carried out in which known carbon and nitrogen sources were used and manipulated in various ways. Firstly, by varying the levels of carbon and nitrogen in the medium on which the fungus was grown prior to inoculation of cotton stems, it was found that the level of both carbon and nitrogen influenced the severity of infection. The effect of different amounts of nitrogen on disease severity is shown in Fig. 1. With 2 gm asparagine/l there is extensive penetration and rotting of the stems, but with 0.5 gm asparagine/l attack is very slight. Similar results were obtained when the nitrogen level was held constant and the carbon level varied. These effects were demonstrated with seedlings inoculated in sand in growth chambers and also in field soil in the greenhouse and in the field.

A   B

Fig. 1. The effect of different levels of nutrients in the culture medium on disease severity on cotton stems. *A,* extensive lesion development with the fungus grown on a medium containing 20gm glucose/l and 2gm aspara-gine/l. *B,* very slight infection with the fungus grown on a medium containing 20gm glucose/l and 0.5gm aspara-gine/l.

One of the important points to consider is the effect of varying nutrition on the amount of fungus in each piece of inoculum and also the effect on the ability of the inoculum to grow. It has been found that levels of carbon and nitrogen which caused a marked reduction in disease severity do not significantly reduce the dry weight of inoculum pieces. Similarly, growth, as estimated by colony diameter on water agar and the density of hyphal tips per unit area, is not markedly restricted by lowering the amounts of carbon and nitrogen in the medium on which the fungus is grown.

There is, however, one important effect of altering the nutritional status which appears to be of some significance. With inoculum grown on high levels of carbon and nitrogen, growth of the fungus on cotton stems was rapid and extensive and large numbers of infection cushions formed within 24 hours. Penetration occurred rapidly from most cushions, resulting in the formation of large necrotic lesions which often coalesced. On the other hand, with inoculum from media with reduced levels of nutrients, growth on stems was reduced in both rate and density. Furthermore, almost no infection cushions were formed, and those that did form either failed to penetrate or produced only small lesions which failed to enlarge. It appears that with the lower levels of nutrition the fungus is unable to accumulate the amounts of reserves which are required for the vigorous vegetative growth and intensive metabolic activity associated with cushion formation and infection. These results demonstrated the importance of the nutritional status of the inoculum and indicated possible ways in which nutrition influenced the ability of the fungus to infect.

The role of exudates and other external nutrients has also been investigated. The addition of various nitrogenous compounds to sand around seedlings inoculated with fungus grown on high carbon and low nitrogen resulted in severe infection. The fungus was able to use the added nitrogen to make extensive growth and to develop many infection cushions, resulting in the formation of large, necrotic lesions. Also, when grown on media with low levels of nutrients and even on water agar the fungus could cause damping-off if the inoculum was placed adjacent to seed at the time of planting. It is thus clear that the fungus can use nutrients from external sources such as seed exudates and organic matter to alter its ability to infect.

This work and that discussed by Hayman (this volume) point out the importance of seed exudates in this disease of cotton in California and also provide some idea of the role that they play. In examining the development of the disease in the field, it appears that the fungus survives in the soil over the winter months either in small pieces of organic matter or as free hyphae. In both forms the fungus possesses low levels of reserves at planting and is unable to produce severe infection on the emerging stems. It is therefore dependent on seed and root exudates and other external nutrients in the soil for a source of utilizable carbon and nitrogen. The greatest part of this requirement is probably derived from seed exudates, and only the

inoculum within the zone of influence of these exudates is able to improve its nutritional status sufficiently to initiate infection. It is clear that in this disease, and perhaps in others, the virulence of the fungus and the severity of the disease are very dependent on nutrients from exudates and other sources in the soil. Further studies of this type should lead to a better understanding of the requirements of the fungus to initiate disease, and of the role that nutrients from different sources play in meeting these requirements. With this knowledge we should then be in a better position to determine means of disease control.

MORPHOGENETIC FACTORS INVOLVED IN PENETRATION.—It is now known that different isolates of *R. solani* may penetrate in different ways (Dodman, Barker, and Walker, 1968). There appears to be some correlation between the mode of penetration and the type of disease produced. Isolates attacking the aerial parts of plants usually penetrate by means of lobate appressoria and through stomata, while isolates causing damping-off and other seedling diseases commonly penetrate from dome-shaped infection cushions.

Although it is clear that a marked morphogenesis occurs during the formation of lobate appressoria, there is no information at present on the factors which control the initiation of these structures. On the other hand, there is now strong evidence that the various morphological changes associated with the formation of dome-shaped infection cushions are controlled by host materials.

In the formation of infection cushions it has been possible to distinguish several stages although it is not clear that all of these are essential for successful penetration. However, the sequence that is usually observed on a susceptible host is as follows: (1) growth of hyphae on to the plant; (2) attachment of hyphae to the cuticle; (3) growth of many hyphae along the lines of junction of the underlying epidermal cell walls; (4) formation of numerous, short, swollen side branches which aggregate into cushions; (5) penetration either from pegs or hyphal tips at the base of the cushion.

Flentje and co-workers (this volume) have suggested that these stages are genetically controlled. From a range of natural and induced mutants of *R. solani* many have been found which are nonpathogenic and which fail to establish successful infection because of a breakdown at one or other of the steps listed above. In fact, an examination of several mutants which all fail at the same step has shown that they are not genetically identical. It is thus possible that each of the steps in the sequence may be further subdivided or that each step is controlled by a number of different genes. It is hoped that further studies with these and other mutants will lead to a better understanding of the stages in the infection process.

Although it has been shown that there is a characteristic sequence of events preceding successful infection, little is known about the factors controlling most of the stages in this sequence. However, it has been demonstrated that host factors initiate the marked morphogenesis associated with the development and aggregation of side branches to form infection cushions.

This hypothesis was first put forward by Flentje (1957) and was supported by the finding that exudates from radish-seedling roots and stems induced a crucifer-attacking isolate to form typical infection cushions on pieces of washed epidermis from radish stems (Kerr and Flentje, 1957). It was then demonstrated by Flentje, Dodman, and Kerr (1963) that structures morphologically identical to infection cushions were formed by crucifer-attacking isolates on collodion membranes covering either the stems of radish seedlings or blocks of agar containing radish-stem exudate (Fig. 2). No cushions were formed on collodion over agar which did not contain exudate, indicating that host factors are required to initiate cushion development.

Other workers have also found that infection-cushion formation is induced by host materials. Wyllie (1962) observed that structures resembling the early stages of cushion development were formed on the outside of cellophane tubing adjacent to enclosed seedling roots. Later, Martinson (1965) found that infection cushions were formed on the surface of a number of synthetic films contiguous to radish and bean seed or seedlings. He also found that materials collected from seed and seedlings were able to diffuse through cellophane and nylon film and stimulate cushion formation.

The evidence from these different sources shows clearly that the change in morphology from the usual pattern of hyphal elongation and branching to the production and aggregation of hyphal branches into cushions is initiated by host materials. At present there is no information regarding the chemical nature of the stimulant, although Martinson and Holder (1967) found that several carbohydrates induced cushion formation when placed in bags of nylon film in soil infested with *R. solani*. Some of the materials tested are known to be present in exudates, and it may be that these are the active constituents released by the plant. However, the compounds used failed to stimulate cushion formation in the absence of soil, suggesting a possible interaction with other microorganisms.

Since the development of infection cushions is the result of a marked morphogenesis, it is not unrealistic to think of fungal hormones being involved. Furthermore, some of the fungal cellular changes which occur are in many ways analogous to the reactions in higher plants produced by plant hormones. During cushion formation the rate of nuclear division in the fungal side branches appears to be increased, while the rate of cell elongation is apparently suppressed. Further cytological investigations of this type, together with chemical studies of the active materials in exudates, should enable the characterization and identification of the specific factor or factors controlling infection-cushion formation. When this is achieved, and if it is

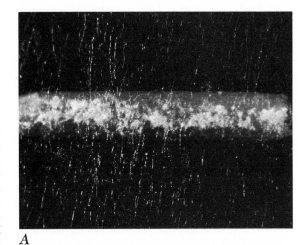

A

B

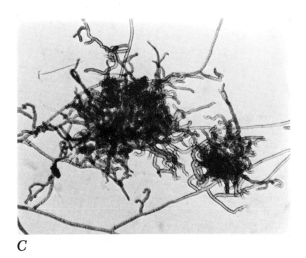

C

Fig. 2.   Infection cushions formed on collodion membranes. *A,* over radish stem tissue (× 6.5) *B* and *C* over exudate from the stems of radish seedlings (× 52.5 and × 96, respectively).

found that the stimulatory factors do act as a type of fungal hormone, it may be possible to synthesize analogues of the active materials, which would interfere with the growth of the fungus in much the same way as analogues of plant hormones affect plant growth. It may then be possible to use such analogues to prevent cushion formation and thereby provide a means of controlling diseases caused by some isolates of *R. solani*.

THE RELATION BETWEEN THE PREPENETRATION STAGES OF INFECTION AND SPECIFICITY.—There are many stages in the process of infection where the ability of an isolate to cause disease may be influenced. Perhaps the earliest stage is during the germination of sclerotia or resting hyphae in soil and the continued growth of mycelium through the soil and in the rhizosphere. It is possible that exudates influence these fungal activities and thereby play some part in determining specificity. Kerr (1956) found that seedling roots enclosed in cellophane caused a selective stimulation of growth of different isolates of *R. solani*; this stimulation was correlated with pathogenicity, suggesting a possible selective effect of exudates at this early stage of infection. Similarly, Nour El Dein and Sharkas (1964) reported a direct correlation between the stimulation of growth induced by exudates from three tomato varieties and the level of disease due to *R. solani*. On the other hand, Flentje, Dodman, and Kerr (1963) and de Silva and Wood (1964) have shown that there is no correlation between *in vitro* stimulation of growth by root exudates and pathogenic specificity of isolates of *R. solani*. It is thus clear that further investigations are needed before it can be concluded that stimulation of growth by root exudates plays a part in determining specificity in the field.

Considering later stages of infection in relation to specificity, it has been pointed out by Flentje (1957) that there are several stages in the formation of infection cushions where successful infection may be halted. At present it is not known if all of these stages are essential for penetration or if all of the stages are controlled by host factors. However, there is considerable evidence that the ability of an isolate to infect a plant is determined by its ability to form infection cushions on that plant and that this is the important factor governing specificity. It has been shown in some cases that exudates from the stems of susceptible seedlings initiate the formation of infection cushions by isolates which are able to infect these seedlings (Flentje, Dodman, and Kerr, 1963). However, at present it cannot be concluded that each host exudes a specific material stimulating cushion formation by isolates specific to the different hosts. An alternative explanation for specificity may be that all plants exude the same stimulatory material and that different hosts release specific materials which suppress infection-cushion formation. It is difficult to demonstrate which of these explanations is correct until more is known about the nature of the factors stimulating cushion development.

De Silva and Wood (1964) have suggested that the nature of the epidermis is involved in the initiation of infection cushions and that this may play a part in determining specificity. There are some other observations on the effects of surface on cushion formation. Flentje, Dodman, and Kerr (1963) found differences in cushion formation on cellophane and collodion, although the source of exudate was similar in both cases. In addition, Dodman (unpublished data) observed that cushions form in drops of exudate on slides, but only in contact with the glass. In other experiments it was found that some isolates of *R. solani* formed cushions on collodion membranes containing a trace of paraffin, but did not respond in the absence of paraffin. These results indicate that the surface is an important factor in infection-cushion formation, and it is possible that differences in the surfaces of different hosts may be involved in specificity.

It is clear that, at present, the factors controlling specificity are poorly understood. Our knowledge is limited to experiments with a few different hosts and isolates, and applies only to penetration from dome-shaped infection cushions. Nothing is known about the specificity of isolates which penetrate by other means or if, in fact, host factors play any part in influencing penetration by these isolates. Investigations of these basic problems are required to determine the importance of plant exudates in both penetration and specificity of a wide range of isolates of *R. solani*.

LITERATURE CITED

BARKER, K. R. 1961. Factors affecting the pathogenicty of *Pellicularia filamentosa*. Ph.D. thesis, University of Wisconsin, Madison.

DE SILVA, R. L., and R. K. S. WOOD. 1964. Infection of plants by *Corticium solani* and *C. praticola*—effect of plant exudates. Trans. Brit. Mycol. Soc. 47:15–24.

DODMAN, R. L., K. R. BARKER, and J. C. WALKER. 1968. Modes of penetration by different isolates of *Rhizoctonia solani*. Phytopathology 58:31–33.

FLENTJE, N. T. 1957. Studies on *Pellicularia filamentosa* (Pat.) Rogers. III. Host penetration and resistance, and strain specialization. Trans. Brit. Mycol. Soc. 40:322–336.

FLENTJE, N. T., R. L. DODMAN, and A. KERR. 1963. The mechanism of host penetration by *Thanatephorus cucumeris*. Australian J. Biol. Sci. 16:784–799.

KAMAL, M., and A. R. WEINHOLD. 1967. Virulence of *Rhizoctonia solani* as influenced by age of inoculum in soil. Can. J. Botany 45:1761–1765.

KERR, A. 1956. Some interactions between plant roots and pathogenic soil fungi. Australian J. Biol. Sci. 9:45–52.

KERR, A., and N. T. FLENTJE. 1957. Host infection in *Pellicularia filamentosa* controlled by chemical stimuli. Nature (London) 179:204–205.

MARTINSON, C. A. 1965. Formation of infection cushions by *Rhizoctonia solani* on synthetic films in soils. Phytopathology 55:129.

MARTINSON, C. A., and R. R. BAKER. 1962. Increasing relative frequency of specific fungus isolations with soil microbiological sampling tubes. Phytopathology 52:619–621.

MARTINSON, C. A., and ELIZABETH HOLDER. 1967. Stimulation of infection cushion formation by *Rhizoctonia solani* in soil. Phytopathology 57:342.

NOUR EL DEIN, M. S., and M. S. SHARKAS. 1964. The pathogenicity of *Rhizoctonia solani* in relation to different tomato root exudates. Phytopathol. Z. 51:285–290.

SCHROTH, M. N., and R. J. COOK. 1964. Seed exudation and its influence on pre-emergence damping-off of bean. Phytopathology 54:670–673.

SCHROTH, M. N., and W. C. SNYDER. 1961. Effect of host exudates on chlamydospore germination of the bean root rot fungus, *Fusarium solani* f. *phaseoli*. Phytopathology 51:389–393.

WEINHOLD, A. R., and T. BOWMAN. 1967. Virulence of *Rhizoctonia solani* as influenced by nutritional status of inoculum. Phytopathology 57:335.

WYLLIE, T. D. 1962. Effects of metabolic by-products of *Rhizoctonia solani* on the roots of Chippewa soybean seedlings. Phytopathology 52:202–206.

Armillaria mellea *Infection Structures: Rhizomorphs*

M. O. GARRAWAY and A. R. WEINHOLD—*Department of Plant Pathology, The Ohio State University, Columbus, and Department of Plant Pathology, University of California, Berkeley.*

Rhizormorphs are the structures by means of which *Armillaria mellea* infects its many hosts. Because of their importance in infection (Leach, 1937; Thomas, 1934), spread (Marsh, 1952), and survival (Darley and Wilbur, 1954), attention has been given to nutritional factors affecting rhizomorph formation and development (Benton and Erlich, 1941). The value of many of these studies has been limited by the fact that nonsynthetic media have been used. Nonsynthetic media contain one or more natural but unknown growth-promoting substances which stimulate rhizomorph formation. The discovery that these substances could be replaced by low molecular weight alcohols (Weinhold, 1963) has made possible a more detailed investigation of nutritional factors affecting growth and rhizomorph formation. In this paper, data are presented on some of the effects of ethanol and of physiological and biochemical changes associated with ethanol-induced rhizomorph formation.

INFLUENCE OF ETHANOL ON GROWTH AND RHIZO-MORPH FORMATION.—*A. mellea* mycelia grew poorly after 21 days of incubation on a chemically defined synthetic medium containing glucose (5 gm/liter) as the source of carbon. If, however, an ethanol supplement (500 ppm) was added to the medium, there was good mycelial growth and extensive rhizomorph development after 20 days of incubation (Weinhold, 1963; Weinhold and Garraway, 1966). In view of this, cultures of the fungus were examined at different times of incubation to determine more precisely the effect of ethanol on the pattern of development of mycelia and rhizomorphs.

Stages of growth.—The fungus was grown on 20 ml of a basal medium consisting of 5 g D-glucose, 2 g L-asparagine, 1.75 g KH_2PO_4, 0.75 g $MgSO_4 \cdot 7H_2O$, 1 mg thiamine, and 0.4 g (500 ppm) ethanol per liter of distilled water, unless otherwise indicated (Weinhold and Garraway, 1966). At different times of incubation up to 21 days, mycelia and rhizomorphs from replicate cultures were separated, placed on weighed pieces of tinfoil, and dried overnight in an oven at 100° C; dry weights were then determined.

During the first 8 days of incubation, cultures consisted mainly of undifferentiated mycelia (Fig. 1),

with rhizomorph primordia localized as dark zones on the underside. After 10 days the growth rate of rhizomorphs (Fig. 1, line R) exceeded that of mycelia (Fig. 1, line M). The stimulation of growth (Fig. 1, line R + M) was due to the ethanol supplement and was in contrast to the check (Fig. 1, line Ck), which had poor mycelial growth and no rhizomorphs.

Analysis of the medium for ethanol and glucose (Garraway and Weinhold, 1968a) showed that by 13 days, with the dry weight at 36 mg/culture, all of the ethanol and 35% of the glucose were consumed. At 21 days, the dry weight was over 120 mg and 80% of the glucose was consumed. This showed that most of the growth occurred after the ethanol supplement had been depleted from the medium, and indicated that glucose was a suitable carbon source for growth after rhizomorph initiation and early development on an ethanol-supplemented medium.

Ethanol concentration.—Cultures of *A. mellea* which were incubated for 6 to 8 days on an ethanol-supplemented medium showed a marked decrease in mycelial growth and a cessation of rhizomorph growth following transfer to, and subsequent incubation on, a nonsupplemented medium (Garraway, 1966; Garraway and Weinhold, 1968a). This was studied further by measuring the growth response of young mycelial

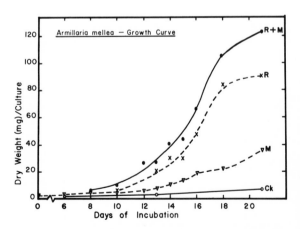

Fig. 1. Growth of *Armillaria mellea* in relation to time.

cultures following transfer to media containing different concentrations of ethanol supplement.

The fungus was grown for 7 days on a supplemented medium, as described previously (Weinhold and Garraway, 1966), then transferred to fresh media containing different concentrations of ethanol (Table 1), and incubated for an additional 14 days. Growth

TABLE 1. Growth and rhizomorph production by young cultures* of *Armillaria mellea* following 14 days of incubation on a glucose medium supplemented with different concentrations of ethanol

Ethanol concentration (ppm)	Growth: dry wt/culture		
	Rhizomorphs	Mycelium	Total
	mg	mg	mg
Control†	—	5.0	5.0
0	—	9.0	9.0
50	6.0	12.0	18.0
100	18.0	14.0	32.0
250	49.0	11.0	60.0
500	49.0	15.0	64.0
1,000	67.0	14.0	81.0

* Cultures grown on glucose medium supplemented with ethanol (500 pm) for 7 days then transferred to media indicated.
† Dry wt of representative colonies at time of transfer.

on the different media was compared with that of 7-day colonies weighed at the time of transfer.

Increase in growth of transfers was correlated with the increase in the concentration of ethanol supplement, being lowest for cultures transferred to the non-supplemented medium and highest for those transferred to a medium with 1,000 ppm ethanol (Table 1). The effect of added ethanol was mainly on rhizomorph development. The growth of transfers was poor (final dry weight/culture 6.0 mg after 14 days of incubation) with 500 ppm ethanol as the sole source of carbon. This indicated that ethanol at the concentration used to stimulate growth did not supply the carbon needed.

INFLUENCE OF ETHANOL ON UPTAKE AND METABOLISM OF GLUCOSE.—Since the ethanol supplement was insufficient to support the amount of growth which occurred, it appeared that ethanol must have stimulated fungal growth by influencing the utilization of the glucose present in the medium. In view of this, the influence of ethanol on uptake and metabolism of glucose was studied.

Time course.—Replicate batches of 7-day cultures (about 100 mg fresh wt/replicate) were incubated for different times at 24° C in sealed jars on 5 ml of basal medium containing glucose (5 g/liter) and glucose-U-C^{14} (1.0 uc/ml, specific activity 2.4 mc/m Mole) with and without a supplement of unlabeled ethanol (500 ppm). After incubation, the cultures were assayed for radioactivity (Garraway and Weinhold, 1968*b*). Uptake of radioactivity and of $C^{14}O_2$ evolved after different incubation times are presented in Fig. 2. There was less glucose uptake in the presence of ethanol than in its absence, and considerably less radioactivity in

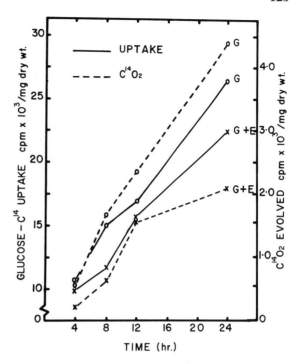

Fig. 2.   Glucose-U-C^{14} uptake and $C^{14}O_2$ evolved by 7-day cultures of *Armillaria mellea* after different incubation times on a medium with ($G+E$) and without (G) an ethanol supplement (500 ppm).

the CO_2 evolved. This indicated that less of the glucose taken up in the presence of ethanol was utilized for energy. Although ethanol appeared to suppress glucose metabolism, a previous study showed that it had no significant effect on fungal respiration (Garraway, 1966).

Ethanol concentration.—Seven-day cultures were incubated for 12 hr in jars containing media with cold glucose (5 g/liter), glucose-U-C^{14} (1 uc/ml), and different concentrations of ethanol supplement as used previously (Table 1). The cultures were then extracted with hot 80% ethanol, and the radioactivity in each fraction (Fig. 3) was determined (Garraway and Weinhold, 1968*b*). Radioactivity was expressed as a percentage of that present in corresponding fractions of the controls, which were incubated on a glucose medium without ethanol.

Ethanol caused a decrease in glucose-C^{14} uptake. It also caused a reduction in the radioactivity from glucose-C^{14} which entered the CO_2, basic, acidic, and lipid fractions (Fig. 3). This was accompanied by an increase in the level of soluble (intracellular) glucose-C^{14}. Thus, the concentrations of ethanol which stimulated growth and rhizomorph formation suppressed glucose uptake and its conversion to other metabolites (Garraway and Weinhold, 1968*b*).

DISCUSSION.—Since ethanol behaves like a growth-promoting substance, the studies reported here and

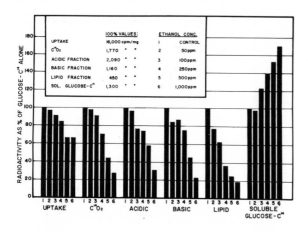

Fig. 3. Effect of different concentrations of ethanol supplement on uptake and metabolism of glucose-U-C^{14} by 7-day cultures of *Armillaria mellea*, following 12 hr of incubation.

elsewhere (Garraway, 1966; Garraway and Weinhold, 1968a) may relate to the significance of growth factors in the biology of *A. mellea*. Thus, growth-promoting substances may be required for initiation of rhizomorphs from undifferentiated mycelia. Once rhizomorphs have developed, however, they may be able to utilize the carbohydrates derived from a food base, such as woody substrates (Bliss, 1946; Garrett, 1956) or infected roots (Marsh, 1952) for growth through soil and infection of other roots.

The finding that ethanol suppressed uptake and metabolism of glucose may relate to the growth-promoting effects of ethanol. Thus, glucose which is utilized in the absence of ethanol could possibly be converted to metabolites which selectively inhibit growth of young mycelia and initiation of rhizomorphs. Possibly, ethanol releases this inhibition of growth by suppressing glucose uptake and subsequent conversion to these inhibitory metabolites. Concurrently, ethanol could enhance the utilization of glucose for growth by redirecting a greater proportion of that which is taken up into other metabolic pathways, e.g., cell-wall polysaccharide synthesis (Garraway and Weinhold, 1968b).

The work presented here is of interest because ethanol is produced by soil microorganisms (Pentland, 1967), and under certain conditions it is produced and secreted by roots (Bolton, 1966). Therefore, ethanol itself may be important as a growth factor in nature. Knowledge of the role of ethanol in morphogenesis in *A. mellea* may give clues concerning the nature of action of other natural-growth substances in initiation and development of structures associated with infection, survival, and reproduction of plant pathogenic fungi.

LITERATURE CITED

BENTON, V. L., and J. EHRLICH. 1941. Variation in culture of several isolates of *Armillaria mellea* from western white pine. Phytopathology 31:803–811.

BLISS, D. E. 1946. Relation of soil temperature to the development of *Armillaria* root rot. Phytopathology 36: 302–318.

BOLTON, E. F. 1966. Effects of soil flooding on ethanol content of tomato plants related to certain environmental conditions. Ph.D. thesis, Michigan State University (Dissertation Abstr. 28, 1–B).

DARLEY, E. F., and W. D. WILBUR. 1954. Some relationships of carbon disulfide and *Trichoderma viride* in control of *Armillaria mellea*. Phytopathology 44: 485. (Abstr.)

GARRAWAY, M. O. 1966. Nutrition and metabolism of *Amillaria mellea* (Vahl ex Fr.) Quél. in relation to growth and rhizomorph formation. Ph.D. thesis, University of California, Berkeley.

GARRAWAY, M. O., and A. R. WEINHOLD. 1968a. Period of access to ethanol in relation to carbon utilization and rhizomorph initiation and growth in *Armillaria mellea*. Phytopathology 58:1190–1191.

GARRAWAY, M. O., and A. R. WEINHOLD. 1968b. Influence of ethanol on the distribution of glucose-C^{14} assimilated by *Armillaria mellea*. Phytopathology 58: 1652–1657.

GARRETT, S. D. 1956. Rhizomorph behavior in *Armillaria mellea* (Vahl) Quél. II. Logistics of infection. Ann. Botany (London) [N.S.] 20:193–209.

LEACH, R. 1937. Observations on the parasitism and control of *Armillaria mellea*. Proc. Royal Soc. B. 121: 561–573.

MARSH, R. W. 1952. Field observations on the spread of *Armillaria mellea* in apple orchards and in a black current plantation. Trans. Brit. Mycol. Soc. 35:201–207.

PENTLAND, GERTRUDE D. 1967. Ethanol produced by *Aureobasidium pullulans* and its effect on the growth of *Armillaria mellea*. Can. J. Microbiol. 13:1631–1639.

THOMAS, H. E. 1934. Studies on *Armillaria mellea* (Vahl) Quél., infection parasitism, and host resistance. J. Agr. Res. 48:187–218.

WEINHOLD, A. R. 1963. Rhizomorph production by *Armillaria mellea* induced by ethanol and related compounds. Science 142:1065–1066.

WEINHOLD, A. R., and M. O. GARRAWAY. 1966. Nitrogen and carbon nutrition of *Armillaria mellea* in relation to growth-promoting effects of ethanol. Phytopathology 56:108–112.

The Significance of Vesicular-arbuscular Mycorrhizae in Plant Nutrition

J. W. GERDEMANN—*Department of Plant Pathology, University of Illinois, Urbana.*

The majority of plant species growing under natural conditions are dual organisms in that the organs through which water and nutrients are absorbed consist of root and fungus tissue. These organs are called mycorrhizae.

The ectotrophic mycorrhizae of certain forest trees are familiar to most plant scientists, and the importance of this type of symbiosis in plant nutrition has long been recognized by foresters. However, the endotrophic, vesicular-arbuscular (VA) mycorrhizae, which occur on a much greater number of plant species, including most agricultural crops, are less familiar. In fact, agronomists, horticulturists, plant physiologists, and plant pathologists often do not realize that the plant species that they study are normally mycorrhizal. Since VA mycorrhiza also has an effect on the physiology of the host, it is important that it receive greater recognition.

The tendency to disregard VA mycorrhiza has probably resulted from the fact that such infections produce very little change in external root morphology, and the fungi involved cannot be readily isolated and grown in pure culture.

METHODS OF STUDY.—Slight morphological changes have been reported to result from VA mycorrhizal infections (Gerdemann, 1968), but they have not proven useful in distinguishing mycorrhizae from nonmycorrhizal roots. VA mycorrhizae can be recognized on plant species with relatively thin unsuberized roots by their bright yellow color; however, this color disappears rapidly if they are exposed to bright light. VA mycorrhizae are also surrounded by an extensive hyphal growth that extends for considerable distances into the soil. However, unless roots are very carefully removed from soil and washed with great care, this hyphal growth is generally lost. The best way to diagnose VA infections is with a clearing and staining technique. This simple process makes the coarse irregular hyphae, arbuscules, and vesicles easily visible under relatively low magnification. There are many excellent descriptions and illustrations of VA mycorrhizae in the literature. Several recent reviews of this subject provide comprehensive lists of references (Mosse, 1963; Harley, 1965; Nicolson, 1967; Gerdemann, 1968).

There have been many attempts to isolate and grow the VA endophytes in pure culture. With one exception, however, investigators have obtained only limited growth of the endophyte on artificial media. Barrett (1961), using a complex hemp-seed baiting technique, obtained a number of isolates which he termed *Rhizophagus*, a name given by Dangeard to the fungus he observed in a VA mycorrhizae. With such isolates, Barrett was able to synthesize typical VA mycorrhizae. I have successfully repeated Barrett's isolation, but I have not been able to obtain infection with my isolates or with one obtained from Barrett. The bulk of our recently acquired knowledge concerning VA mycorrhizae has not been obtained with the use of pure cultures.

Mosse (1956), using spores and sporocarps that she found attached to mycorrhizal strawberry roots, was the first to demonstrate experimentally that a species of *Endogone* could form VA mycorrhizae. She produced "pot cultures" of this species by growing it on the roots of living plants in partially sterilized soil. In this way she produced inoculum that was used in many important experiments.

The process of wet-sieving and decanting has been used to demonstrate the abundance of *Endogone* spores in most soils (Gerdemann 1955, 1961; Gerdemann and Nicolson, 1963; Mosse and Bowen, 1968). A number of *Endogone* species obtained in this way have been shown to be mycorrhizal, and "pot cultures" of them have been established. Inoculum, consisting of spores (Fig. 1), sporocarps, or infected roots from such cultures, has proven useful in studying the taxonomy of species (Nicolson and Gerdemann, 1968), their host ranges, effect on nutrient uptake and growth of their hosts. Spores sieved directly from field soil have also been used as inoculum (Clark, in press).

Baylis (1959, 1967) inoculated plants by growing seedlings in natural soils containing VA mycorrhizal fungi. Plants that failed to become infected when grown under the same conditions were used as controls.

In most experiments designed to determine the effect of VA mycorrhiza on nutrient absorption and plant growth, pure inoculum has not been used. Therefore, two criticisms can be directed against them: (1) Organisms other than mycorrhizal fungi were undoubtedly present on the inoculum and it is conceivable that they may have made nutrients more available or stimulated growth of the test plants in some other way (Nicholas, 1965; Bowen and Rovira, 1966). (2) Partially sterilized soil has been used in many experiments. Soil sterilization, especially with heat, sometimes results in soil toxicity (Warcup, 1957). Since microorganisms can neutralize such toxins (Rovira and Bowen, 1966), it is possible that the observed growth stimulation could be caused by neutralization of toxins by either the contaminating microorganisms or the mycorrhizal fungi.

Various workers have tried to eliminate these criticisms. In an attempt to introduce contaminating microorganisms into the soil of control pots, the inoculum has been washed and the wash-water either decanted or filtered from the inoculum and then added to control soil. Murdoch, Jackobs, and Gerdemann (1967) tested the effect of such a filtrate on the growth of sudan grass. The inoculum, consisting of maize roots infected with *E. mosseae*, was washed and the water was passed through filter paper. This filtrate, which probably contained many microorganisms, significantly stimulated growth; however, this stimulation was much less than that obtained from inoculation with living mycorrhizae and only the mycorrhizal plants had a significantly higher percentage of phosphorus (Table 1). Clark (in press), who inoculated tree seedlings with *E. gigantea* spores that he sieved from maize fields, went so far as to "inoculate" his control pots with an equal number of *E. gigantea* spores that he had killed by piercing them with a needle.

Fig. 1. Chlamydospore of *E. mosseae*. Such spores or sporocarps containing several spores are produced on the roots of living plants and they can be used for inoculum. (× 275)

TABLE 1. The effect of mycorrhiza and of a water-filtrate, from the mycorrhizae used as inoculum, on the growth of sudan grass*

Treatment	Dry matter (g)	Phosphorus %
No treatment	8.3	.061
Filtrate from mycorrhizae	12.2	.056
Heat-killed mycorrhizae	9.6	.059
Living mycorrhizae	20.1	.071

*From C. L. Murdoch, J. A. Jackobs, and J. W. Gerdemann. 1967.

This laborious procedure should supply both mycorrhizal and nonmycorrhizal plants with the same microbial contaminants. The inoculation procedure used by Baylis (1959) should also result in the introduction of a similar microbial flora in both the mycorrhizal and nonmycorrhizal treatments.

The addition of contaminating microorganisms on the inoculum to the control soil also helps to counter the criticism directed against the use of partially sterilized soil. This criticism, of course, does not apply to experiments in which plants were grown in sand and watered with a nutrient solution, or to experiments in which nonsterile subsoil was used.

The fact that similar results have been obtained from many experiments using diverse inoculation techniques constitutes good evidence that the VA mycorrhizal infections, rather than contaminating microorganisms, are primarily responsible for the observed stimulation in growth.

THE EFFECT ON GROWTH.—Most investigators have found that VA mycorrhizae stimulate plant growth. Increases in growth of mycorrhizal plants have been obtained under the following conditions: infected roots as inoculum (Winter and Meloh, 1958; Meloh, 1963), or spores (Daft and Nicolson, 1966), with the plants grown in sand culture; mycorrhizae "implanted" in roots, with plants grown in water culture (Peuss, 1958); infected roots as inoculum, with plants grown in partially sterilized soil (Clark, 1963; Gerdemann, 1965; Murdoch, Jackobs, and Gerdemann, 1967); sporocarps or spores as inoculum, with plants grown in partially sterilized soil (Mosse, 1957; Gerdemann, 1964, 1965; Holevas, 1966; Clark, in press); infected roots as inoculum, with plants grown in nonsterile fallow soil or subsoil (Peuss, 1958); sporocarps as inoculum, with plants grown in nonsterile subsoil (Gerdemann, 1964); plants inoculated by seeding them in nonsterile soil, with the effect on growth tested in steamed soil (Baylis, 1959, 1967).

Increases in growth have been obtained with the following *Endogone* species: unknown species (Peuss, 1958; Winter and Meloh, 1958; Meloh, 1963; Baylis, 1959, 1967); *E. mosseae* (Mosse, 1957; Gerdemann, 1964; Holevas, 1966; Daft and Nicolson, 1966; Murdoch, Jackobs, and Gerdemann, 1967); *E. macrocarpa* var. *caledonia* and *E. macrocarpa* var. *geospora* (Daft and Nicolson, 1966); *E. fasciculata* (Gerdemann, 1965); and *E. gigantea* (Clark, in press). Meloh (1963), who inoculated plants with a form of *E. mos-*

seae that produces large chlamydospores within the roots, obtained a slight reduction in growth of mycorrhizal plants.

Differences in growth between mycorrhizal and nonmycorrhizal plants are often very striking, especially when the plants are grown in infertile soil. Maize inoculated with sporocarps of *E. mosseae* with "washwater" decanted from the sporocarps added to the controls was grown in a steamed soil deficient in phosphorus (Gerdemann, 1964). The mean dry weight of tops of mycorrhizal plants was 10.4g while that of nonmycorrhizal plants was only 2.7g. These differences were so consistent that they were significant at the 1% level (Fig. 2). One uninoculated plant grew as well as the inoculated plants, and it proved to be the only control that had become mycorrhizal.

At higher nutrient levels there is less difference in growth between mycorrhizal and nonmycorrhizal plants. Only one investigator (Peuss, 1958) has reported that mycorrhizal plants grew much larger than the nonmycorrhizal at both low and high fertility levels. Holevas (1966), Daft and Nicolson (1966), Baylis (1967), and Murdoch, Jackobs, and Gerdemann (1967) grew plants at low and at high phosphorus levels. At the low phosphorus levels mycorrhizal infection greatly stimulated growth of the test plants, whereas at the high phosphorus levels the experiments were not sensitive enough to determine if the slight differences in growth obtained were statistically significant.

Information is needed on the effect of VA mycorrhiza on plants growing in fertile soil under the best of conditions. Crop plants in Illinois may be highly mycorrhizal when growing in very fertile soils. Even a slight increase or decrease in yield resulting from mycorrhizal infection would be of considerable importance. At the higher nutrient levels, experiments will need to be replicated many times in order to determine whether slight differences in growth are statistically significant. Ideally such experiments should be done under field conditions; however, the nearly universal distribution of *Endogone* species in natural soils will make it extremely difficult to obtain nonmycorrhizal controls.

THE EFFECT ON NUTRIENT ABSORPTION.—There is evidence that VA mycorrhizae increase the ability of plants to absorb nutrients. The quantity of phosphorus and potassium in the soil is reduced more by mycorrhizal plants than by nonmycorrhizal plants (Gerdemann, 1964, 1965) (Table 2). Since mycorrhizal plants

TABLE 2. Phosphorus and potassium content of soils in which mycorrhizal and nonmycorrhizal maize had grown for 93 days[*]

	P-1[†] lb per acre	P-2[†] lb per acre	K[†] lb per acre
Soil from pots of mycorrhizal maize	7.4	12.5	167
Soil from pots of nonmycorrhizal maize	8.8[§]	14.9[§]	181[‡]

[*] From Gerdemann. 1965.
[†] P-1, available phosphorus; P-2, available and acid-soluble phosphorus; K, exchangeable potassium.
[‡] Difference significant at the 5% level.
[§] Difference significant at the 1% level.

are often much larger than nonmycorrhizal plants, they may contain larger quantities of most essential elements. In addition, mycorrhizal plants may contain a higher concentration of certain elements (Mosse, 1957; Baylis, 1959). When grown in phosphorus-deficient soils, mycorrhizal plants generally have higher percentages of phosphorus than nonmycorrhizal controls (Baylis, 1959, 1967; Gerdemann, 1964; Holevas, 1966).

Mycorrhizal plants are better able to utilize less available forms of phosphorus. Daft and Nicolson (1966) obtained the largest increases in growth of mycorrhizal plants under conditions of low phosphorus availability. Murdoch, Jackobs, and Gerdemann (1967) found that maize inoculated with *E. mosseae* did not grow significantly better than uninoculated maize when both were given superphosphate or monocalcium phosphate. When the phosphorus source was rock phosphate or tricalcium phosphate, inoculated maize grew much larger and had a higher phosphorus content than nonmycorrhizal maize (Table 3).

The enhanced ability of plants with VA mycorrhizae to absorb phosphorus has also been demonstrated with p32. Roots of mycorrhizal tuliptrees and foliage of mycorrhizal sweet gum accumulated more p32 than did the nonmycorrhizal controls (Gray and Gerdemann, 1967). The roots and tops of mycorrhizal onion also accumulated much larger quantities of p32 than did nonmycorrhizal onions (Gray and Gerdemann, 1969). Bowen and Mosse (Bowen and Theodorou, 1967) found that mycorrhizal onion and clover plants absorbed larger quantities of p32 than

Fig. 2. Maize plants growing in soil low in phosphorus. Left to right, mycorrhizal (*Endogone mosseae*), nonmycorrhizal.

TABLE 3. Dry matter yield and phosphorus content of mycorrhizal and nonmycorrhizal maize grown in a steamed soil low in phosphorus with added phosphorus sources of different availability*

	Dry matter (g)	Phosphorus (%)
Check	7.5	.061
Check mycorrhizal	12.4	.065
Tricalcium phosphate	8.2	.067
Tricalcium phosphate mycorrhizal	23.1	.088
Monocalcium phosphate	40.4	.075
Monocalcium phosphate mycorrhizal	42.0	.080
Rock phosphate	8.7	.068
Rock phosphate mycorrhizal	20.8	.094
Superphosphate	40.2	.070
Superphosphate mycorrhizal	40.8	.079

* From Murdoch, Jackobs, Gerdemann. 1967.

nonmycorrhizal plants. Detached mycorrhizal nodules of *Agathis australis* absorbed more phosphate than did nonmycorrhizal nodules (Morrison and English, 1967).

There is evidence that the fungus is directly responsible for the increased uptake of phosphorus. Mycorrhizae of onion accumulated larger quantities of p32 than did comparable nonmycorrhizal root segments of root tips (Gray and Gerdemann, 1969) (Table 4). Application of a fungicide, known to effectively suppress

TABLE 4. Radioactivity in mycorrhizal and nonmycorrhizal root segments of onion supplied with p32*

Plants and plant parts	Count/Min/Segment
Mycorrhizal plants	
Mycorrhizal segments	5,635†
Uninfected segments	214
Root tips	212
Nonmycorrhizal plants	
Root segments	10
Root tips	10

* From Gray and Gerdemann. Plant Soil (in press). 1969.
† Mean of mycorrhizal segments significantly different from other treatment means at greater than the 5% level.

mycorrhizal formation of *Endogone*, destroyed the enhanced ability of mycorrhizae to accumulate p32. Bowen and Mosse (Bowen and Theodorou, 1967) showed by means of autoradiography that mycorrhizal portions of roots exposed to p32 were most radioactive and that the radioactivity was concentrated in fungal structures.

Pathologists have studied root exudates, root infection, and root pathology. I believe it is time we studied exudates produced by mycorrhizae, infection of mycorrhizae by pathogens, and pathology of mycorrhizae. If we wish to understand plants as they grow in nature, we cannot ignore the most common and most highly specialized fungus-root relationship, the VA mycorrhizae.

LITERATURE CITED

BARRETT, J. T. 1961. Isolation, culture, and host relation of the phycomycetoid vesicular arbuscular mycorrhizal endophyte *Rhizophagus*, p. 1725–1727. *In* Recent Advances in Botany. Univ. of Toronto Press, Toronto.

BAYLIS, G. T. S. 1959. The effect of vesicular-arbuscular mycorrhizas on growth of *Griselinia littoralis* (Cornaceae). New Phytologist 58:274–280.

BAYLIS, G. T. S. 1967. Experiments on the ecological significance of phycomycetous mycorrhizas. New Phytologist 66:231–243.

BOWEN, G. D., and A. D. ROVIRA. 1966. Microbial factor in short-term phosphate uptake studies with plant roots. Nature (London) 211:665–666.

BOWEN, G. D., and C. THEODOROU. 1967. Studies on phosphate uptake by mycorrhizas. Proc. 14th IUFRO, Munich 5:116–138.

CLARK, F. B. 1963. Endotrophic mycorrhizae influence yellow poplar seedling growth. Science 140:1220–1221.

CLARK, F. B. Endotrophic mycorrhizal infection of tree seedlings with *Endogone* spores. Forest Sci. (In press.)

DAFT, M. J., and T. H. NICOLSON. 1966. Effect of *Endogone* mycorrhiza on plant growth. New Phytologist 65:343–350.

GERDEMANN, J. W. 1955. Relation of a large soil-borne spore to phycomycetous mycorrhizal infections. Mycologia 47:619–632.

GERDEMANN, J. W. 1961. A species of *Endogone* from corn causing vesicular-arbuscular mycorrhiza. Mycologia 53:254–261.

GERDEMANN, J. W. 1964. The effect of mycorrhiza on the growth of maize. Mycologia 56:342–349.

GERDEMANN, J. W. 1965. Vesicular-arbuscular mycorrhizae formed on maize and tuliptree by *Endogone fasciculata*. Mycologia 57:562–575.

GERDEMANN, J. W. 1968. Vesicular-arbuscular mycorrhiza and plant growth. Ann. Rev. Phytopathol. 6: (In press.)

GERDEMANN, J. W., and T. H. NICOLSON. 1963. Spores of mycorrhizal *Endogone* species extracted from soil by wet sieving and decanting. Trans. Brit. Mycol. Soc. 46:235–244.

GRAY, L. E., and J. W. GERDEMANN. 1967. Influence of vesicular-arbuscular mycorrhizas on the uptake of phosphorus-32 by *Liriodendron tulipifera* and *Liquidambar styraciflua*. Nature (London) 213:106–107.

GRAY, L. E., and J. W. GERDEMANN. 1969. Uptake of phosphorus-32 by vesicular-arbuscular mycorrhizae. Plant Soil. (In press.)

HARLEY, J. L. 1965. Mycorrhiza, p. 218–230. *In* K. F. Baker and W. C. Snyder [ed.], Ecology of soil-borne plant pathogens. Univ. of California Press, Berkeley and Los Angeles.

HOLEVAS, C. D. 1966. The effect of a vesicular-arbuscular mycorrhiza on the uptake of soil phosphorus by strawberry (Fragaria sp. var. Cambridge Favorite) J. Hort. Sci. 41:57–64.

MELOH, K. A. 1963. Untersuchungen zur Biologie der endotrophen Mycorrhiza bei *Zea mays* L. und *Avena sativa* L. Arch. Mikrobiol. 46:369–381.

MORRISON, T. M., and D. A. ENGLISH. 1967. The significance of mycorrhizal nodules of *Agathis australis*. New Phytologist 66:245–250.

MOSSE, B. 1956. Fructifications of an *Endogone* species causing endotrophic mycorrhiza on fruit plants. Ann. Botany (London) [N.S.] 20:349–362.

MOSSE, B. 1957. Growth and chemical composition of mycorrhizal and non-mycorrhizal apples. Nature (London) 179:922–924.

MOSSE, B. 1963. Vesicular-arbuscular mycorrhiza: An extreme form of fungal adaptation, p. 146–170. *In* P. S. Nutman and B. Mosse [ed.], Symbiotic Associations. 13th Symp. of the Soc. for Gen. Microbiol. Cambridge Univ. Press, London.

MOSSE, B., and G. D. BOWEN. 1968. The distribution of *Endogone* spores in some Australian and New Zealand soils, and in an experimental field soil at Rothamsted. Trans. Brit. Mycol. Soc. 51. (In press.)

MURDOCH, C. L., J. A. JACKOBS, and J. W. GERDEMANN. 1967. Utilization of phosphorus sources of different availability by mycorrhizal and nonmycorrhizal maize. Plant Soil 27:329–334.

NICHOLAS, D. J. D. 1965. Influence of the rhizosphere on the mineral nutrition of the plant, p. 210–217. *In* K. F. Baker and W. C. Snyder [ed.], Ecology of soil-borne plant pathogens. Univ. of California Press, Berkeley and Los Angeles.

NICOLSON, T. H. 1967. Vesicular-arbuscular mycorrhiza —a universal plant symbiosis. Sci. Progr. (Oxford) 55: 561–581.

NICOLSON, T. H., and J. W. GERDEMANN. 1968. Mycorrhizal *Endogone* species. Mycologia 60:313–325.

PEUSS, H. 1958. Untersuchungen zur ökologie und Bedeutung der Tabakmycorrhiza. Arch. Mikrobiol. 29: 112–142.

ROVIRA, A. D., and G. D. BOWEN. 1966. The effects of micro-organisms upon plant growth. II. Detoxication of heat-sterilized soils by fungi and bacteria. Plant Soil 25: 129–142.

WARCUP, J. H. 1957. Chemical and biological aspects of soil sterilization. Soils fertilizers 20:1-5.

WINTER, A. G., and K. A. MELOH. 1958. Untersuchungen über den Einfluss der endotrophen Mycorrhiza auf die Entwicklung von *Zea mays* L. Naturwissenschaften. 45:319.

Response of Resting Structures of Root-Infecting Fungi to Host Exudates: An Example of Specificity

J. R. COLEY-SMITH and J. E. KING—*Department of Botany, University of Hull, England.*

Responses of fungal resting structures to the presence of plants or to exudates from these have been recorded on numerous occasions (Schroth and Hildebrand, 1964). For the most part the responses appear to be nonspecific in that germination occurred equally well in the presence of host and nonhost species. There are, nevertheless, a few cases where specific responses have been claimed, and the account which follows will describe one such example.

Sclerotium cepivorum is a typical root-infecting fungus (Garrett, 1956). It is a poor competitor with soil saprophytes and will only grow in soil from an infected food base (Scott, 1956). When the food supply is exhausted, growth ceases. In the absence of the host, *S. cepivorum* survives in the form of sclerotia which are produced in large numbers on the bulbs of infected plants. These sclerotia are extremely resistant structures and can persist for many years either on or beneath the soil surface (Coley-Smith, 1959).

EFFECT OF ALLIUM SPECIES ON GERMINATION OF SCLEROTIA.—During the last few years a study has been made of the conditions under which sclerotia of *Sclerotium cepivorum* germinate. The fungus has a narrow host range and only infects members of the genus *Allium*. Using a glass-tube-and-nylon-strip technique (Fig. 1) it was found, with a limited number of plant types (Coley-Smith, 1960), that in soil sclerotia reacted by mass germination (Fig. 2) to the presence of *Allium* species, whereas in the absence of these plants little or no germination occurred. In spite of a report to the contrary by Tichelaar (1961), the specific effect of *Allium* on germination has since been confirmed, using a much more extensive range of plants (Coley-Smith and Holt, 1966). This stimulatory effect has been shown to occur under field conditions as well as in the laboratory (Coley-Smith, 1960).

From early experiments with seedlings it was suspected that a chemical effect was involved. Sclerotia frequently germinated at distances of up to a centimetre from the nearest *Allium* root. It was also found that stimulatory effects could be obtained with water extracts of *Allium* roots, bulbs, or even leaves (Coley-Smith, 1960; Coley-Smith and Holt, 1966), thus confirming that a chemical response was involved. Water

extracts of non-*Allium* species or simple sugar and amino-acid solutions had no effect on germination in soil.

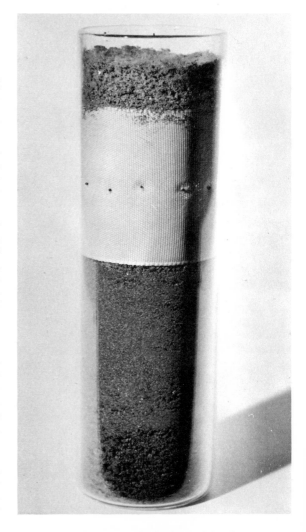

Fig. 1. Nylon strip-soil tube method for observing germination of sclerotia of *Sclerotium cepivorum*.

130

Fig. 2. Sclerotium of *Sclerotium cepivorum* germinating on a strip of nylon fabric in a soil tube.

By placing sclerotia on different parts of seedling roots of *Allium*, it was found that there was a marked positional effect on germination. The highest levels were recorded at the root tip region. On the other hand, when the root surface was injured, not only was germination considerably faster but the positional effect disappeared. It is quite clear that the root tip is the main site of exudation of the stimulatory principle, at least as far as seedlings are concerned.

GERMINATION OF SCLEROTIA UNDER ASEPTIC CONDITIONS.—The soil-tube method was used in the initial stages of this work in order to be certain that the stimulatory effect occurred under conditions similar to those which might be expected in nature. For recent biochemical work, attempts have been made to find alternative methods so that smaller quantities of test solutions might be used. The response of sclerotia to a number of plant extracts and nutrient solutions has been examined, aseptically, on agars, silica gels,

filter-paper discs and on columns of soil (Coley-Smith et al., 1967). It was found that under aseptic conditions there was no specific response to *Allium* extracts. Sclerotia were shown to be subject to soil mycostasis (Dobbs and Hinson, 1953). There was no obligate requirement for materials in *Alliums* for sclerotial germination, but in some way only species of this genus appeared to be capable of annulling the mycostatic effect of soil on sclerotia. Attempts have been made to discover how this reversal of mycostasis is effected (Coley-Smith et al., 1968). *Allium* species are known to contain substances, the alkythiolsulphinates (Cavallito and Bailey, 1944; Virtanen and Matikkala, 1959), which exhibit marked antibiotic properties. It was thought that these materials might be exuded in sufficient quantities to affect the microflora of the soil and bring about a consequent reduction of its fungistatic effect. However, in the first place, no evidence has been found of any reduction in bacterial numbers in soil in the presence of *Allium* species. Secondly, extracts of the latter were shown to stimulate germination of sclerotia at concentrations which were far too dilute to cause antibiotic effects *in vitro* against bacteria. Finally, methylmethanethiolsulphinate, one of the main antibiotic constituents of onions, was synthesized and shown to be without appreciable effect on germination of sclerotia (Coley-Smith et al., 1968). It is clear that however *Allium* species annul mycostasis, they do not do so by exuding antibiotic compounds.

Unfortunately the lack of a specific response under aseptic conditions means that at present the soil-tube method, although cumbersome, is the only technique which is sufficiently reliable for a biochemical investigation of the nature of the stimulatory principle.

VOLATILE STIMULANTS IN ALLIUM SPECIES.—Some years ago (Coley-Smith, 1960) slight evidence was reported which indicated that volatile components given off by onion seedlings might be important in stimulating germination of sclerotia. Since this original report additional evidence of the importance of volatile materials has been obtained (King and Coley-Smith, 1968). By using a modified soil-tube technique (Fig. 3) it was shown that actively growing onion and leek seedlings gave off volatile materials which induced germination of sclerotia. Air which was bubbled through onion and garlic extracts (Fig. 4) also stimulated sclerotial germination. The vapours evolved at room temperature by garlic extract were collected as a condensate in a tube immersed in liquid air, and this condensate induced germination of sclerotia. Stimulatory activity has also been found in steam distillates of garlic juice.

THE EFFECT OF ORGANIC SULPHIDES ON GERMINATION.—Many of the volatile components of *Allium* species have been identified by other workers (Carson and Wong, 1961; Oaks, Hartmann, and Dimick, 1964). The majority, though not all, of these are organic sulphides, methyl and propyl sulphides being the prin-

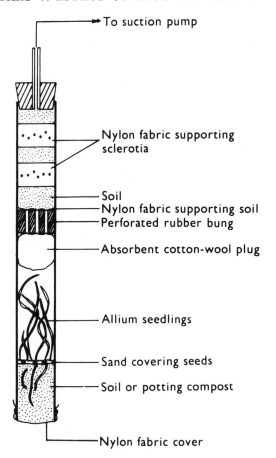

Fig. 3. Method of testing effect on germination of sclerotia of volatile compounds arising from seedlings. (J. E. King and J. R. Coley-Smith. 1968.)

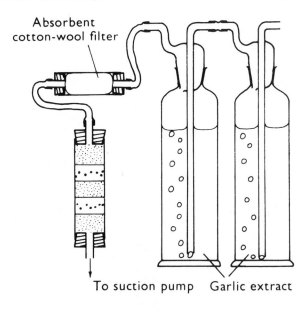

Fig. 4. Method of testing effect on germination of sclerotia of volatile compounds arising from plant extracts. (J. E. King and J. R. Coley-Smith. 1968.)

cipal constituents of onion, and methyl and allyl sulphides of garlic.

A number of these organic sulphides, which are available commercially, have been tested for their effects on germination of sclerotia, using, so far, a single isolate of *Sclerotium cepivorum* (A.T.C.C. 11793). Preliminary results which have been obtained

are summarized in Table 1. These results should at the moment be treated with caution since S. *cepivorum* A.T.C.C.11793 is a somewhat atypical isolate in that it normally germinates much more rapidly and shows a greater response to *Allium* species than a number of other isolates (Coley-Smith and Holt, 1966). In addition, the majority of the compounds which have been tested have been shown by gas-liquid chromatography and mass spectrometry to contain substantial quantities of impurities. Experiments are now in progress with a more extensive range of sulphides, and ultimately, those which are active will be purified by preparative gas-liquid chromatography and retested. Attempts will then be made to discover whether these sulphides are given off as gases by undamaged *Allium* plants and whether they occur in solution in root exudates. At present it seems likely, however, that alkyl sulphides are at least part of the source of stimulation in *Allium* species.

TABLE 1. Effect of sulphur compounds on germination in soil of sclerotia of *Sclerotium cepivorum*

Compound	Formula	Effect on germination of sclerotia	Main impurities in compounds
Methyl mercaptan	CH_3SH	Inactive	Not tested
Dimethyl sulphide	$(CH_3)_2S$	Inactive	Not tested
Dimethyl disulphide	$(CH_3S)_2$	Inactive	Not tested
Methyl-n-propyl sulphide	$CH_3SC_3H_7$	Active down to 1 μg/ml	Traces of unknown compounds
Di-n-propyl sulphide	$(C_3H_7)_2S$	Active down to 10 μg/ml	$(C_3H_7S)_2$
Di-n-propyl disulphide	$(C_3H_7S)_2$	Active down to 10 μg/ml	Traces of unknown compounds
Di allyl sulphide	$(C_3H_5)_2S$	Active down to 1-10 μg/ml	$(C_3H_5S)_2$
Di allyl disulphide	$(C_3H_5S)_2$	Active down to 1-10 μg/ml	$(C_3H_5)_2S$

GERMINATION OF SCLEROTIA OF OTHER FUNGI.—It is clear that germination of sclerotia of *Sclerotium cepivorum* is an unusual example of specificity in the response of fungal sclerotia to the presence of host plants and it is interesting to consider why this should be so.

In the first place, *Sclerotium cepivorum* has a very narrow host range, being confined to plants of a single genus. This situation itself is not always the case. With many sclerotial fungi the host range is extremely wide. *Sclerotinia sclerotiorum*, *Sclerotium rolfsii*, and *Verticillium dahliae* are well-known examples of wide-host-range types. Specific responses to hosts would hardly be expected amongst organisms of this type. Other sclerotial fungi have alternative mechanisms by which survival is ensured. Some, unlike *S. cepivorum*, possess sclerotia which can germinate many times, e.g. *Botrytis cinerea* (Vanev, 1966), *Phymatotrichum omnivorum* (Taubenhaus and Ezekiel, 1930), and *Rhizoctonia solani* (Pitt, 1964). A host-specific response is not perhaps as important where there is more than a single chance of germination. *S. cepivorum* usually germinates only once (Coley-Smith, 1960), the whole of the sclerotial contents being used up in the process. Another type of survival mechanism is the rapid production of secondary sclerotia following germination, e.g. *Sclerotinia sclerotiorum* (Williams and Western, 1965) and *Verticillium dahliae* (Wilhelm, 1954). The sclerotia of many fungi produce fruiting bodies, e.g. *Claviceps purpurea*, *Sclerotinia sclerotiorum*, and *Typhula* spp. (Potatosova, 1960), or spores e.g. *Botrytis cinerea* (Hockey, 1950), *Helminthosporium sigmoideum* (Nonaka and Yoshii, 1956), and *Gloeocercospora sorghi* (Dean, 1968). With these there is obviously a greater chance of reaching suitable plants in the absence of host-stimulated germination than with a mycelium-producing type like *S. cepivorum*.

Lastly, it should be emphasized that specific host stimulation of *Sclerotium cepivorum* is only apparent under unsterile conditions in soil. It is possible that other host-specific reactions may yet be discovered by a search amongst appropriate fungal types using soil techniques similar to those described here.

ACKNOWLEDGEMENT.—The authors wish to thank the Agricultural Research Council for a grant supporting this work.

LITERATURE CITED

CARSON, J. F., and F. F. WONG. 1961. The volatile flavor components of onions. J. Agr. Food Chem. 9:140–143.

CAVALLITO, C. J., and J. H. BAILEY. 1944. Allicin, the antibacterial principle of *Allium sativum*. I. Isolation, physical properties and antibacterial action. J. Am. Chem. Soc. 66:1950–1951.

COLEY-SMITH, J. R. 1959. Studies of the biology of *Sclerotium cepivorum* Berk. III. Host range: persistence and viability of sclerotia. Ann. Appl. Biol. 47:511–518.

COLEY-SMITH, J. R. 1960. Studies of the biology of *Sclerotium cepivorum* Berk. IV. Germination of sclerotia. Ann. Appl. Biol. 48:8–18.

COLEY-SMITH, J. R., D. J. DICKINSON, J. E. KING, and R. W. HOLT. 1968. The effect of species of *Allium* on soil bacteria in relation to germination of sclerotia of *Sclerotium cepivorum* Berk. Ann. Appl. Biol. 62:103–111.

COLEY-SMITH, J. R., and R. W. HOLT. 1966. The effect of species of *Allium* on germination in soil of sclerotia of *Sclerotium cepivorum* Berk. Ann. Appl. Biol. 58:273–278.

COLEY-SMITH, J. R., J. E. KING, D. J. DICKINSON, and R. W. HOLT. 1967. Germination of sclerotia of *Sclerotium cepivorum* Berk. under aseptic conditions. Ann. Appl. Biol. 60:109–115.

DEAN, J. L. 1968. Germination and overwintering of sclerotia of *Gloeocercospora sorghi*. Phytopathology 58:113–114.

DOBBS, C. G., and W. H. HINSON. 1953. A widespread fungistasis in the soil. Nature (London) 172:197.

GARRETT, S. D. 1956. Biology of root-infecting fungi. Cambridge University Press, London and New York. 293 p.

HOCKEY, J. F. 1950. A *Botrytis* cane blight of raspberry. Proc. Can. Phytopathol. Soc. 18:11.

KING, J. E., and J. R. COLEY-SMITH. 1968. Effects of volatile products of *Allium* species and their extracts on germination of sclerotia of *Sclerotium cepivorum* Berk. Ann. Appl. Biol. 61:407–414.

NONAKA, F., and H. YOSHII. 1956. Observations of the conidia of rice stem rot fungus (*Helminthosporium sigmoideum* Cav.). I. Dissemination of conidia in fields and their sporulation from sclerotia. Sci. Bull. Fac. Agric. Kyushu Univ. 15:435–440.

OAKS, D. M., H. HARTMANN, and K. P. DIMICK. 1964. Analysis of sulfur compounds with electron capture/hydrogen flame dual channel gas chromatography. Anal. Chem. 36:1560–1565.

PITT, D. 1964. Studies on sharp eyespot disease of cereals. II. Viability of sclerotia: persistence of the causal fungus, *Rhizoctonia solani* Kühn. Ann. Appl. Biol. 54:231–240.

POTATOSOVA, E. G. 1960. Conditions of germination of the sclerotia of the fungi of the genus *Typhula*. Zashch. Rast. Bol. 5:40.

SCHROTH, M. N., and D. C. HILDEBRAND. 1964. Influence of plant exudates on root infecting fungi. Ann. Rev. Phytopathol. 2:101–132.

SCOTT, M. R. 1956. Studies of the biology of *Sclerotium cepivorum* Berk. I. Growth of the mycelium in soil. Ann. Appl. Biol. 44:576–583.

TAUBENHAUS, J. J., and W. N. EZEKIEL. 1930. Studies on the overwintering of *Phymatotrichum* root rot. Phytopathology 20:761–785.

TICHELAAR, G. M. 1961. De invloed van *Gladiolus* op de Kieming van sclerotiën van *Sclerotium cepivorum*. Tijdschr. PlZiekt. 67:290–295.

VANEV, S. 1966. Bioekologichni prouchvaniya vŭrkhu *Botrytis cinerea* Pers.—prichinitel na sivoto gniene na Grozdeto. II. Formirane i prorastvane na sklerotsiite. (Bioecological studies on *B. cinerea*, the causal agent of grey mould of vine. II. Formation and germination of sclerotia). Isv. botan. Inst., Sof. 16:183–204.

VIRTANEN, A. I., and E. J. MATIKKALA. 1959. The isolation of S-methylcysteine-sulphoxide and S-n-propyl-cysteine sulphoxide from onion (*Allium cepa*) and the antibiotic activity of crushed onion. Acta. Chem. Scand. 13:1898–1900.

WILHELM, S. 1954. Aerial microsclerotia of *Verticillium* resulting from conidial anastomosis. Phytopathology 44:609–610.

WILLIAMS, G. H., and J. H. WESTERN. 1965. The biology of *Sclerotinia trifoliorum* Erikss. and other species of sclerotium forming fungi. II. The survival of sclerotia in soil. Ann. Appl. Biol. 56:261–268.

The Influence of Root Exudates on the Activity of Some Plant-parasitic Nematodes

AUDREY M. SHEPHERD—*Nematology Department, Rothamsted Experimental Station, Harpenden, Herts, England.*

Nematodes are organisms equipped with a neuro-sensory system, and their behaviour patterns are influenced by their environment. Most plant-parasitic nematodes spend part of their lives in pore spaces in the soil near roots, where they are exposed to dissolved salts, to exudates from roots, and to substances produced by soil microorganisms. These all influence the physiology, behaviour, and distribution of some nematode species, as has been demonstrated in the hatching of eggs, in moulting, and in host-finding. Root exudates are suspected of playing a part in the selection of feeding or entry sites and in activating free-living stages in the soil.

From recent studies using ciné and time-lapse photography, much has been learnt about the behaviour of nematodes during different stages in their life-cycles. Improved methods, especially microtechniques, should tell us more of their biochemistry and physiology, and improvement in preparing nematodes for electron microscopy will give further details of their histology and cytology.

In the meantime we can only speculate on the sites of action and the nature of the reactions involved between nematodes and root exudates.

HATCHING.—Cyst-forming nematodes (*Heterodera* spp.) demonstrate the survival value of a quiescent stage that becomes active only when near a source of food. The eggs are retained and protected inside the dead female, of which only the tanned cuticle remains, forming the so-called cyst, a horny envelope containing 200–600 eggs. The egg within the cyst develops until it contains a larva that has undergone the first of its four moults. This larva remains quiescent and does not hatch from the egg and emerge from the cyst into the soil unless stimulated by exudates from plant roots, usually, but not invariably, roots of host plants. In the field some larvae hatch "spontaneously" without the stimulus of a root exudate, but with most species this "water hatch" is much smaller than with an active root exudate.

How the hatching stimulus works is unknown, although filming eggs in hatching agents has shown the sequence of events when they are stimulated (Doncaster and Shepherd, 1967). Soon after exposure to root exudate, the larva begins to move around intermittently inside the eggshell. Periods of movement increase in duration and frequency until, after some days, the larva begins to thrust its mouth spear at one end of the eggshell. Soon it does this more precisely and more purposefully, so that a neat slit is cut in the shell, through which the larva emerges when the slit is long enough. When kept in distilled water, larvae make occasional slow movements, but never become really active. After stimulation the dorsal pharyngeal glands become active even before the larvae move, but there is no evidence that their secretions, whose main function is in feeding, assist the larva to escape.

Many organic and inorganic compounds hatch larvae of *Heterodera* species, but the chemical composition of natural hatching factors for the various species has not been elucidated. Concentrating and separating the relatively unstable active material from the many other constituents of root exudates in amounts large enough for analysis is beset with problems. Purification is slow and laborious, the bio-assay of resultant fractions is lengthy and tedious, and interpreting results is difficult when the concentration of hatch inhibitors fluctuates simultaneously with that of hatch stimulants. In the meantime, the mode of action of hatch stimulants has been approached empirically by studying artificial hatching agents. Similarities and differences between natural and artificial hatching agents are clearer when those of several species are studied. For example, *H. schachtii* (beet cyst-nematode) has a wide host range and is stimulated by root exudates of many plant hosts and a few nonhosts (Shepherd, 1962*a*). Many synthetic and inorganic compounds hatch eggs of this species as well as, or even better than, root exudates (Winner, 1957; Shepherd, 1962*b*; Clarke and Shepherd, 1964, 1965, 1966*a*, *b*, 1967; Viglierchio and Yu, 1966). Of 455 compounds tested by Clarke and Shepherd, 63 were very active and 156 were moderately active. The host range of *H. rostochiensis* (potato cyst-nematode) is more limited, and only root exudates from a few solanaceous species cause its eggs to hatch. Very few artificial hatching agents are known for this species (Marrian et al., 1947; Clarke and Shepherd, 1966*b*, *c*, 1968). Of 513 compounds tested, only 3 were very active and 20 were moderately active.

A wide range of organic compounds hatch *H.*

schachtii well. Some dyes are very active stimulants, especially redox dyes; so, too, are some quinones, esters and lactones, phenols, and aldehydes. No amino acids, sugars, or Krebs-cycle acids hatched *H. schachtii* or *H. rostochiensis* well, which suggests that hatching is not associated with the availability to the larva of a common metabolite (Table 1). Some inorganic

TABLE 1. Hatching activity of some organic compounds with *Heterodera schachtii*, beet cyst-nematode

Class of compound	No. tested	No. moderately active*	No. very active†
Acids (excluding Krebs-cycle)	67	21	6
Acids (Krebs-cycle)	10	1	0
Alcohols	11	1	0
Aldehydes	10	2	1
Amines	22	8	1
Amino acids	10	1	0
Dyes (excluding redox)	63	29	12
Dyes (redox)	31	11	15
Esters and lactones	24	6	3
Heterocyclic compounds (excluding pyridine derivatives)	34	14	2
Ketones	14	4	1
Phenols	32	11	5
Pyridine derivatives	11	3	1
Quinones	15	3	5
Miscellaneous organic compounds	58	23	7
Sugars	5	0	0

* Hatched more larvae than water but fewer than the appropriate root exudate.

† Hatched as many larvae as the appropriate root exudate, or more.

ions hatch *Heterodera* eggs as well as root exudates (Clarke and Shepherd, 1966*b*). Most are highly specific, but Zn^{2+} is an exception in being nonspecific. Zinc salts are the only compounds yet found that hatch almost all the species tested.

Clarke and Shepherd (1964) thought that hatching factors might act as electron acceptors, but this cannot be generally so because some known organic and inorganic hatching agents, e.g. Zn^{2+}, are unlikely to act in this way (Clarke and Shepherd, 1968). It seems more probable that many hatching agents are taken up by some constituent of the egg or larva and that this alters the structure and function of the binding material. A feature common to many of the more active organic compounds is a pair of polarisable atoms or groups connected by a conjugated unsaturated chain, such as O—C=C—C=O or O—C=C—O=N. Clarke and Shepherd (1968) speculated that specifity of hatching factors might be related to the length of this chain, *H. rostochiensis*, for example, needing a chain length of about 6.7Å, and *H. schachtii* 4Å. Protein has been suggested as a likely substrate (Marrian et al., 1947) and this is supported by the fact that the most effective hatching agents include compounds such as picric, picrolonic, and flavianic acids, which are amine and amino-acid precipitants.

Natural hatching factors in root exudates may not act in the same way as the artificial hatching agents,

because there may be more than one way in which quiescence can be broken. However, studies of stimulated larvae of *H. rostochiensis* show that the behaviour patterns and the sequence of events during hatching are essentially similar whether stimulation is by root exudate or by sodium metavanadate, one of the few compounds that hatch this species as well as root exudate. Mouth-spear movements are less well coordinated in metavanadate, which also seems to soften the eggshell (Doncaster and Shepherd, 1967), but from the nature, range, and specificity of artificial hatching agents, it is unlikely that they affect the eggshell alone. The very small concentrations of natural factor (down to 1 part in 10,000,000) that stimulate hatch bring it within the limits of nervous perception, with a sensitivity on the part of the nematode approaching that in organs of smell. Artificial hatching agents need to be more concentrated and most have a threshold at about 20 ppm.

MOULTING.—The migratory, root-feeding nematode genus *Paratylenchus* is the first in which the influence of root exudates on moulting has been demonstrated. In this genus, the fourth-stage larva, that is the pre-adult, is the survival stage, although not necessarily quiescent. The fourth-stage larvae of *P. projectus* and *P. dianthus* survive in moist soil for long periods without feeding. Rhoades and Linford (1959) showed that in water 1–2% *P. dianthus* and 16% *P. projectus* larvae moulted to the adult stage, whereas in host-root exudates 85% *P. dianthus* and 100% *P. projectus* larvae moulted. Exudates from roots of some plants that are not hosts also stimulated moulting. Fisher (1966) found that root exudates trigger the moulting of *P. nanus*. Even when only briefly exposed and then transferred to water, the larvae moulted. The stimulus seems to be received in the anterior half of the nematode because, when ligatured, only this half of the larva moulted. Boiling exudate for 10 minutes removed the moulting stimulus; hydrogen-ion concentration had no effect between pH 4–8.

HOST-FINDING.—Attraction of nematodes to roots was reviewed by Jones (1960), Seinhorst (1961), Klingler (1965), and Flegg (1965). Several authors observed nematodes grouped around root tips and at the points of origin of lateral roots. The question is whether they arrive there by random movements, and once there remain, or whether they respond to a concentration gradient of some constituent of the exudates from the roots (Wallace, 1960). Response to a gradient could be either by taxis (direct alignment of the body in the gradient) or by kinesis (a change in rate of linear movement or of frequency of turning moments), or both. Accumulation at the source of the stimulant by kinetic responses could occur (a) if response to increased concentration was slowed movement (this has not been observed), or (b) if the nematode at first turned less as concentration increased but adapted quickly to the new concentration. This would eventually bring it to the source and retain it there if any movement down-gradient (i.e. away from the source) caused random turning until movement was up-gradient again (i.e. back to the source).

Kinetic response has been recorded from work on the attraction of male *Heterodera* to females (Green, personal communication), by observing nematode tracks on agar and by time-lapse cinematography. There seems every reason to suppose that similar behaviour patterns might occur in root-exudate gradients, although this has yet to be demonstrated. With *Heterodera* males, kinetic response to the female sex attractant starts with stimulation into activity basically similar to the initial stages of hatching stimulation, when quiescent larvae begin to move.

Klingler (1961, 1963, 1965) favoured the view that CO_2 concentration plays a major part in attracting nematodes to roots in soil, but Peacock (1961) showed that, whereas removing root exudates by charcoal greatly lessened the number of larvae of *Meloidogyne incognita* invading host roots, removing CO_2 by basic ion-exchange resins had little effect. There is, however, little doubt that many factors associated with the presence of plant roots can influence soil organisms, and it is probably unrealistic to consider these factors in isolation.

FEEDING AND ENTRY SITES.—Having arrived at the root, different species choose different sites to feed or penetrate. *Heterodera* larvae tend to accumulate around and penetrate the zone of elongation just behind the root tip, or at the points of origin of lateral roots (Widdowson, Doncaster, and Fenwick, 1958). Whether selection of these sites is determined chemically or physically is unknown. Where a larva of *Heterodera* or *Meloidogyne* has entered the root, others cluster around the point of entry, probably in response to sap leaking from the damaged cells (Godfrey and Oliviera, 1932; Linford, 1939, 1942; Peacock, 1959; Doncaster, personal communication).

Using time-lapse cinematography through windows in an underground root-observation chamber, Pitcher (1967) demonstrated the massing of the ectoparasitic nematode *Trichodorus viruliferus* around the zone of elongation of young apple roots (Fig. 1). Only white, rapidly extending roots were attractive to nematodes, not older roots or the fine feeder roots. The mass of nematodes, containing often 100–300, moved forward with the tip as it elongated, and dispersed when root growth stopped. Nematodes aggregated within four days of the formation of an attractive root tip, so the mass was not the result of multiplication, and knowing the mean population densities of the nematodes, Pitcher calculated they must have accumulated by moving to the tip from a radius of at least 8 cm, so that a chemical stimulus exuded by the root may have had a greater range than is usually assumed (Wallace, 1958; Blake, 1962). Luc (1961) also found that soil taken from up to 40 cm away from millet roots attracted *Hemicycliophora paradoxa*.

GENERAL AMOUNT OF NEMATODE ACTIVITY.—That activity of free-living stages of plant-parasitic nematodes fluctuates not only with temperature changes

Fig. 1. *Trichodorus viruliferus* massed around the zone of elongation of an extending apple root. (From Pitcher, 1967.)

but also with chemical changes in their environment is suggested by several recent observations. For example, *Heterodera* males first react to a sex attractant released by females by becoming more active (Green, personal communication). There is no direct evidence to confirm that root exudates excite nematodes, but several observations suggest they may. Seinhorst (1968) reported that populations of *Pratylenchus penetrans* and *Tylenchorhynchus dubius* decreased faster during the first month after a host crop was removed than afterwards, and concluded that nematodes need more food when stimulated to activity by host roots. Van Gundy (1966) reported that larvae of *Meloidogyne javanica* and *Tylenchulus semipenetrans* lost their body contents more rapidly in soil with host roots than without. Away from roots nematodes seem quiescent, need less food, and survive longer. Harrison and Hooper (1963) found that *Longidorus elongatus* and several other species survived for more than a year without feeding in soil stored in polythene bags, and there are many other examples of long life without a source of food. Weischer (1959) reported that *Heterodera* larvae moved faster and further in soil when in contact with root exudate. The hatching stimulus (see earlier) acts by stimulating larvae to move, after which the behaviour pattern leads to hatching.

Seinhorst (1968) suggested that activation of free-living stages in the soil is similar to hatching of *Heterodera* eggs. This may well be so, with the stimulus to move being followed by the stimulus to feed. However, more direct observations are needed to test such a hypothesis.

In brief, the action of root exudates on plant-parasitic nematodes is two-fold. (1) Quiescent individuals are activated and a behaviour pattern is set in train, accompanied by physiological changes such as secretion by the salivary glands. (2) The behaviour of activated individuals probably helps them to find their host by responding to gradients of stimulants in exudates, which arise from their food source. The response to root exudates has survival value, for quiescence away from roots avoids squandering food reserves in useless activity, whereas response to them increases the chance of finding a feeding site on or in the host plant.

LITERATURE CITED

BLAKE, C. D. 1962. Some observations on the orientation of *Ditylenchus dipsaci* and the invasion of oat seedlings. Nematologica 8:177–192.

CLARKE, A. J., and A. M. SHEPHERD. 1964. Synthetic hatching agents for *Heterodera schachtii* Schm. and their mode of action. Nematologica 10:431–453.

CLARKE, A. J., and A. M. SHEPHERD. 1965. Zinc and other metallic ions as hatching agents for the beet cyst nematode, *Heterodera schachtii* Schm. Nature (London) 208:502.

CLARKE, A. J., and A. M. SHEPHERD. 1966a. The action of nabam, metham-sodium and other sulphur compounds on *Heterodera schachtii* cysts. Ann. Appl. Biol. 57:241–255.

CLARKE, A. J., and A. M. SHEPHERD. 1966b. Inorganic ions and the hatching of *Heterodera* spp. Ann. Appl. Biol. 58:497–508.

CLARKE, A. J., and A. M. SHEPHERD. 1966c. Picrolonic acid as a hatching agent for the potato cyst nematode, *Heterodera rostochiensis* Woll. Nature (London) 211:546.

CLARKE, A. J., and A. M. SHEPHERD. 1967. Flavianic acid as a hatching agent for *Heterodera cruciferae* Franklin and other cyst nematodes. Nature (London) 213:419–420.

CLARKE, A. J., and A. M. SHEPHERD. 1968. Hatching agents for the potato cyst-nematode, *Heterodera rostochiensis* Woll. Ann. Appl. Biol. 61:139–149.

DONCASTER, C. C., and A. M. SHEPHERD. 1967. The behaviour of second-stage *Heterodera rostochiensis* larvae leading to their emergence from the egg. Nematologica 13:476–478.

FISHER, J. M. 1966. Observations on moulting of fourth stage larvae of *Paratylenchus nanus*. Australian J. Biol. Sci. 19:1073;1079.

FLEGG, J. J. M. 1965. The plant-pathogen relationship (Blackman Essay). Rept. E. Malling Res. Sta. Year 1964, p. 62–70.

GODFREY, G. H., and J. OLIVIERA. 1932. The development of root knot nematode in relation to root tissues of pineapple and cowpea. Phytopathology 22:325–348.

HARRISON, B. D., and D. J. HOOPER. 1963. Longevity of *Longidorus elongatus* (de Man) and other nematodes in soil kept in polythene bags. Nematologica 9:159–160.

JONES, F. G. W. 1960. Some observations and reflections on host finding by plant nematodes. Meded. LandbHoogesch. OpzoekStns Gent 25:1009–1024.

KLINGLER, J. 1961. Anziehungsversuche mit *Ditylenchus dipsaci* unter Berücksichtigung der Wirkung des Kohlendioxyds, des Redoxpotentials und ander Faktoren. Nematologica 6:69–84.

KLINGLER, J. 1963. Die Orientierung von *Ditylenchus dipsaci* in gemessenen künstlichen und biologischen CO_2-Gradienten. Nematologica 9:185–199.

KLINGLER, J. 1965. On the orientation of plant nematodes and of some other soil animals. Nematologica 11:4–18.

LINFORD, M. B. 1939. Attractiveness of roots and excised shoot tissues to certain nematodes. Proc. Helminthol. Soc. Wash. D.C. 6:11–18.

LINFORD, M. B. 1942. The transient feeding of root-knot nematode larvae. Phytopathology 32:580–589.

LUC, M. 1961. Note préliminaire sur le déplacement de *Hemicycliophora paradoxa* Luc (Nematoda-Criconematidae) dans le sol. Nematologica 6:95–106.

MARRIAN, D. H., P. B. RUSSELL, A. R. TODD, and W. S. WARING. 1947. The structure of anhydrotetronic acid. J. Chem. Soc., Year 1947, p. 1365–1369.

PEACOCK, F. C. 1959. The development of a technique for studying the host-parasite relationship of the root knot nematode *Meloidogyne incognita* under controlled conditions. Nematologica 4:43–55.

PEACOCK, F. C. 1961. A note on the attractiveness of roots to plant parasitic nematodes. Nematologica 6:85–86.

PITCHER, R. S. 1967. The host-parasite relations and ecology of *Trichodorus viruliferus* on apple roots, as observed from an underground laboratory. Nematologica 13:547–557.

RHOADES, H. L., and M. B. LINFORD. 1959. Molting of preadult nematodes of the genus *Paratylenchus* stimulated by root diffusates. Science 130:1476–1477.

SEINHORST, J. W. 1961. Plant-nematode inter-relationships. Ann. Rev. Microbiol. 15:177–196.

SEINHORST, J. W. 1968. Mortality of plant-parasitic nematodes shortly after removal of the host plant. Compt. Rend. 8eme Symp. Int. Nématologie, Antibes, 1965, p. 41–42.

SHEPHERD, A. M. 1962a. The emergence of larvae from cysts in the genus *Heterodera*. Tech. Comm. No. 32, Commonwealth Bureau of Helminthology, St. Albans, Herts. Commonwealth Agric. Bureaux. 90 p.

SHEPHERD, A. M. 1962b. Dyes as artificial hatching agents for beet eelworm, *Heterodera schachtii* Schm. Nature (London) 196:391–392.

Van Gundy, S. D. 1966. Survival of nematodes in fallow soil. Proc. Symp. Nematode Control, Univ. Hawaii, Sept. 1966. p. 6–12.

VIGLIERCHIO, D. R., and P. K. YU. 1966. On the nature of hatching of *Heterodera schachtii*. III. Principles of hatching activity. J. Am. Soc. Sugar Beet Technol. 13:698–715.

WALLACE, H. R. 1958. Observations on the emergence from cysts and the orientation of larvae of three species of the genus *Heterodera* in the presence of host plant roots. Nematologica 3:236–243.

WALLACE, H. R. 1960. Movement of eelworms. VI. The influence of soil type, moisture gradients and host plant roots on the migration of the potato-root eelworm *Heterodera rostochiensis* Wollenweber. Ann. Appl. Biol. 48:107–120.

WEISCHER, B. 1959. Experimentelle Untersuchungen über die Wanderung von Nematoden. Nematologica 4:172–186.

WIDDOWSON, E., C. C. DONCASTER, and D. W. FENWICK. 1958. Observations on the development of *Heterodera rostochiensis* Woll. in sterile root cultures. Nematologica 3:308–314.

WINNER, C. 1957. Über die aktivierende Wirkung von Aminoacridinen auf *Heterodera schachtii*. Nematologica 2:126–130.

The Role of Basidiospores in Stump Infection by Armillaria mellea

J. RISHBETH—*Botany School, University of Cambridge, England.*

In Britain *Armillaria mellea* is very common in old deciduous and mixed woodland, and the great majority of stump infections which it causes there result from penetration by rhizomorphs or, perhaps more rarely, direct mycelial transfer, below ground (Greig, 1962; Redfern, 1966). The suggestion of Hiley (1919) that spores are the common means by which the fungus attacks dead, but uninfected, stumps has received little support from more recent investigations. However, even if stump colonization by spores were relatively rare, it could still play a significant part in establishing new infection foci. This would have important implications for forestry in the damage caused, for instance, by root attacks on conifers, whilst the extreme longevity of *A. mellea*, once established, in the presence of hardwoods might well cause concern. The creation of new infection centres in orchards, parks, and gardens is also potentially important. The investigation described here was started at Cambridge in 1960; a brief summary of the initial work has been given earlier (Rishbeth, 1964).

At sites where *A. mellea* is already well established it would be very difficult to detect natural stump infection by basidiospores owing to rapid invasion of stumps by rhizomorphs. Circumstantial evidence of spore infection was discovered, however, in a first-rotation, 35-year-old stand of Douglas fir, *Pseudotsuga menziesii*, on former arable at Thetford Chase, East Anglia, which also contained a few Scots pine, *Pinus sylvestris*. A stump of the latter species had been colonized by *A. mellea* some 50 m from a prolific spore source, and the fungus had spread to an adjacent stump of Douglas fir; there was convincing evidence that it had not spread underground from the spore source. Several recent attacks on trees or shrubs in Cambridge were investigated and found to be associated with a hardwood stump containing the fungus. These were discrete areas of infection, and the circumstances again suggested stump infection by spores, although the possibility that the trees were infected before felling could not be excluded altogether.

In East Anglia the maximum development of *A. mellea* fruit bodies usually occurs during October and November. Over this period spore production is so profuse that in fairly calm conditions deposition rates of up to 1,000 viable spores/dm²/min may be re-

corded close to fruit bodies, and spores drifting away from their sources may be detected by trapping on muslin screens (Rishbeth, 1959; Swift, 1964). Sporophores may be produced sporadically in smaller numbers from late July onwards and have also been observed to survive moderate ground frost ($-10°C$) in early December: the total period over which spores may be liberated in this part of Britain is thus nearly 6 months. In Sweden, Molin and Rennerfelt (1959) recorded *A. mellea* spores in the air over the period July to September by exposing freshly cut wood discs. Such evidence, taken in conjunction with reports of sporophore production in other countries, suggests that spores are readily available for colonizing stumps at the appropriate season in forests throughout the temperate zone, and also in some parts of the tropics.

In practice, however, it has been hard to establish that such colonization occurs. One difficulty is that the most suitable trees for experiments are often located at sites where *A. mellea* is already abundant, so that great care is needed for interpreting the results of stump inoculation. During the early part of the investigation many hundreds of trees, varying widely in age and species, were felled and the stumps inoculated with viable basidiospores of *A. mellea*, without obtaining a single infection. This behaviour, strikingly different from that of *Fomes annosus*, might be due to a wide variety of factors. It could be postulated, for instance, that physical conditions within the stump, such as moisture content or temperature, are often unsuitable, or that nutrient conditions are seldom satisfied. Important differences, in the latter respect, might occur between tree species and even between individuals within a species. Changing nutrient levels in woody tissues might determine that a certain period within the wider one of spore availability was particularly suitable for colonization of the stump created by felling. Again, stumps might differ considerably in resistance to invasion by *A. mellea*, especially as a result of meristematic activity. Whereas with some exceptions conifer stumps enter a phase of decline from the time of felling so that the broad outline of fungal succession is generally predictable, hardwood stumps often have great potentialities for callus formation and regrowth; the final outcome of fungal infections is correspondingly more open. Competitive or synergistic effects between *A. mellea* and other stump colonizers

141

Fig. 1. A stem length of *Prunus spinosa* which had been inoculated with basidiospores of *Armillaria mellea* and then incubated in moist sand for 4 months at 20°C. The bark has been folded back (left) to show colonization common with hardwoods involving broad, fanlike mycelial sheets. (× 3/4)

Fig. 2. A stem length of *Pseudotsuga menziesii* inoculated with basidiospores of *A. mellea* and incubated as in Fig. 1. The bark has been removed to show the characteristic pattern of colonization in the cambial region. The woody tissues were appreciably darker in the vicinity of the narrow mycelial strands. (× 1/2)

must also be borne in mind. Since results from stump experiments are often obtainable only after a period of 12 to 18 months, there is clearly a good case for trying to learn something about the operation of such factors in the laboratory. An account will therefore be given of one such line of research before considering the present outcome of field experiments.

LABORATORY STUDIES.—The following method was devised to test the ability of *A. mellea* to colonize woody stems after inoculation with basidiospores. Deposits from sporophores of local origin, obtained in petri dishes and stored at about 2°C until required, served as inoculum. Freshly cut stems 3–4 cm in diameter were thoroughly washed and cut into 6 cm lengths. After being inoculated at one end with about 0.2 ml suspension containing at least 5×10^3 viable basidiospores, lengths were put into clean 1 lb jars and surrounded with steamed quarry sand having a moisture content of 8–10%. Jars were covered with polythene sheeting and incubated in polythene bags for 4 months at 20°C. For any given treatment six or more replicates were prepared. Lengths were examined by separating off the bark and looking for mycelium of *A. mellea*. Mycelial growth was scored 0–3 on the basis of the proportion of cambial surface in-

vaded. Patterns of colonization varied: in one common type flattened, branching mycelial strands invaded the cambial zone irregularly from the inoculated end (Fig. 1); more rarely the strands were narrow and grew vertically (Fig. 2). In later stages strands expanded into broader sheets, which ramified extensively in the inner and middle bark. The outermost wood was soon colonized, but progress further into the wood was relatively slow.

Rhizomorphs were generally formed when the stem lengths had been extensively colonized, and these were scored 0–3 on the basis of abundance. However, since initiation and subsequent growth of rhizomorphs are not necessarily governed by the same factors as initial colonization, which is the immediate concern here, no further reference to the former will be made.

Variation with different tree species.—First to be tested was the ability of *A. mellea* to colonize stem samples from different species of tree, which proved to be very variable. Of the hardwoods, sycamore, *Acer pseudoplatanus*, for example, gave high mean ratings of 1.5–3.0, whereas many others gave generally lower mean ratings; examples were oak, *Quercus robur* (0—

1.2), birch, *Betula verrucosa* (0–1.5), and beech, *Fagus sylvatica* (0–1.8). By contrast lime, *Tilia vulgaris*, and aspen, *Populus tremula*, were not colonized at all. With stem lengths of some species, such as willow, *Salix alba*, and elm, *Ulmus carpinifolia*, this failure of *A. mellea* to become established was associated with the production of shoots and sometimes also roots. Conifers often gave high mean ratings, for example, Norway spruce, *Picea abies* (2.0–3.0) and Scots pine (0.7–2.5). Such results, obtained with random samples from young stems, do not necessarily reflect the suitability of stump tissues for colonization by *A. mellea*. Samples were therefore also taken from stumps of older trees used in field experiments: when prepared for tests in jars they comprised the outer wood and the bark, the surface of which was removed to limit contamination. The mean values for hardwoods thus obtained were somewhat lower, as with sycamore (0.6–1.7), oak (0.2–0.5), and beech (0.2–0.6). Variation within a species was well illustrated by field maple, *Acer campestre*, for which ratings of 0.25, 0.25, 2.5, 0, and 0 were obtained with five individual stumps (mean of four samples from each). Mean ratings for conifers were again high, examples being Norway spruce (2.2) and Scots pine (2.7). Five individual stumps of the former species gave mean ratings of 3.0, 3.0, 0.5, 2.0, and 2.0. Hence, even with species having stumps generally favourable for colonization by *A. mellea*, unsuitable individuals apparently occur; the converse may well be true. On the basis of these tests, stumps of conifers in general appear to be more readily colonized than those of hardwoods.

Inoculum dosage.—The effect on colonization of varying the dosage of basidiospores was studied by inoculating stem lengths of sycamore and Norway spruce. For the latter species, the mean ratings (ten replicates) obtained with successive ten-fold dilutions of a heavy spore suspension were as follows, the dosage quoted being the number of viable spores applied to each stem length: 8×10^4, 2.7; 8×10^3, 2.6; 8×10^2, 2.0; 80, 0.9; 8, 0.3. Colonization was nearly maximal at the two highest dosages, and thereafter fell off markedly with further dilution of the inoculum. With sycamore a rating of 0.1 was recorded with a dosage of five basidiospores. Under these conditions *A. mellea* becomes established with very low dosages. With some other tree species, by contrast, even a huge dosage provides insufficient inoculum for colonization to proceed. Thus in one experiment with hazel, *Corylus avellana*, no growth occurred after inoculating stem lengths with 4×10^4 viable spores, whereas complete colonization was obtained when similar lengths were inoculated by attaching a small wood disc permeated by mycelium of *A. mellea*.

Moisture content.—It seemed likely that availability of moisture might limit the ability of *A. mellea* spores to colonize wood lengths, especially since Benton and Ehrlich (1941) found the optimum moisture content for growth of the fungus in wood of *Pinus monticola*

to be as high as 150%. In an experiment with holly, *Ilex aquifolium*, no growth of *A. mellea* was recorded when inoculated lengths were stored at 20°C for 4 months in sand initially containing 6% moisture, but a mean rating of 2.1 was obtained when similar lengths were held in sand with a moisture content of 11%. With lengths of chestnut, *Castanea sativa*, no colonization occurred with sand having initial moisture contents of 4, 8, or 13%, but with sand at 20% moisture content a mean rating of 2.5 was recorded. It was found subsequently that during storage in jars, wood lengths, initially at field moisture content (40–55%), steadily withdrew water from the sand. Unless the sand was sufficiently moist at the outset its water content fell to a critical value, probably about 1%, at which little or no colonization by *A. mellea* occurred. Such failure occurred when the aggregate moisture content of the wood was as high as 56%. By contrast complete colonization occurred when wood containing a similar amount of moisture was surrounded by sand whose moisture content did not drop much below 3% during storage. Two related effects might be expected to follow desiccation of the sand. Firstly, the outermost woody tissues might themselves lose sufficient moisture to prevent or restrict colonization by *A. mellea*: this loss would hardly be detectable by determining aggregate moisture content. Secondly, drying of the wood surface might stimulate meristematic activity, leading in some instances to formation of callus. This in fact often happens with hardwood lengths, in which case *A. mellea* rarely becomes established. It is noteworthy that callus formation tends to be maximal in the very region initially colonized by the fungus. It seems probable that under natural conditions desiccation of the stump surface limits colonization by spores of *A. mellea*.

Temperature.—Stem lengths of sycamore and Norway spruce were inoculated with a heavy spore suspension to determine the effect of temperature on colonization. After incubation for a standard period of 4 months the following mean ratings (six replicates) were obtained with the former species at temperatures (°C) of: 5°, 0.3; 10°, 3.0; 15°, 1.5; 20°, 0.5; 25°, 0.5; 27°, 0.2; 30°, 0. Only one replicate was colonized at 5°. The rating at 20° was unusually low. With Norway spruce, on which growth was meagre throughout, the lowest temperature at which colonization occurred was 10° and the highest again 27°. Since there was little or no growth at 5°, it was thought desirable to find out whether *A. mellea* could develop on inoculated wood lengths held for a considerable period at a low temperature and then incubated at 20°. This would partially simulate natural conditions where spores are deposited on stumps in late autumn and temperatures are subsequently too low to permit appreciable growth of the fungus until the spring.

In late October wood lengths of two species were selected—sycamore, which is usually colonized by *A. mellea* with ease, and willow, generally colonized only with difficulty. As a basis for comparison such lengths were inoculated as usual and incubated for 4 months at 20°, when the mean ratings (ten replicates) for growth of *A. mellea* were 2.1 and 0 respectively. In the main experiment similar lengths were inoculated and kept for 6 months at 5°, after which period no growth by *A. mellea* was detected in representative samples. The remainder were then transferred to 20° for another 4 months, when the mean ratings were 0.9 for sycamore and 0.8 for willow. Other inoculated lengths were stored first in a shallow, covered soil pit for 6 months, where they were subjected to a fluctuating low temperature (minimum −5°) and then transferred to 20° as before. The respective ratings for the two species after this period were 0.4 and 0.8. Apart from showing that under these conditions colonization may indeed take place in the spring from an inoculum of spores applied the previous autumn, the experiment reveals in the case of willow a reduced resistance to infection associated with prolonged storage at low temperatures. It is well to remember that under natural conditions the surface tissues of stumps may dry out considerably during the winter: moisture content might then become limiting for growth of *A. mellea*.

Nutrient status.—Autoclaving renders stem lengths of all woody species so far tested readily colonizable by spores of *A. mellea*, so that the failures described earlier are probably due to the inaccessibility of nutrients in fresh stems rather than to actual deficiency. Nonetheless, in view of earlier work with the fungus it was thought desirable to determine whether its ability to colonize was related to starch content of the tissues. This was estimated visually after applying a solution of iodine to a freshly cut surface. No such relationship was found. The tissue regions preferentially colonized by *A. mellea* strongly suggest that nitrogen content may be important. Thus Anderson and Pigman (1947) give values for percentage nitrogen (Kjeldahl) in tissues of black spruce, *Picea mariana*, as follows: outer bark, 0.32; inner bark, 0.60; cambial zone, 1.09; young sapwood, 0.26; sapwood, 0.05; heartwood, 0.06. More recently Merrill and Cowling (1966) have recorded similar distributions of nitrogen in tissues of conifers and hardwoods and have discussed their significance for wood-inhabiting fungi.

In the current investigation plum trees of the variety Victoria were liberally fertilized with nitrogen during the growing season and then felled in November. The total nitrogen in the bark of such trees was about 12% higher than that in the bark of unfertilized ones. Samples from the stem base were then tested for colonization by *A. mellea* when those from fertilized and unfertilized trees gave mean ratings (twelve replicates) of 1.2 and 0.3 respectively. This difference was not significant at the 5% level. In each series significantly higher ratings were obtained with samples from the upper branches than with those from the stem base. This aspect of the problem needs further investigation.

Availability of nutrients is linked with resistance of the woody tissues to invasion, which resistance can be modified by application of tissue poisons. Perhaps especially interesting, in the light of forestry practice, is the effect of 2,4,5-trichlorophenoxyacetic acid. When a 1.5% solution of the butyl ester in oil was applied to the bark surface of stem lengths before inoculation, species such as willow, lime, and hawthorn, *Crataegus monogyna*, normally colonized little if at all, were rapidly invaded by *A. mellea*. In the case of willow, production of shoots and roots by stem lengths during incubation was correspondingly inhibited (Fig. 3). Under natural conditions the ability of stumps to produce regrowth may similarly be correlated with resistance to infection by spores of *A. mellea*: thus at one experimental site three chestnut stumps, samples from which gave low mean ratings (0.2, 0, and 0.2), subsequently produced much regrowth, whereas another, samples from which gave a high mean rating (2.5), produced none.

Interactions with other organisms.—Lastly may be considered the effect of inoculating stem lengths simultaneously with *A. mellea* and other microorganisms. With hazel, a species not very readily colonized by *A. mellea*, more rapid and abundant growth of the fungus was obtained by joint inoculation with bacteria such as *Bacillus polymyxa* and certain strains of *Pectobacterium carotovorum*. All bacteria showing this synergistic effect were proteolytic, although not all the proteolytic bacteria tested were effective. It seems likely that the enhanced growth of *A. mellea* resulted

Fig. 3. Stem lengths of *Salix alba* similarly inoculated and incubated. Left, an otherwise untreated length which produced a sprout and was not colonized by the fungus. Right, a length which had been treated with 2,4,5-T at the time of inoculation. Colonization and subsequent rhizomorph production were extensive; no sprouting occurred. (× 3/4)

from rapid killing of the woody tissues and a subsequent release of nutrients. By contrast, colonization by *A. mellea* was reduced when the fungus *Xylosphaera polymorpha* was also inoculated, although it was later found that competition from this fungus was reduced where moisture content of the sand was high and the type of wood was more favourable for growth of *A. mellea*. Colonization of sycamore lengths, for example, was less affected by joint inoculation with *X. polymorpha* and was virtually unaffected by joint inoculation with primary stump colonizers such as *Polyporus adustus*, *Polystictus versicolor*, and *Stereum hirsutum*.

The growth rates of *A. mellea*, *P. versicolor*, and *S. hirsutum* in lengths of sycamore stem were estimated by attaching a small wood inoculum at one end and placing them in jars of moist sand. At suitable intervals the lengths were cut serially into discs 0.5 cm thick, and these were reincubated on moist sand to determine the depth to which the fungus had penetrated. At 20°C, for instance, the three species all grew in the wood at a rate of 0.4 mm/day, whereas *A. mellea* alone grew in the cambial zone at a rate of 2.4 mm/day. If this finding is applicable to stump colonization by spores, fungi such as *P. versicolor* and *S. hirsutum* might have an initial advantage in the central wood over *A. mellea*, which penetrates to this region relatively late. Below the stump surface, however, such an advantage might be progressively lost through the ability of *A. mellea* to outflank the wood-colonizing fungi in the bark.

STUMP EXPERIMENTS.—Inoculations of conifer stumps with spores of *A. mellea* were generally carried out in first-rotation stands; stumps were examined at intervals after a period of 12 months. In an experiment with 12-year-old Scots pines, two stumps out of twelve became infected (Fig. 4) after inoculation in late September, the weather which followed being unusually mild and with well distributed rainfall. More commonly, however, such inoculations entirely failed, perhaps because stumps tended to dry out fairly rapidly in the sandy soils. Later experiments were performed at much moister sites in a fen valley during late July. In 9-year-old Norway spruce having a generally open canopy, six stumps out of ten became infected by *A. mellea* when they were covered with moist peat and polythene sheeting after inoculation, whereas only one out of ten, in a well-shaded position, was infected when the stump surface was left exposed. In 24-year-old Scots pines having a closed canopy the same proportion of stumps, two out of ten, became infected in covered and uncovered series. There is thus reasonably strong support for the contention that colonization of stumps by spores of *A. mellea* is often prevented by their being too dry. Another limitation is suggested by the result of inoculating similar Norway spruce stumps in November, rather than July, and then covering them: here *A. mellea* failed to appear the following spring. Conceivably with some types of stump *A. mellea* overwinters only if spore

Fig. 4. A stump of *Pinus sylvestris* which had been inoculated with basidiospores of *A. mellea* 8 months before. The bark has been partially removed to show the branching mycelial sheets. (× 2/5)

infection occurs early enough to permit some prior mycelial growth.

In the experiment with Scots pine just described some stumps, subsequently covered, were inoculated both with *A. mellea* and *Peniophora gigantea*. One out of ten such stumps was colonized by *A. mellea*, a result almost certainly reflecting its facility for rapid growth in the bark. Under comparable conditions of joint inoculation with *P. gigantea*, *Fomes annosus* is not to be found in pine stumps after a year has elapsed (Rishbeth, 1963). Examination of the conifer stumps inoculated with spores of *A. mellea* revealed that after a year the fungus had occupied the cambial zone in the stump and proximal portions of stump roots. After 3 years it had extensively invaded the wood both within the stump and far out into the lateral roots. This type of colonization, first seen in laboratory experiments, seems effective under natural conditions.

Inoculation of hardwood stumps has been much less successful. With stumps of young trees vigorous regrowth regularly occurred and no colonization by *A. mellea,* or indeed most other stump fungi, took place. Treatment of such stumps with 2.4.5-T did not promote colonization by *A. mellea* from spores, though if rhizomorphs were already present, invasion by these was greatly enhanced. With stumps of much older trees, on which little or no regrowth developed, inoculations at dry sites failed, possibly because the stump surfaces were not kept sufficiently moist. The only success was obtained at a wet site on heavy clay where the stump of an ash, *Fraxinus excelsior,* became infected after inoculation in spring; it produced minimal regrowth.

Experimental evidence thus tends to confirm that stump infection by basidiospores of *A. mellea* is rare. The association of callus formation and regrowth with resistance to colonization by *A. mellea* basidiospores accords well with the belief that it is not a very strong parasite. Under the formerly widespread system of hardwood management involving coppice and standards, felling of standards might well have created occasional opportunities for establishment of new infection centres, whereas coppicing was almost ideally suited for perpetuating them. Although caution is required for interpreting the results obtained under artificial conditions in the laboratory, some of them appear to be relevant to field conditions; further work is required before a more realistic evaluation can be made.

LITERATURE CITED

ANDERSON, E., and W. W. PIGMAN. 1947. A study of the inner bark and cambial zone of black spruce (*Picea mariana* B.S.P.). Science 105:601–602.

BENTON, V. L., and J. EHRLICH. 1941. Variation in culture of several isolates of *Armillaria mellea* from western white pine. Phytopathology 31:803–811.

GREIG, B. J. W. 1962. *Fomes annosus* (Fr.) Cke. and other root-rotting fungi in conifers on ex-hardwood sites. Forest. 35:164–182.

HILEY, W. E. 1919. The fungal diseases of the common larch. Clarendon Press, Oxford. 204 p.

MERRILL, W., and E. B. COWLING. 1966. Role of nitrogen in wood deterioration: amounts and distribution of nitrogen in tree stems. Can. J. Botany 44:1555–1580.

MOLIN, N., and E. RENNERFELT. 1959. Honungsskivlingen, *Armillaria mellea* (Vahl) Quél., som parasit på barrträd. Meddn. St. Skogsforsk Inst. 48 (10):1–26.

REDFERN, D. B. 1966. Root infection by *Armillaria mellea*. Ph.D. thesis, University of Cambridge.

RISHBETH, J. 1959. Dispersal of *Fomes annosus* Fr. and *Peniophora gigantea* (Fr.) Massee. Trans. Brit. Mycol. Soc. 42:243–260.

RISHBETH, J. 1963. Stump protection against *Fomes annosus*. III. Inoculation with *Peniophora gigantea*. Trans. Brit. Mycol. Soc. 52:63–77.

RISHBETH, J. 1964. Stump infection by basidiospores of *Armillaria mellea*. Trans. Brit. Mycol. Soc. 47:460. (Abstr.)

SWIFT, M. J. 1964. The biology of *Armillaria mellea* in Central Africa. Ph.D. thesis, University of London.

The Ecology of Armillaria mellea: *Rhizomorph Growth Through Soil*

D. B. REDFERN—*Department of Forestry and Rural Development, Forest Research Laboratory, Fredericton, New Brunswick, Canada.*

The rhizomorphs and mycelial strands of most specialized root-infecting fungi are not usually found growing through soil other than on root systems (Garrett, 1960), where soil conditions are modified by root exudates and bark scales. The rhizomorphs of *Armillaria mellea* are less closely associated with root systems: they grow out from a food base into the soil, which can thus exert a direct influence. The effects of food base and soil type on rhizomorph growth have, with a few notable exceptions, been little studied. This paper mainly describes work on this topic begun recently in England at the Botany School, University of Cambridge, and continued in Canada.

SOIL TYPE AND RHIZOMORPH GROWTH.—In contrast to temperate countries, rhizomorph production is sparse or nonexistent in parts of West and Central Africa (Leach, 1939; Fox, 1964). While investigating this phenomenon, Swift (1964) demonstrated the presence in a Rhodesian soil of a substance which totally inhibits rhizomorph growth. In Britain, rhizo-morphs have been observed so widely that if any such inhibitor is present, it either occurs very locally or does not cause total inhibition.

Field observations regarding the influence of soil type on rhizomorph growth and disease development are of limited value because they are so affected by host resistance, and by the type and distribution of food bases. However, the abundance of such observations, together with experiments by Bliss (1941), indicate that in temperate countries rhizomorphs develop in soils differing widely in type and reaction. Even so, some soils are more favorable for growth than others.

In one experiment by the author, segments of oak (*Quercus* sp.) stem, 4.0 cm long and 1.5 cm in diameter, were autoclaved and colonized by *A. mellea* under pure-culture conditions in the manner devised and described by Garrett (1956) and buried in soils in glass jars capped with plastic sheet. After 4 months incubation at 25°C segments were carefully washed free of soil and the rhizomorph systems counted, dried, and weighed (Table 1). Alkaline soils were

TABLE 1. The number and dry weight of rhizomorphs produced in a variety of soils from oak stem segments colonized by *A. mellea*

Soil type	% saturation	Segments* with living *A. mellea*	Segments* producing rhizomorphs	Rhizomorphs		
				Mean no. systems per segment*	Mean d.w. per segment (mg)	Mean d.w. per system (mg)
Ac. sand	25	10	7	1.5	5.9	3.9
" "	50	10	2	0.2	1.6	8.0
Al. sand	25	10	10	4.9	57.6	11.7
" "	50	10	9	4.0	48.9	12.2
Ac. loam	50	10	0	0.0	0.0	0.0
Al. clay	50	10	6	1.7	34.1	20.1
Ac. peat (amorphous)	50	6	0	0.0	0.0	0.0
Ac. peat (fibrous)	50	2	0	0.0	0.0	0.0
Al. loam	25	10	10	5.6	124.5	22.3
" "	50	10	10	4.9	167.7	34.3
" "	75	10	10	6.1	144.9	23.8
" "	100	0	0	0.0	0.0	0.0
Significant difference (P = 0.05)†				2.7	50.5	

* Out of 10.
† Snedecor's (1956) modification of Tukey's test.

more favorable to rhizomorph initiation and growth than were acid ones. Indeed, *A. mellea* died in some segments buried in the acid-fibrous and acid-amorphous peat soils. Moisture content, in those soils in which it was varied, had no significant effect on either rhizomorph initiation or growth, at least over the range 25-75% saturation. Garrett (1956) also found this to be true for an alkaline loam over the range 40-80% saturation. The few rhizomorphs initiated in alkaline clay grew well (Table 1, last column), which stresses the need to assess both initiation and subsequent growth in such experiments.

Although the above experiment provides a useful comparison of rhizomorph growth in several soils, total inhibition cannot be assumed in those soils in which growth did not occur: the acid loam tested was from a site heavily infested by *A. mellea*! Strains of this fungus vary in their ability to form rhizomorphs *in vitro* (Benton and Ehrlich, 1941; Raabe, 1966, 1967), and similar variations may be expected *in vivo*. The method of inoculum preparation may also be important in experiments of this nature. Inocula of willow (*Salix* sp.) prepared by Garrett's method (1956) produced fewer rhizomorphs in acid sand, a mean of 4.9 per segment, than inocula prepared by infecting live segments under unsterile conditions, which produced 16.4 rhizomorphs per segment. The latter technique is described more fully below.

Recent pure-culture experiments by Weinhold (1963) have shown that ethanol and other short-chain alcohols greatly stimulate rhizomorph production *in vitro*. *Aureobasidium pullulans* was found to have a similar effect through production of ethanol (Pentland, 1967). Pentland commented that ethanol may be continuously produced in soil and may be encountered by *A. mellea*, and that soils may differ in the amount available to a fungus.

FOOD BASE AND RHIZOMORPH GROWTH.—Although substances in soil may vitally affect the growth of *A. mellea*, probably the bulk of its growth requirements are obtained from root tissues, the quality of which may greatly influence rhizomorph production. Day (1929) commented that "In natural or semi-natural forest, it [*A. mellea*] seems to be confined to broadleaved forest, pure or mixed with conifers. In pure natural conifer forest it seems at least to be rare, but in pure conifer plantations established on the site of old broadleaved forest it is extremely common at least during the first rotation." Later authors apparently agreed with these observations, but made no attempt to study their implications. Greig (1962) made the equally significant observation that in conifer plantations established on hardwood sites, "the vigour of *A. mellea* appears to decline" as the hardwood stumps disintegrate. Conifer stumps apparently do not act as suitable food bases, although they are readily invaded: a point of considerable importance in plantation forestry. If killing occurs, can it be expected to continue unabated in succeeding rotations, or to decline?

A preliminary experiment was performed in England to assess the value of the food base provided by roots of various tree species. Segments of conifer roots infected by *A. mellea* produced the same total length and dry weight of rhizomorphs as hardwood segments of the same volume, but the latter produced significantly more rhizomorph systems. Thus, surprisingly, the conifer roots apparently provided as effective a food base for *A. mellea* as did the hardwood roots. However, because rhizomorph growing tips may be considered the "infective propagules" of *A. mellea*, the different numbers of rhizomorph systems produced from the two substrates may be of considerable importance. These results raise the question of what criteria should be adopted as a measure of food base quality. Although it seems reasonable to suppose that any superiority of hardwood over conifer roots as a food base would be expressed in the total amount of fungal material produced, this is not necessarily so. Even though *A. mellea* can compete as a saprophyte, probably more successfully than most fungi considered to be specialized parasites (Garrett, 1960), long-term survival depends on its ability to infect living trees. This ability may or may not be positively correlated with either the number, growth rate, or duration of growth of rhizomorphs produced from a given food base.

The experiment just mentioned was repeated in Canada to allow more detailed examination of the rhizomorphs produced. Freshly excised and undamaged roots of *Picea rubens* and *Acer rubrum*, 1.5 cm in diameter and without side roots, were cut into segments approximately 5.0 cm long and of equal volume (8.0 cm³). Autoclaved maple inoculum segments of the same diameter, prepared 3 months earlier, were held firmly to the end of each living root segment by two rubber bands and the resulting joined segments were incubated in moist, nonsterile sand. When infection was complete, a process requiring approximately 6 weeks, the joined segments were separated and each newly infected root segment was placed in dry soil (a sandy loam of pH 4.1) in a glass jar. Water was added to bring the soil to 50% saturation and the jars were loosely capped with metal covers. After 6 months incubation at 25°C, during which time the soil was maintained at 50% saturation by addition of water at monthly intervals, the 37 replicate segments of each substrate, and the rhizomorphs produced from them, were carefully washed from the jars. As before, significantly (5% level) more rhizomorph systems were formed from the hardwood substrates (13.9) than from the conifer substrates (9.3), but this time the hardwood material also produced a significantly greater total length and dry weight of rhizomorphs: 109.6 cm and 29.2 mg respectively for maple, in contrast to 75.8 cm and 19.5 mg respectively for spruce. However, the two types of substrate segment, although of equal volume, were of different initial densities: that of the spruce being 0.407 g cm^{-3} and the maple 0.465 g cm^{-3}. After correcting on this basis, the length and weight of rhizomorphs produced by maple no longer differed significantly from that of rhizomorphs produced by the spruce. However, the numbers of rhizomorph systems and rhizomorph growing tips remained significantly different even when corrected: 12.2 and 3.1 respectively for maple and 9.3 and 1.2 respectively for spruce. In the light of Garrett's (1956) work on inoculum potential, this correction was considered not only justified but necessary. Losses in dry weight of the two substrates, 33% for the maple and 35% for the spruce, suggest that both species would have become exhausted as food bases at about the same time if rotting had been allowed to continue.

It is unlikely that the postulated superiority of hardwood roots over conifer roots as food bases for *A. mellea* can be fully explained by the fact that hardwood root segments apparently permit growth of a slightly greater amount of fungal material than conifer segments of equal volume. The association of disease caused by this fungus with hardwood sites alone is too consistent for the reason to lie only in small growth differences. Either the postulate is incorrect and there is another explanation for the field observations or, more probably, the parameters chosen to measure food base quality are not related to the parasitic ability of *A. mellea;* although the number of rhizomorph growing tips may be an exception. Because of these shortcomings additional experiments are being conducted on food base quality and the host-parasite relationship.

ACKNOWLEDGEMENTS.—That portion of the work carried out in England forms part of a thesis submitted for the degree of Ph.D. at the University of Cambridge. I am indebted to the U.K. Science Research Council for the award of a Research Studentship which enabled me to carry out this work. I would like to thank my supervisor, Dr. J. Rishbeth, for his advice and encouragement and also for helpful criticism of this manuscript.

LITERATURE CITED

BENTON, V. L., and J. EHRLICH. 1941. Variation in culture of several isolates of *Armillaria mellea* from western white pine. Phytopathology 31:803–811.

BLISS, D. E. 1941. Artificial inoculation of plants with *Armillaria mellea.* Phytopathology 31:859.

DAY, W. R. 1929. Environment and disease. A discussion on the parasitism of *Armillaria mellea* (Vahl). Fr. Forest. 3:94–103.

FOX, R. A. 1964. A report on a visit to Nigeria (9–30 May, 1963) undertaken to make a preliminary study of root diseases of Rubber. Docum. Res. Arch. Rubber Res. Inst. Malaya 27. 34 p.

GARRETT, S. D. 1956. Rhizomorph behaviour in *Armillaria mellea* (Vahl) Quél. II. Logistics of infection. Ann. Botany (London) (N.S.) 20:193–209.

GARRETT, S. D. 1960. Rhizomorph behaviour in *Armillaria mellea* (Fr.) Quél. III. Saprophytic colonization of woody substrates in soil. Ann. Botany (London) (N.S.) 24:275–285.

GREIG, B. J. W. 1962. *Fomes annosus* (Fr.) Cke. and other root-rotting fungi in conifers on ex-hardwood sites. Forest. 35:164–182.

LEACH, R. 1939. Biological control and ecology of *Armillaria mellea* (Vahl). Fr. Trans. Brit. Mycol. Soc. 23:320–329.

PENTLAND, G. D. 1967. Ethanol produced by *Aureobasidium pullulans* and its effect on the growth of *Armillaria mellea.* Can. J. Microbiol. 13:1631–1639.

RAABE, R. D. 1966. Variation of *Armillaria mellea* in culture. Phytopathology 56:1241–1244.

RAABE, R. D. 1967. Variation in pathogenicity and virulence in *Armillaria mellea.* Phytopathology 57:73–75.

SNEDECOR, G. W. 1956. Statistical methods. Iowa State University Press, Ames. 534 p.

SWIFT, M. J. 1964. The biology of *Armillaria mellea* in Central Africa. Ph.D. thesis, University of London.

WEINHOLD, A. R. 1963. Rhizomorph production by *Armillaria mellea* induced by ethanol and related compounds. Science 142:1065–1066.

Armillaria mellea *(Vahl ex Fries) Kummer in Central Africa: Studies on Substrate Colonisation Relating to the Mechanism of Biological Control by Ring-barking*

M. J. SWIFT—*Botany Department, Birkbeck College, University of London, England (formerly at University College of Rhodesia).*

Interest in *Armillaria mellea* in Central Africa has mainly centred round the pioneering work of Leach (1937, 1939) in Malawi, which resulted in the development of his method of biological control. The use of this method has successfully reduced the incidence of *A. mellea* deaths in tea gardens throughout the area to negligible proportions compared with the very heavy losses reported by Leach and others in prewar years (Wiehe, 1952). In other crops where application of the method is uneconomical, *Armillaria* continues to be a serious danger. A situation of this kind has recently been reported for conifer crops in Rhodesia (Swift, 1964, 1968). In this instance the disease follows the pattern first described by Leach and since confirmed elsewhere (Gibson and Goodchild, 1960); attacks in plantation crops result from a buildup of a large inoculum in the stumps of felled indigenous hardwoods e.g. *Brachystegia spiciformis*. *Armillaria* is present on a low percentage of these trees prior to felling, though generally in a nonpathogenic form. When the tree is felled, the resistance to invasion is lowered and a rapid growth of the fungus in the stump ensues.

Subsequent to penetration of a root, hyphae of *A. mellea* appear to follow the path of high starch concentration within the root tissues (Leach, 1937, 1939). Leach therefore suggested that the critical factor for invasion might be the concentration of stored carbohydrate within the roots. Thus if the quantity of carbohydrate in the roots can be lowered before felling, buildup of *Armillaria* in stumps may be avoided. Ring-barking the trees two years before felling is a way of achieving such a change in the carbohydrate balance of the root and provides the basis of Leach's successful method of control. Recent observations, however, suggest that *A. mellea* may persist within ring-barked areas. Second plantings may show a recurrence of Armillaria disease; in other instances shade trees (e.g. *Grevillea robusta*) in tea gardens, which had been ring-barked prior to felling, acted as foci for outbreaks in adjacent tea bushes. Redfern (1968) has described situations in temperate areas where ring-barking appears to enhance, rather than inhibit, invasion of hardwood root systems by *A. mellea*.

INFECTION BEHAVIOUR OF A. MELLEA IN THE FIELD. —*Natural Infection.*—A sample of 150 trees from *Brachystegia spiciformis* woodland was felled and examined after eighteen months for *A. mellea*; 14% of the stumps showed colonisation by the fungus. Close examination revealed that in the majority of cases colonisation had originated from localised infections present on tap or lateral roots prior to felling.

Artificial Infection.—One hundred replicates comprising four different treatments of *Brachystegia* were inoculated with *A. mellea* by tying infected woodblocks to several roots and were examined six months later. No *Armillaria* was found on the roots of trees left standing nor on stumps of trees that had been ring-barked twelve months or more prior to felling. Successful colonisation by *A. mellea* was found in eleven of the stumps that were felled at the same time as inoculation, and in three of those that were left standing but were ring-barked at the time of inoculation. Colonisation of the latter, however, originated from infections already present on the roots prior to treatment rather than from the artificial inoculum.

Conclusions.—Although ring-barking is effective in preventing invasion of stumps from an external (saprophytic) inoculum source of *A. mellea,* the spread of the fungus from parasitic lesions is not inhibited. In the latter case ring-barking has an effect comparable to that of felling in lowering host resistance to invasion.

GROWTH ACTIVITY OF A. MELLEA IN RELATION TO SAWDUST SUBSTRATES.—*Breakdown of root tissue by A. mellea.*—The fungus was grown on 20 gm samples of sterile sawdust from the root-wood of *B. spiciformis*. Cultures were sampled in duplicate every four weeks over a period of fifteen months and the loss in dry weight of the sawdust was determined. The sawdust was then chemically analysed for a number of fractions. Starch is rapidly utilised; 46% of the original starch content disappears during the initial 9% loss in the dry weight, and 98% by the time the sawdust has been 53% decomposed. Over the same periods losses of lignin were 8% and 67%, and of cellulose, 8% and 56%. Thus, although cellulose breakdown is the most important component of weight loss in the sawdust, on a weight-for-weight basis lignin and starch are more extensively degraded.

The growth of A. mellea *in sawdust substrates of varying composition.*—Growth media for *A. mellea* were prepared by suspending 0.5 gm of sterile sawdusts of different compositions (Table 1), in 15 ml of 2% water agar in a petri plate. The plates were centrally inoculated with *A. mellea* and incubated in the dark at 25°C. Growth was estimated after ten days as total rhizomorph length.

A series of control plates was established in which 0.05% yeast extract was added to the suspensions. This was to ascertain whether alterations in the content of materials other than carbohydrates was affecting the growth of the fungus. In all cases but lignin, the mean growth rate on these control plates was the same as that shown in Table 1, indicating that the results may be interpreted in terms of the altered carbohydrate constitution of the media.

Gross changes in the content of sugars and starch significantly affect the rate of growth of the fungus (Table 1). Growth on lignin alone was minimal but was stimulated to the level of that on cellulose plus lignin by the addition of 0.05% yeast extract. The yeast extract used contains roughly 50% amino acids, 10% carbohydrates, and a wide range of growth factors and inorganic salts. The stimulatory component was found to be sugar; when grown in liquid medium with a basal medium of salts *A. mellea* showed significantly greater growth in glucose *and* lignin (63.6 ± 1.77 mg) when supplied at concentrations of 0.05% and 0.5% respectively, than on glucose alone (28.7 ± 0.81 mg), or lignin alone (4.9 ± 0.31 mg). This synergistic effect of glucose and lignin is similar to that found by Gottlieb, Day, and Pelczar (1950) for three species of *Polyporus*, and is suggestive of an increased efficiency of lignin utilisation in the presence of sugars.

Conclusions.—*A. mellea* is able to utilise all the major structural and storage compounds of wood tissue as growth substrates, but growth is most markedly related to the availability of sugars in a readily metabolised form.

Competitive Ability of A. mellea on Sawdust Substrates.—The success of *Armillaria* in occupying natural substrates such as stump tissues will depend not only on its ability to utilise such tissues as growth substrates but on its ability to do so in a competitive situation.

Four unidentified fungi, two of which (*C* and *D* in Table 2) produced clamp connections in agar culture, were isolated from the tissues of ring-barked stumps and tested as antagonists to the growth of *A. mellea*. These fungi showed no depression of growth rate on starch-free sawdust in direct contrast to that recorded for *Armillaria*. The fungi were co-inoculated with *Armillaria* on plates of sawdust media, the inocula being initially 2 cm apart. The subsequent interaction of the fungi was noted and scored according to the relative dominance of *Armillaria* (Table 2). Interactions were compared on two media differing in their starch content and in situations where *A. mellea* was inoculated either at the same time or five days prior to the antagonist.

The results confirm that the starch concentration of the substrate strongly affects the competitive ability of *A. mellea*. Prior possession of the substrate, however, can modify the effect of low starch. This effect is limited, for after twenty days the *Armillaria* scores in row 3 of Table 2 were lower.

Conclusions.—The competitive ability of *A. mellea* on sawdust substrates is enhanced by the presence of starch. In the absence of starch *Armillaria* may be replaced by other fungi.

Discussion.—It is important to distinguish between the two types of effect ring-barking might have: (a) on the rate of spread of *A. mellea* within stumps already parasitised, and (b) on the rate of spread between stumps. In the latter case Leach's conclusions concerning the effect of starch concentration are borne out by this work. The former case is more difficult to interpret. The initial effect of ring-barking is to stimulate invasion, as has also been demonstrated by

Table 1. Growth of *A. mellea* on sawdust media of different composition[*]

	Growth as total rhizomorph length in mm					
	Type of sawdust substrate					
	Sawdust 3.8% sugars 17.8% starch	Sawdust 11.5% starch	Cellulose Hemicellulose Lignin	Cellulose Lignin	Lignin Yeast Extract	Lignin alone
Mean[†]	28.5	17.2	10.6	11.1	11.1	1.4
Differences[‡]	—	11.3	6.6	0	0	9.2

[*] Media prepared from *Brachystegia* root-wood sawdust by extraction with diastase (starch), alkali (hemicellulose); lignin by method of Brauns (1939).

[†] Ten replicates for each treatment.

[‡] Analysis of variance gives D (= S x̄ Q) = 6.6 at p = 0.05, i.e. differences greater than this are significant.

TABLE 2. Growth of *A. mellea* on sawdust media in mixed culture

		Relative dominance of *A. mellea* on 10 plates*			
		Antagonist fungus			
Medium	Inoculation	A	B	C	D
Starch present	Lapsed	11	16	10	12
	Simultaneous	9	7	8	10
Starch absent	Lapsed	2	6	7	8
	Simultaneous	0	0	0	0

* The dominance of *Armillaria* is expressed on a scale related to three types of observed reaction between the fungi on the plates (a), *A. mellea* overgrows antagonist (score = 2); (b) antagonist overgrows *A. mellea* (score = 0); (c) neither overgrown (score = 1). Assessment made after ten days for ten replicates in each series (i.e. maximum score = 20).

Redfern (1968). However, as Redfern indicates, it is probable that infectivity is lowered. This is probably both due to the extended period between invasion and contact with the planted crop, and to the very rapid exhaustion of starch reserves. This view is compatible with the observations quoted earlier of survival of *A. mellea* in plantations apparently free of it and also with the results of the mixed-culture experiments, where saprophytic replacement is enhanced by time and starch concentration.

Although *A. mellea* can degrade both cellulose and lignin in pure culture, its ability to occupy substrates largely composed of these substances in competition with fungi of similar capability is dependent on the additional presence of sugars or starch. This evidence indicates that *A. mellea* belongs to the physiological category of "sugar fungi," whereas the four fungi tested as antagonists resemble the typical late colonisers, i.e. cellulose or lignin fungi, of Burges (1939) and Garrett (1956).

Woody substrates, however, may prove to be specialised materials compared with the relatively transient plant residues that have been more commonly studied. Recent evidence indicates that in pine stumps (Meredith, 1960) and in woody beech litter (Swift, unpublished data) the succession of fungi is the reverse of that shown by Burges and Garrett; lignin decomposers provide the primary flora and are succeeded by those of the "sugar-fungus" type. This may explain the observations of Garrett (1960), who demonstrated the ability of *A. mellea* to colonise autoclaved willow segments even when they had been buried in soil for three weeks. As he points out, the conditions of Garrett's experiments were such as to make it probable that the competing flora was largely composed of sugar fungi rather than of types capable of causing extensive decay of wood.

In this situation it is the lignin- and cellulose-decomposing ability which confers high competitive ability on *Armillaria* in contrast with the situation described above. In general the distribution of *A. mellea* is most closely related to its parasitic ability. Indeed, in Garrett's experiments the colonisation of living willow segments was always quantitatively more successful than that of autoclaved segments. All too little is known of pathogenic mechanisms in the wood-rotting fungi. One thing is certain, however—they must relate to features distinct from those of saprophytic importance such as lignin- or cellulose-degrading ability.

ACKNOWLEDGEMENT.—The work described in this paper represents part of a thesis presented for the Ph.D. degree of the University of London. The research was financed by the Rockefeller Foundation. I am also grateful to Dr. A. R. Loveless for supervision and to the British South Africa Company for providing facilities in the field.

LITERATURE CITED

BRAUNS, F. E. 1939. Native Lignin. Its isolation and methylation. J. Am. Chem. Soc. 61:2120–2127.

BURGES, A. 1939. Soil fungi and root infection. Broteria 8:64–81.

GARRETT, S. D. 1956. Biology of root-infecting fungi. Cambridge Univ. Press, London and New York. 293 p.

GARRETT, S. D. 1960. Rhizomorph behaviour in *Armillaria mellea* (Vahl) Quél. III. Saprophytic colonisation of woody substrates in the soil. Ann. Botany (London) [N.S.] 24:275–285.

GIBSON, I. A. S., and N. A. GOODCHILD. 1960. *Armillaria mellea* in Kenya forests. E. Afr. Agr. J. 26:142–143.

GOTTLIEB, S., W. C. DAY, and M. J. PELCZAR. 1950. The biological degradation of lignin. II. The adaptation of white-rot fungi to growth on lignin media. Phytopathology 40:926–935.

LEACH, R. 1937. Observations on the parasitism and control of *Armillaria mellea*. Proc. Roy. Soc. Series B 121:561–575.

LEACH, R. 1939. Biological control and ecology of *Armillaria mellea*. Trans. Brit. Mycol. Soc. 23:320–329.

MEREDITH, D. S. 1960. Further observations on fungi inhabiting pine stumps. Ann. Botany (London) [N.S.] 24:65–78.

REDFERN, D. B. 1968. The ecology of *Armillaria mellea* in Britain: biological control. Ann. Botany (London) [N.S.] 32:293–300.

SWIFT, M. J. 1964. The biology of *Armillaria mellea* in Central Africa. Ph.D. thesis, University of London.

SWIFT, M. J. 1968. Inhibition of rhizomorph development by *Armillaria mellea* in Rhodesian forest soils. Trans. Brit. Mycol. Soc. 51:241–247.

WIEHE, P. O. 1952. The spread of *Armillaria mellea* (Fr.) Quél. in tung orchards. E. Afr. Agr. J. 18:67–72.

Fomes annosus *in the Southern United States*

C. S. HODGES—*Southeastern Forest Experiment Station, Forest Service, U.S. Department of Agriculture, Research Triangle Park, North Carolina.*

Fomes annosus was first collected in the southern United States from a pine log in Alabama in 1897 (Underwood), and first reported as a pathogen on white pine *(Pinus strobus)* in North Carolina in 1923 (Haasis). Subsequently the fungus has been found killing slash *(Pinus elliottii* var. *elliottii),* loblolly *(P. taeda),* longleaf *(P. palustris),* shortleaf *(P. echinata),* and pitch *(P. rigida)* pines, eastern redcedar *(Juniperus virginiana),* Atlantic white cedar *(Chamaecyparis thyoides),* and Fraser fir *(Abies fraseri).*

With respect to symptomatology and factors which influence disease development, the hosts attacked by *F. annosus* fall into three distinct categories: (1) eastern redcedar, (2) white pine, and (3) slash, loblolly, and other hard pine species. In the Piedmont of the southeastern United States, eastern redcedar is primarily an understory species in stands of loblolly and shortleaf pines. Most of these stands are on moderately heavy clay soils of low pH. *Fomes annosus* often colonizes a high percentage of pine stumps resulting from logging operations, but the residual pine is rarely attacked. Redcedar mortality, however, may be 50% or more in some stands. Miller (1943) attributes the susceptibility of redcedar under these conditions to general unthriftiness and suppressed growth. However, loblolly and shortleaf pines of suppressed and intermediate crown classes growing adjacent to killed redcedar are not affected. No explanation presently accounts for the differential susceptibility of pines and redcedar on Piedmont sites.

Fomes annosus infection on redcedar causes a typical root rot, seldom growing more than a few centimeters above the root collar. The sapwood at the base of infected trees may be badly decayed but the heartwood is resistant to decay. This has been attributed to a higher content of cedarwood oil in heartwood (Dwyer, 1951). Occasionally, infected redcedar trees are windthrown and remain alive on the ground for a considerable time. Usually, however, they die upright and are readily recognized by the brick-red foliage.

In the southeastern United States, eastern white pine is restricted to the upper Piedmont and mountain regions, where it is the most economically important conifer. *F. annosus* occurs in both natural and planted stands but causes much greater losses in the latter. As with other hosts, *F. annosus* in most white pine stands is related to the presence of stumps that result from thinning operations. The fungus also may be found in unthinned stands, but it is uncommon (Boyce, 1962).

White pine differs from other pine hosts in that heart rot is an important part of the disease syndrome. Heart rot can be found several meters high in trees that are less than 50 years of age. In this respect the disease on white pine is more similar to that on spruce and fir. The reason for the occurrence of heart rot in white pine is not known, but Platt, Cowling, and Hodges (1965) showed that the relative decay susceptibility of root wood and stem wood of white pine was approximately the same, whereas stem wood of southern pine species was considerably more resistant to decay than root wood.

Although similar to spruce and fir with respect to heart rot, white pine roots are also extensively rotted. Both tap and lateral roots may be affected. Standing trees rarely die, however, even when most of the roots are decayed. Trees generally die after windthrow.

White pine infected by *F. annosus* seldom exhibit crown symptoms, other than a slight yellowing and thinning. Some known infected trees under observation for 11 years have exhibited no crown symptoms and no detectable decrease in diameter increment. Production of sporophores on these infected trees was highly variable, ranging from new sporophores almost yearly to one or two during the observation period.

Slash and loblolly pines are the most economically important hosts of *F. annosus* in the southern United States. During the last 15 years, over one-half million acres of pines have been planted annually. Slash and loblolly pines account for more than 95% of this total. These species are planted principally in the Coastal Plain under optimum conditions for development of annosus root rot.

In 1960, a survey of thinned and unthinned natural and planted stands in the southern United States (Powers and Verrall, 1962) confirmed what has long been observed in Europe—that the fungus is more prevalent and more damaging in thinned than in unthinned plantations and thinned or unthinned natural stands. The fungus was found in 59 and 44% of the thinned plantations of loblolly and slash pine, respectively. Overall losses were small, with only 2.8 and 2.2% of the trees in loblolly and slash pine plantations, respectively, killed or infected by *F. annosus*. Infection in individual plantations, however, ranged up to 30%, which emphasizes the potential threat in thinned plantations by this disease. Only 0.07% of the trees in natural slash pine were dead or dying. The survey also revealed that light sandy soils, deep forest litter, stands on slopes, and increasing depth of the A horizon were correlated with increased damage from *F. annosus* (Powers and Verrall, 1962).

The disease caused by *F. annosus* on loblolly and slash pines, as well as other southern pines, is a typical root rot. Heart rot, except for some discoloration several centimeters above the root collar, has not been observed. Infected loblolly and slash pines usually die upright. Windthrow of trees with badly decayed roots but no crown symptoms is not uncommon, however, especially on certain sites.

Detailed observations of several hundred infected loblolly and slash pines over the past 10 years have revealed no distinct patterns of symptomatology. Infected trees sometimes show progressive crown symptoms over a period of years, consisting of needle yellowing, decrease in needle length and leader growth, and loss of all but the current year's needles. Usually, however, trees die without any outward indication of infection. Lack of symptoms, plus the fact that trees usually die singly and scattered over a large area, makes salvage impractical.

As on white pine, production of sporophores on loblolly and slash pines is quite variable. New sporophores are sometimes produced on infected trees for several years before and after death, whereas on some trees they are never produced. Because of their location in or beneath the litter, or in some cases below ground level in rodent tunnels, they seldom persist for more than one year because of the activity of fungi and insects.

An interesting symptom sometimes associated with infected slash and loblolly pines has been termed "bark resin soaking" by the author. The bark of dead trees showing this symptom may be partially or completely soaked with resin for up to 3 meters above ground level, giving the base of the tree the appearance of being saturated with oil. Resin exudation or "bleeding," similar to that associated with canker diseases, does not occur. The wood beneath the bark is infiltrated with resin for a depth of several centimeters. *F. annosus* can always be isolated from the infiltrated wood. This wood remains free from blue stain and other fungi for several months after death of the tree. Bark resin soaking has not been observed on living trees. Of the small number of dead trees exhibiting resin soaking in permanent observation plots, most showed advanced crown symptoms before death. Trees with resin-soaked bark are visible for some distance in the woods and are helpful in detecting infection centers.

In infection centers created by inoculation of stumps in unthinned stands, some trees within 2 meters of inoculated stumps died 10 months following stump inoculation. In one case, a tree situated 5 meters from the nearest infected stump died after 2 years. *F. annosus* can grow about 2 meters per year through the roots of pine stumps (Hodges, 1968). Growth rate through the roots of living trees is considerably less, averaging from 1.3 cm per month (Towers and Stambaugh, 1968) to 4.3 cm per month (Miller and Kelman, 1966) in inoculated roots of loblolly pine. These growth rates of *F. annosus* through stump and tree roots cannot account for the death of trees 2 meters away from stumps whose surfaces were infected only 10 months before.

Rishbeth (1951) observed external growth of *F. annosus* on roots of trees growing in alkaline soil up to 1 meter beyond the point of invasion in living tissue, but no superficial growth was noted on roots in acid soil. Observations in slash pine plantations on soils of about pH 5 disclosed no discernible superficial mycelial growth on the surface of roots or under bark scales, either at or beyond the area where *F. annosus* had invaded the wood. On many roots, however, small isolated patches of heavy resin exudation developed along the root for a considerable distance ahead of wood colonized by *F. annosus*. When bark scales from beneath the exuded resin and from points between the areas of resin exudation were plated on a selective medium (Kuhlman and Hendrix, 1962), *F. annosus* was recovered. The fungus was apparently growing on or between the bark scales but had not formed strands of mycelium visible to the unaided eye. No data are available on the rate of growth of the fungus in the bark scales, but it is obviously considerably faster than through roots of living trees. This faster growth, plus multiple infection, could account for the relatively rapid death of residual trees following stump infection.

Examination of the isolated areas on roots exhibiting resin exudation revealed that both resin exudation and some resin infiltration of the xylem occurred before the fungus was present in the xylem. This suggests that a toxin produced by the fungus may be involved in pathogenesis. Whether it is the toxin reported by Persson (1957) or Bassett et al. (1967) is not known.

Although most infection centers are in thinned stands, centers in unthinned stands are not uncommon. Observations made in a number of these centers indicate that initial infection took place in trees broken near the ground at cankers caused by *Cronartium fusiforme* Hedgc. and Hunt ex Cumm. In others, intact standing trees of suppressed or intermediate crown class were apparently the first to become infected. It is possible that the roots of these trees became infected directly, as has been reported by Hendrix and Kuhlman (1964).

Spread of the fungus in unthinned stands is often rapid because of the large number of root contacts. In one slash pine plantation, 70 trees were killed or infected in one center in 8 years. Spread is more rapid along the rows than across the rows.

As in other areas where *F. annosus* is found, reforestation of infested sites is a management problem. In the southern United States, seedling losses have been observed on sites clear-cut because of heavy infestation by *F. annosus* and where *F. annosus* was believed to have become established after the residual stand was cleared (Hendrix et al., 1964). In both cases, seedling losses have not averaged more than about 1% per year. The fungus is still active after periods of up to 8 years, however, and some root-to-root contact between infected and healthy trees is taking place. This may result in continued losses. Mortality of natural regeneration in infection centers is seldom noted; this suggests that direct seeding of infested sites is a possible means of reducing losses on these sites.

Two control measures for annosus root rot are currently recommended in the southern United States. One involves sprinkling dry, powdered borax on the surface of the stump immediately after felling (Driver, 1963). This material gives excellent protection against *F. annosus* at the stump surface, is easy to apply, cheap, and safe to use. One disadvantage is that stumps remain alive longer, which may result in more direct infection of the stump roots by *F. annosus* (Hodges, 1968).

The second recommended control measure is to thin during the summer months when temperatures at the stump surface reach levels that are lethal to *F. annosus* (40°C for 2 hours) (Ross, 1967). In the more southern areas of the region, no or very little stump infection takes place from April through September; in the northern portion of the region, it is safe to thin only from mid-June through August. In addition to unfavorable temperatures for stump infection, spore production by the fungus is lower during these periods.

LITERATURE CITED

Bassett, C., R. T. Sherwood, J. A. Kepler, and P. B. Hamilton. 1967. Production and biological activity of fomannosin, a toxic sesquiterpene metabolite of *Fomes annosus*. Phytopathology 57:1046–1052.

Boyce, J. S. 1962. *Fomes annosus* in white pine in North Carolina. J. Forest. 60:553–557.

Driver, C. H. 1963. Further data on borax as a control of surface infection of slash pine stumps by *Fomes annosus*. Plant Disease Reptr. 47:1006–1009.

Dwyer, W. W., Jr. 1951. *Fomes annosus* on eastern red cedar in two Piedmont forests. J. Forest. 49:259–262.

Haasis, F. W. 1923. Root rot as a factor in survival. J. Forest. 21:506.

Hendrix, F. F., and E. G. Kuhlman. 1964. Root infection of *Pinus elliottii* by *Fomes annosus*. Nature (London) 201:55–56.

Hendrix, F. F., Jr., E. G. Kuhlman, Charles S. Hodges, Jr., and Eldon W. Ross. 1964. *Fomes annosus*—a serious threat to regeneration of pine. U. S. Forest Serv., S. E. Forest Exp. Sta., Res. Note SE–24. 4 p.

Hodges, C. S. 1968. Evaluation of stump treatment chemicals for control of *Fomes annosus*. (In press.)

Kuhlman, E. G., and F. F. Hendrix. 1962. A selective medium for the isolation of *Fomes annosus*. Phytopathology 52:1310–1312.

Miller, J. K. 1943. *Fomes annosus* and red cedar. J. Forest. 41:37–40.

Miller, Thomas, and A. Kelman. 1966. Growth of *Fomes annosus* in roots of suppressed and dominant loblolly pines. Forest Sci. 12:225–233.

Persson, E. 1957. Über den Stoffwechsel und eine antibiotisch wirksame substanz von *Polyporus annosus* Fr. Phytopathol. Z. 30:45–86.

Platt, W. D., E. B. Cowling, and C. S. Hodges. 1965. Comparative resistance of coniferous root wood and stem wood to decay by isolates of *Fomes annosus*. Phytopathology 55:1347–1353.

Powers, H. R., and A. F. Verrall. 1962. A closer look at *Fomes annosus*. Forest Farmer 21:8–9, 16–17.

Rishbeth, J. 1951. Observations on the biology of *Fomes annosus*, with particular reference to East Anglian pine plantations. III. Natural and experimental infection of pines, and some factors affecting severity of the disease. Ann. Botany (London), [N.S.] 15:221–246.

Ross, E. W. 1967. Practical control of *Fomes annosus* in southern pines. XIV IUFRO Congress, Sect. 24, p. 321–324.

Towers, Barry, and W. J. Stambaugh. 1968. The influence of induced soil moisture stress upon *Fomes annosus* root rot of loblolly pine. Phytopathology 58:269–272.

Underwood, L. M. 1897. Some new fungi, chiefly from Alabama. Bull. Torrey Botan. Club 24:81–86.

Fomes annosus *in Eastern Canada*

DAVID PUNTER—*Department of Forestry and Rural Development, Sault Ste. Marie, Ontario, Canada.*

Fomes annosus is found in most regions of the northern hemisphere on a wide range of vascular plants. Its significance as a pathogen is greatest in areas of coniferous forest (Anonymous, 1954, 1962; Bega, 1962, 1963; Sinclair, 1964). Collections from the northeastern states of the U.S.A. date back to the early years of forest mycology; however, there are no corresponding records from the adjacent provinces of Canada. A specimen from Nova Scotia (Somers, 1880) was almost certainly misidentified as *Polyporus annosus,* and Dr. Pomerleau (personal communication) now considers that the basis for the single Quebec record (Pomerleau, 1963) is a collection of *Trametes serialis* and not *F. annosus.*

F. annosus was first reported in Ontario by Jorgensen (1956) and is now recorded from eight principal locations (Fig. 1). Although red pine *(Pinus resinosa)* has been the commonest host species, other pines and two native hardwoods (Punter and Cafley, 1968) also have been attacked. The concentration of reliable records in the plantations of southern Ontario has led to speculation that *F. annosus* may be a recent introduction to eastern Canada. Since *F. annosus* is a serious forest pathogen, it is important to know if it is native to the area and to understand the factors which control its spread.

NATURAL LONG-RANGE DISPERSAL.—The only process which warrants serious consideration is aerial dissemination of spores. Between 1962 and 1967 more than 500 muslin-square spore traps were exposed throughout Ontario in the manner described by Rishbeth (1959). At first, the trapping was in summer at various locations, but from June 1966 to October 1967 muslins were exposed at two-week intervals at ten fixed sampling stations (Fig. 2). Samples of red pine foliage also were collected at some locations. Spores washed from the muslins and foliage were plated on fresh sections of red pine stem. The sixteen occurrences of *F. annosus* detailed in Fig. 2 were widely distributed in space but confined in time to the period between August and January.

In addition to this province-wide program, a regular schedule of trapping was followed from 1963 to 1967 near St. Williams, Ontario. Four fresh red pine stem sections, 8 cm in diameter, and one muslin square were exposed concurrently for three to four hours during the last week of each month. The average counts of viable *F. annosus* spores from three locations are presented in Fig. 3. They indicate that the numbers of airborne spores may undergo intense seasonal fluctuations even in an area where *F. annosus* is abundant. The peak values recorded during the fall agreed with those obtained by Rishbeth (1959) in East Anglia, and similar seasonal fluctuations occurred in New York State (Sinclair, 1964). These seasonal changes at St. Williams reflected the effects of climate (Fig. 4) upon sporophore activity. The sporophores remain frozen during most of January and February; those which are still capable of spore liberation after the spring thaw are quickly inactivated by hot, dry weather, and it is only with the production of new pore layers in the fall that spore discharge is resumed. Rishbeth's (1959) studies indicate that spores of *F. annosus* can travel hundreds of miles without loss of viability, but since spores were trapped in northern Ontario only during the fall, it is probable that their source was limited by climate. The large catch from near Geraldton (location b, Fig. 2) suggested a local source but none was revealed by careful searching.

The infectivity of these airborne spores was checked by providing a natural substrate for colonization.

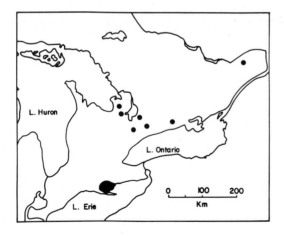

Fig. 1. Distribution of *Fomes annosus* in Ontario showing the eight principal locations on record in June 1968.

156

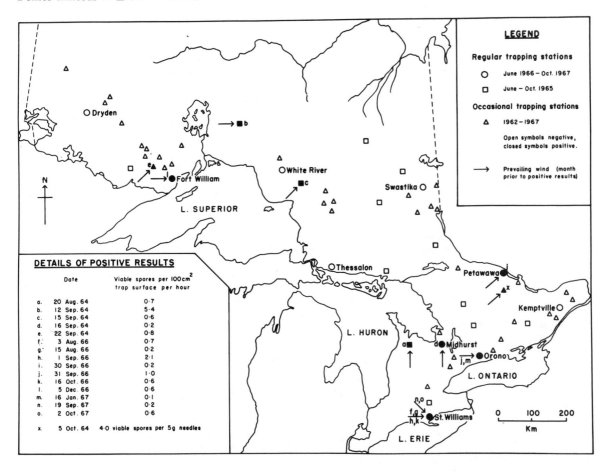

Fig. 2. Summary of trapping for airborne spores of *Fomes annosus* in Ontario between 1963 and 1967 with details of positive results (solid symbols).

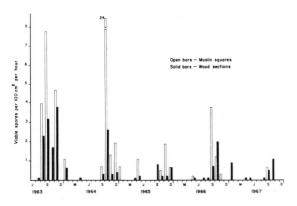

Fig. 3. Average counts of viable *Fomes annosus* spores trapped at three locations near St. Williams, Ontario. Muslin squares and wood sections were exposed at monthly intervals.

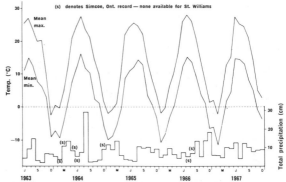

Fig. 4. Monthly mean maximum and minimum temperatures and total precipitation at St. Williams, Ontario. (Canada Department of Transport)

Each month, while the traps were being exposed at St. Williams, 100 red pines were felled in a nearby

plantation. Six months later, sections were cut from the tops of the resulting stumps and examined, after incubation, for evidence of *F. annosus*. The percentage frequencies of colonization are given in Fig. 5.

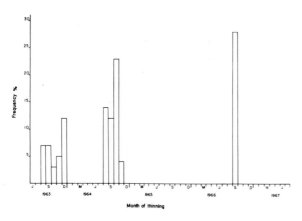

Fig. 5. Seasonal variation in frequency of colonization of *Pinus resinosa* thinning stumps by *Fomes annosus* at St. Williams, Ontario.

The availability of airborne inoculum determined the frequency of colonization, but successful stump infection was rare when spores were trapped at rates lower than 1/100 cm²/hour.

DISSEMINATION BY MAN.—As Jorgensen (1962) has reasoned, the likelihood of infested poles or lumber being brought into Canada is remote. Seedlings have been imported, but *F. annosus* root rot is not normally found in nursery beds. Jorgensen (1961), however, demonstrated that basidiospores deposited on seedling roots may remain viable for eight weeks during transportation. Although basidiospores did not infect wounds on seedling roots, they did colonize stump roots when applied directly to wounds made during a simulated planting operation. He therefore postulated that stumps might become infected during reforestation and serve as foci for new infections. This suggestion has prompted an examination of the susceptibility of severed stump roots to infection by various types of inoculum.

An unthinned block of 120 healthy red pines was felled and the stumps were numbered in six random groups. The treatments and inoculations are summarized in Table 1 *(a)*. Surface treatment prevented interference from airborne *F. annosus*. Sample discs, taken from the tops of Group *A* stumps immediately before inoculation, were free from *F. annosus*. At each stump, one sound lateral root no less than 3 cm in diameter was exposed. A hole was drilled as aseptically as possible into the root of each Group *B* stump, 15 cm from the body; into this was inserted a beech-wood plug permeated by *F. annosus*. The exposed roots were all severed approximately 25 cm from the stump body. Basidiospore deposits were collected during the week prior to inoculation; conidia were obtained by washing colonies on 1.2% malt extract agar. The freshly prepared suspensions were sprayed on the exposed wood surfaces as indicated in Table 1 *(a)*. Germination of the basidiospores (28%) and the coni-

dia (89%) was checked before and after inoculation. A 2–0 red pine seedling was planted beside each inoculated stump, its root system in close juxtaposition to the cut root end. The roots of seedlings used for Group *E* stumps had been placed under sporophores of *F. annosus* until they bore a visible coating of spores. The seedlings were inspected monthly, and as soon as one died, wood chips were isolated from the roots and lower stem on a medium selective for *F. annosus* (Kuhlman and Hendrix, 1962). The associated stump was extracted along with the proximal 50 cm of each lateral root. Isolations were made from (a) the inoculated root at 5 cm intervals, (b) all other lateral roots 5 cm from the proximal end, and (c) the root-collar region of the stump opposite each lateral root. All roots, split longitudinally, and a disc, 5 cm thick, from the root collar of each stump were incubated in moistened newspaper. All samples were examined microscopically for conidia of *F. annosus*. Thirty-two months after inoculation, all the remaining seedlings and stumps were harvested. The distribution of *F. annosus* within the seedlings and stumps is summarized in Table 1 *(b)*. The only types of inoculum which brought about any colonization were the wood plugs *(B)* and the basidiospore suspension *(C)* applied directly to cut root ends. Whenever *F. annosus* was isolated from a seedling, the adjacent stump also was found to be thoroughly colonized. The remaining dead seedlings *(A, D, F)* from which *F. annosus* was not isolated appeared to have died from other causes. A few seedlings beside stumps which contained *F. annosus* had not died by the end of the experiment; in those cases the pathogen had entered the stump body and other roots but had been replaced by other organisms in the inoculated root. This experiment confirmed that red pine stumps may remain susceptible to root infection by *F. annosus* basidiospores for several months after felling and may serve as infection foci in the next rotation. It was surprising that conidia were noninfective under these same conditions although their germination was superior to basidiospores *in vitro*. The failure of the basidiospores on seedlings to infect stump roots suggests that, in the face of saprophytic competitors, intimate contact is essential for the colonization of wood which has lost much of its selectivity for *F. annosus*. On the other hand, this positional advantage is still insufficient to promote a direct parasitic attack on seedling roots. Although the intention was to simulate a normal planting operation, very high doses of inoculum were used deliberately to promote the establishment of the pathogen. Natural contamination of seedlings would never reach such a level even in the fall; moreover, most seedlings are lifted in the spring when airborne spores are relatively scarce. Thus, while infection by basidiospores may be a theoretical hazard for stump roots, it is most unlikely that *F. annosus* has been or will be introduced into new areas on seedling root systems. This, the last significant method by which an introduction could have been effected, is rendered the more improbable since the plantations in which

TABLE 1. Behavior of *F. annosus* in *P. resinosa* stumps and adjacent seedlings following inoculation of stumps by five methods

	Group (20 stumps)					
	A	B	C	D	E	F
(a) *Summary of treatments and inoculations*						
Surface treatment immediately after felling (Jan. 1964); 20% aq. NiSO₄	−	+	+	+	+	+
Position of inoculation (Sept. 1964)	Cut surface	One root	←———————One cut root end———————→			
Type of *F. annosus* inoculum	Basidiospore suspension; 5 x 10⁵ spores per stump	Beechwood plug; 12.5 mm diam.	Basidiospore suspension; 2 x 10⁵ spores per root	Conidial suspension; 2 x 10⁵ conidia per root	Basidiospores on seedling roots	None
(b) *Condition of stumps and seedlings 34 months after inoculation*						
Total seedlings dead	1	9	2	1	0	2
Dead seedlings containing *F. annosus*	0	7	2	0	0	0
Number of stumps in which *F. annosus* present on						
(i) inoculated root	0	9	3	0	0	0
(ii) other roots	0	11	4	0	0	0
(iii) root collar region	0	10	3	0	0	0

F. annosus has occurred to date were established on derelict arable land rather than on cut-over forest sites.

PAST AND PRESENT DISTRIBUTION OF F. ANNOSUS.—Airborne inoculum has been detected over much of Ontario during the fall, and in 1967, although 100 muslin squares exposed in New Brunswick failed to detect *F. annosus,* it was isolated from a sample of spruce foliage collected there in September (Redfern and Van Sickle, 1968).

The ability of such inoculum to colonize suitable substrates has been demonstrated, sources of inoculum have been available in adjacent states for many years, and host species, of which red pine is a highly susceptible example, are native and plentiful in eastern Canada. It is therefore unnecessary to postulate introduction by man to account for the presence of this fungus.

If *F. annosus* was not introduced into eastern Canada, its known distribution must be interpreted in some other way. Several factors may explain why *F. annosus* was detected during decay studies in the west but not in the east. Although stumps have long been available for colonization in eastern forests, most of the felling there has been done in winter when airborne *F. annosus* spores are scarce. On the west coast, however, spores may be trapped throughout the year (Reynolds and Wallis, 1966), and logging wounds on several prominent species may be invaded directly (Wright and Isaac, 1956), thus providing many more opportunities for infection. In addition, many western conifers are susceptible to butt rot by *F. annosus,* whereas the stems of the species most heavily represented in the east are not decayed until after death. Since dead trees were not examined, there was therefore little likelihood that *F. annosus*

would be isolated during eastern decay studies. The proportion of the total tree population of eastern Canada which has received critical pathological examination is minute. Untrained observers may easily confuse *F. annosus* root rot with other diseases and may miss sporophores which often develop incompletely or are hidden from view. Thus it seems probable that *F. annosus* is a native fungus which for many reasons has remained undetected in the natural forests of eastern Canada just as it did in the plantations for a number of years.

Although plantations in eastern Canada have been extended greatly during the past 60 years, they are still almost confined to southern Ontario and parts of Quebec. These artificial monocultures, especially of pines on derelict agricultural land, provide an ideal environment for *F. annosus.* Inspection of plantations has been relatively thorough, and groups of dead trees are more conspicuous than in the natural forest. Considering also that thinning often has been carried out during the fall when airborne spores are most abundant, it is hardly surprising that the recorded distribution of *F. annosus* has paralleled the development of pine plantations in Ontario.

CONCLUSIONS.—Certain conclusions based on the foregoing evidence are of practical significance in forest management.

1. *Fomes annosus* is probably an uncommon component of the native forest flora in eastern Canada.

2. Viable spores occur throughout the region, but wide seasonal fluctuations in their numbers have helped to limit the natural spread of the fungus.

3. Introduction of the pathogen on seedlings is unlikely to have occurred and represents an insignificant hazard by comparison with aerial dissemination.

4. As in other parts of the world, plantations, es-

pecially of pines, offer an environment in which *F. annosus* can become a dangerous pathogen if preventive measures are not applied.

LITERATURE CITED

ANONYMOUS. 1954. Special conference on root and butt rots of forest trees. IUFRO Section 24. Wageningen, July 22–26.

ANONYMOUS. 1962. Conference and study tour on *Fomes annosus*. IUFRO Section 24. Scotland, June, 1960.

BEGA, R. V. 1962. Tree killing by *Fomes annosus* in a genetics arboretum. Plant Disease Reptr. 46:107–110.

BEGA, R. V. 1963. *Fomes annosus*. Phytopathology 53:1120–1123.

JORGENSEN, E. 1956. *Fomes annosus* (Fr.) Cke. on red pine in Ontario. Forest. Chron. 32:86–88.

JORGENSEN, E. 1961. On the spread of *Fomes annosus* (Fr.) Cke. Can. J. Botany 39:1437–1445.

JORGENSEN, E. 1962. *Fomes annosus* (Fr.) Cke. in Canada, with particular references to its occurrence in Ontario. IUFRO Section 24, Conference and study tour on *Fomes annosus*, Scotland, June 1960. p. 37–43.

KUHLMAN, E. G., and F. F. HENDRIX. 1962. A selective medium for the isolation of *Fomes annosus*. Phytopathology 52:1310–1312.

POMERLEAU, R. 1963. A record of *Fomes annosus* (Fr.) Cke. in Quebec. Bi-M. Progr. Rept., Dept. Forest. Can. 19(1):2.

PUNTER, D., and J. D. CAFLEY. 1968. Two new hardwood hosts of *Fomes annosus*. Plant Disease Reptr. 52:692.

REDFERN, D. B., and G. A. VAN SICKLE. 1968. *Fomes annosus* in eastern Canada. Plant Disease Reptr. 52:638.

REYNOLDS, G., and G. W. WALLIS. 1966. Seasonal variation in spore deposition of *Fomes annosus* in coastal forests of British Columbia. Bi-M. Res. Notes, Dept. Forest. Can. 22(4):6–7.

RISHBETH, J. 1959. Dispersal of *Fomes annosus* Fr. and *Peniophora gigantea* (Fr.) Massee. Trans. Brit. Mycol. Soc. 42:243–260.

SINCLAIR, W. A. 1964. Root- and butt-rot of conifers caused by *Fomes annosus*, with special reference to inoculum dispersal and control of the disease in New York. Cornell University Agr. Expt. Sta. Memoir 391. 54 p.

SOMERS, J. 1880. Nova Scotia fungi. Proc. and Trans. Nova Scotian Inst. Nat. Sci. 5:188–192.

WRIGHT, E., and L. A. ISAAC. 1956. Decay following logging injury to western hemlock, Sitka spruce, and true firs. U.S. Dept. Agr. Tech. Bull. 1148. 34 p.

The Role of Resin in the Resistance of Conifers to Fomes annosus

J. N. GIBBS—*Botany School, University of Cambridge, England.*

▶

Fomes annosus is a wood-rotting Basidiomycete which is most serious as a pathogen of conifers. Attack in most conifer genera takes the form of a butt rot, root infection leading to extensive decay of the central wood while the outer sapwood, and consequently the tree itself, remains alive. By contrast such butt-rotting is rare in pines, although killing attacks can be serious under certain conditions, particularly in plantations of young trees on light sandy soils.

This resistance of pines to central rot is reproduced following artificial infection. When woody inoculum was inserted in roots of young trees of a number of conifer species (Gibbs, 1968) central infection developed in Sitka spruce *Picea sitchensis* and Douglas fir *Pseudotsuga menziesii* but not in Scots pine *(Pinus sylvestris)* or Corsican pine. *(Pinus nigra* var. *calabrica).*

THE NATURE OF RESISTANCE.—That this resistance of pine roots to infection is an active process is demonstrated by the rapid development of *F. annosus* in the severed roots of Scots pine (Wallis, 1961; Gibbs, 1968). In the experiments of Wallis where a woody inoculum was inserted between the two cut ends of a root the rate of infection in the distal (severed) portion was fifty times as rapid as in the proximal (unsevered) portion. Moreover, while in the former the fungus had completely invaded the roots, in the latter infection was largely confined to the bark and cambium. Clearly the root portion attached to the tree possesses active resistance to infection. This active resistance is much reduced in suppressed trees, and indeed in very suppressed trees the rate of infection is as fast as in severed roots (Wallis, 1961; Gibbs, 1967).

In spruce a different situation is found. With the same inoculation method on Sitka spruce, infection was found to be as rapid in the proximal portion as in the distal and twenty times the rate found in pine (Gibbs, 1968). Here, while complete invasion of the severed root portion again occurred, in the proximal portion infection was concentrated in the central wood, the outer sapwood remaining healthy.

Thus in vigorous 20-year-old pines a capacity for active resistance exists throughout the sapwood, whereas in spruce of the same age only the outer sapwood possesses equivalent resistance.

THE MECHANISM OF RESISTANCE.—When conifers are wounded or infected the neighbouring tissue frequently becomes impregnated with resin. This resin accumulation consists mainly of preformed oleoresins liberated from the resin ducts, and the extent to which it occurs depends on the structure and activity of the resin-duct system. The systems of the Pineae are more elaborate than those of the Abieteae, and even within the former group there are differences. Indeed, there is a series of decreasing activity from *Pinus* with its fully intercommunicating system of long vertical canals lined with viable secretory cells through *Picea* and *Pseudotsuga* to *Larix* with its short separate ducts which rapidly become thick walled and functionless (Gibbs, 1968). These differences affect resin mobilisation in response to a wound, as the value of *Pinus* for resin-tapping demonstrates. That they also affect the resistance of the different genera to infection by *F. annosus* is indicated by the experiments mentioned above. The series *Pinus–Picea–Pseudotsuga* is paralleled by the rate of infection in these three genera following root-inoculation (Gibbs, 1968). Furthermore, the development of infection in the central sapwood of the latter genera can be explained in the same terms since the central wood is the region where a lower proportion of active resin ducts is to be expected. Observations supported this view of the role of resin in the resistance of the outer sapwood of spruce, since the exterior margin of the infected central wood was typically bounded by a resin-impregnated zone. In addition, in those roots where infection had failed, resin soaking was invariably evident.

In pines resin accumulation in infected roots is profuse, and many workers have concluded that it is important in resistance. Wallis (1961) observed that while resin impregnation of the proximal portion of the severed roots was considerable, it was negligible in the distal portion. One way of strengthening the argument that resin has a function in active resistance would be to establish a correlation between the ability of pines to mobilize resin and factors consistently associated with disease resistance. Such a factor is soil moisture, for, as stated above, killing of young pines has frequently been reported on dry sandy soils both in Europe (Rennerfelt, 1952; Low and Gladman, 1960) and in the United States (Powers and Verrall, 1962). Munch (1919) was the first to provide a basis for such a relationship by showing that the pressure with which resin is forced from a ruptured duct depends on the water regime in the tree. Further evidence comes from the work of Bourdeau and Schopmeyer (1958) on *Pinus elliottii* and Vité (1961) on *P. ponderosa*. They have shown a diurnal cycle in resin pressure with a minimum between 2 and 4 PM and a maximum between midnight and dawn. These changes correspond to those observed in the girths of trees due to variations in water content. Furthermore, resin pressure is modified by factors such as temperature and humidity which affect the rate of transpiration. That pressure is directly related to actual resin yields is demonstrated by the fact that diurnal rhythms have been observed in the rate of resin flow during commercial tapping.

Since Vité (1961) has shown that 95-year-old *P. ponderosa* trees on a dry site have a lower average value than those on a wet one, possibilities exist for an environmental influence on host resistance. In order to study this further, and in particular to investigate trees younger than those studied by Vité, a method involving the measurement of the volume of resin exuded in 24 hours from a standard wound was devised (Gibbs, 1968).

The main comparison made was between 20-year-old Scots pines on three sites in Thetford Chase, East Anglia. Site (1) was in a river valley, the soil (pH 4.2) was moist and contained a high proportion of peat. Site (2) was dry sandy podsol (pH 4.0) with a typical surface layer of undecomposed litter and humus. Site (3) was a shallow calcareous soil (pH 7.5–8.0) with very little organic matter on the surface. Resin yields were obtained every 4 to 6 weeks throughout the year, and a marked seasonal flow was found with complete cessation between November and February. When resin was exuded, however, yields from trees on the wet site were much greater (on average 69%) than yields from the other trees. Also interesting was the discovery that yields from the trees on the dry acidic site were almost invariably higher than those on the dry alkaline site. This finding is supported by numerous observations on the relative amounts of resin-soaked sand accumulated around root wounds at the two types of site and suggests that soil pH may have an important influence on host resistance, presumably through the effect of different amounts of surface organic matter on water availability. It has been shown (Rishbeth, 1951) that in the pine-*F. annosus* relationship disease escape is greater on light acidic soils than on light alkaline ones. It now appears that disease resistance may also be greater on such soils.

The effects of other factors on resin mobilization were studied. Vigorous trees produce far more resin than suppressed ones, and trees on the margin of a stand, known to possess greater resistance than interior trees, similarly yield more resin. Again, resin yields increase with tree age, and it is in young trees that most severe killing attacks develop. Of the two commonly planted pines in East Anglia, Scots pine is more susceptible than Corsican pine, and the yield from 18-year-old trees of the latter species is double that from the former (Gibbs, 1968).

There is evidence that in pines both resin viscosity and exudation pressure are under genetical control. Some trees produce consistently low yields and this is unrelated to age and vigour. Mathre (1964) produced some of the strongest evidence for the role of resin in disease resistance by showing that artificial infection of two blue-stain fungi (*Ceratocystis ips* and *C. minor*) was invariably successful in trees with low resin-exudation pressure but not in those with medium or high pressure.

The results presented here provide evidence for the importance of resin in the resistance of conifers, particularly pines, to *F. annosus*. However, no consideration has so far been made of the actual mechanism of resistance involved. In part this may involve a simple mechanical blocking of the vessels and rays, but the recent discovery by Cobb et al. (1968) that volatile components of the resin of *P. ponderosa* are toxic to *F. annosus* raises the possibility of an important chemical as well as physical barrier. In addition, in pines, recent work has laid emphasis on the accumulation of the fungitoxic pinosylvins in infected sapwood (Jorgensen, 1961; Shain, 1967). It would appear, however, that pinosylvins alone cannot provide a complete explanation. Shain (1967) found that while the rate of infection was twelve times as rapid in dominant trees inoculated in the dormant season as in those inoculated in the growing season, there was no difference in the amount of pinosylvins present. On the other hand, as mentioned above, there is a large difference in the rate of resin flow in the two seasons. Since resin is preformed, it may be that it provides a rapid response to infection while pinosylvins act more slowly. This would seem to be the case in the resistance of *Pinus radiata* to the fungus associated with *Sirex noctilio* (Coutts and Dolezal, 1966). Quite possibly, as suggested by Shain (1967), a resin matrix is necessary for the distribution of the pinosylvins (and other toxins) through the tissues, and that consequently the degree of resin impregnation would be critical. Much more needs to be known before an exact understanding of the resistance processes can be reached, but it would seem probable that it is the ability of a pine to mobilize resin that to a large extent determines its resistance to *F. annosus* and many other diseases.

LITERATURE CITED

BOURDEAU, P., and C. S. SCHOPMEYER. 1958. Oleoresin exudation in slash pine, p. 313–319. In K. V. Thimann [ed.], The physiology of forest trees. Ronald Press, New York.

COBB, F. W. JR., M. KRSTIC, E. ZAVARIN, and H. W. BARBER JR. 1968. Inhibitory effects of volatile oleoresin components on *Fomes annosus* and four *Ceratocystis* species. Phytopathology 58:1327–1335.

COUTTS, M. P., and J. E. DOLEZAL. 1966. Polyphenols and resin in the resistance mechanism of *Pinus radiata* attacked by the wood wasp, *Sirex noctilio,* and its associated fungus. Forest. and Timber Bur. Leaflet 10. 18 p.

GIBBS, J. N. 1967. The role of host vigour in the susceptibility of pines to *Fomes annosus.* Ann. Botany (London) 31:803–815.

GIBBS, J. N. 1968. Resin and the resistance of conifers to *Fomes annosus.* Ann. Botany (London) 32:649–665.

JORGENSEN, E. 1961. The formation of pinosylvin and its monomethyl ether in the sapwood of *Pinus radiata.* Can. J. Botany 39:1765–1772.

LOW, J. D., and R. J. GLADMAN. 1960. *Fomes annosus* in Great Britain, an assessment of the situation in 1959. Forest Rec., London 41. 22 p.

MATHRE, D. E. 1964. Pathogenicity of *Ceratocystis ips* and *Ceratocystis minor* to *Pinus ponderosa.* Contr. Boyce Thomson Inst. Pl. Res. 22:363–388.

MUNCH, E. 1919. Naturwissenschaftliche Grundlagen des Kieferharz-nutzung. Arb. biol. BundAnst. Land-u. Forstw. 10:1–118.

POWERS, H., and A. VERRALL. 1962. A closer look at *Fomes annosus.* Forest Farmer 21:8–9, 16–17.

RENNERFELT, E. 1952. On root-rot attack in Scots pine. Meddn. St. Skogsforsk Inst. 41:1–39.

RISHBETH, J. 1951. Observations on the biology of *Fomes annosus.* III. Natural and experimental infection of pines and some factors affecting severity of the disease. Ann. Botany (London) 15:221–246.

SHAIN, L. 1967. Resistance of sapwood in stems of Loblolly pine to infection by *Fomes annosus.* Phytopathology 57:1034–1045.

VITÉ, J. P. 1961. The influence of water supply on oleoresin exudation and resistance to bark beetle attack in *Pinus ponderosa.* Contr. Boyce Thomson Inst. Pl. Res. 21:37–66.

WALLIS, G. W. 1961. Infection of Scots pine roots by *Fomes annosus.* Can. J. Botany 39:109–121.

Poria Root Rot: Problems and Progress in the Pacific Northwest

KEITH R. SHEA—*Principal Plant Pathologist, Pacific Northwest Forest and Range Experiment Station, Forest Service, U.S. Department of Agriculture, Corvallis, Oregon.*

The fungus *Poria weirii* causes a very destructive root rot of conifers west of the Cascade Range in Oregon and Washington in the United States and adjacent British Columbia in Canada. The disease in young Douglas-fir (*Pseudotsuga menziesii*) stands was first noted in 1929 on Vancouver Island, British Columbia. Disease symptoms and the causal fungus were described by Mounce, Bier, and Nobles (1940).

Most native conifers of all ages are susceptible, but greatest damage occurs in pole and sawtimber stands of Douglas-fir. In western Oregon and Washington, annual growth loss and mortality in Douglas-fir are estimated at 32 million cubic feet (896,000 m³) (Childs and Shea, 1967). Climate, soil, and forest site apparently are not related to disease incidence and severity. Susceptibility of the principal conifers, especially Douglas-fir, to Poria root rot poses a serious threat to the productivity of many forests and creates hazardous conditions in forest recreational areas.

In the forests, Poria root rot infection centers range in size from a few trees to an acre or more. Typical centers contain standing dead or dying trees, often with a "distress" crop of numerous small cones, trees with sparse or off-color foliage, leaning trees, and down trees with the roots broken close to the root collar (Fig. 1). The broken roots reveal the characteristic decay which often extends into the lower portions of the bole.

Incipient decay appears as longitudinal reddish-brown to brown streaks or broad bands and circular or crescent-shaped areas in cross sections. Advanced decay is laminated, the wood separating readily at annual rings, from which the common name, laminated root rot, is derived. The yellowish-brown laminations contain numerous small pockets about 0.5 x 1.0 mm in size. Tufts or thin layers of brownish mycelia usually are present in the decayed wood, and brown setal hyphae are readily visible with a hand lens. On the bark or surfaces of broken roots, thin, brown mycelial crusts often form. In the final stages, the wood decays to a stringy mass which eventually disintegrates.

The cinnamon-brown sporophores develop on the undersides of down trees, uprooted stumps, and occasionally on the boles of dead, standing trees. Sporophores on Douglas-fir are annual, but a second layer

Fig. 1. Poria root rot infection center in a 80-year-old stand of Douglas-fir.

of tubes sometimes arises from the substrate. The importance of spores in dissemination of the fungus and in infection of new hosts is unknown.

The disease spreads primarily through contact of roots of healthy trees with those of adjacent diseased trees or with inocula remaining in roots and wood from the previous forest stand. Mycelia may live in old roots of previous stands for 50 years or more, providing a continuing source of infection for successive forests (Buckland and Wallis, 1956; Childs, 1963).

No practical methods are known for controlling Poria root rot in forest stands. Trenching around centers of infection prevents spread to adjacent trees (Wallis and Buckland, 1955) as does cutting, poisoning, and girdling trees around infection centers (Shea, unpublished). However, none of these methods are practical under forest conditions, and inocula will remain viable within the center to infect future stands. Moreover, isolations of fungus clones from root rot centers (Childs, 1963) suggest that the fungus may be more widespread in the forest than is indicated by above-ground symptoms. For example, the boundaries surrounding infection centers resulting from the same

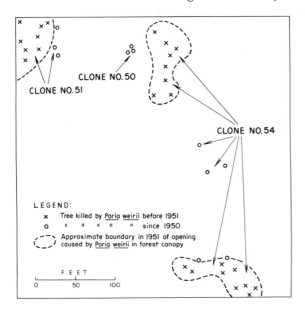

Fig. 2. *Poria weirii* clones in young Douglas-fir saw-timber stand (from Childs, 1963).

fungus clones in Fig. 2 likely are contiguous. Thus, treatment of individual infection centers is of questionable value since the fungus is likely to be present throughout an area is which the same clone is represented. Root-disease-control practices such as the planting of resistant tree varieties, crop rotation, soil fumigation, the planting of mixed forest stands, and other silvicultural practices either have practical or economic limitations, or knowledge is inadequate to permit their use. Present forest practices, therefore, are restricted to salvaging merchantable dead or dying trees.

Because direct approaches now offer little opportunity for controlling Poria root rot, fundamental investigations of fungus biology and soil microbiological relationships are being favored.

Some interesting concepts in survival and biological control of *P. weirii* are being studied by scientists at the U. S. Forest Service Forestry Sciences Laboratory and at Oregon State University in Corvallis, Oregon. Highlights of their research follow:

Nelson (1964) found a positive correlation between survival of the fungus in inoculated, buried wood blocks and zone line formation. In inoculated wood blocks in which zone lines did not form, *P. weirii* was seldom reisolated. Two antagonists to *P. weirii*, *Trichoderma viride* and *T. album*, were commonly found in wood blocks without zone lines. Although the cause-effect relationship needs further study, zone lines in wood invaded by *P. weirii* may exclude antagonistic soil fungi and facilitate survival. Rishbeth (1951) noted a similar situation in root material invaded by *Fomes annosus.*

Li et al. (1967) reported *P. weirii* does not produce nitrate reductase and cannot therefore use nitrate as a nitrogen source. This suggests a hypothetical approach to biological control whereby antagonistic soil microorganisms which utilize nitrate nitrogen might be favored. For example, increases in nitrate nitrogen in forest soils could mitigate against *P. weirii* and favor competing or antagonistic microorganisms which utilize nitrate nitrogen.

The hypothesis further suggests that mixtures of red alder (*Alnus rubra*) with susceptible conifers might reduce Poria root rot damage. Red alder root nodules have been found by Tarrant and Miller (1963) to fix large amounts of total nitrogen in pure alder and in mixed conifer-alder stands. Bollen et al. (1967) report higher levels of nitrate nitrogen in mixed conifer-red alder stands than in pure conifer stands. Thus, incorporating red alder with susceptible conifers or a rotation of red alder may be a promising control method since increases in nitrate nitrogen could favor many of the antagonistic fungi. Furthermore, mixed alder-conifer forests would tend to reduce root contacts among susceptible trees, thus reducing opportunities for root-to-root spread.

U. S. Forest Service scientists are continuing to investigate the alder-conifer approach to biological control of Poria root rot. Preliminary results by Nelson (unpublished) show that survival of *P. weirii* in inoculated wood blocks buried in soil under red alder stands is reduced as compared with that in buried blocks under conifer stands. Additional tests are being made to determine effect of adding nitrate nitrogen to forest soils on survival of *P. weirii* in buried wood.

Disease-resistant conifers have not been found, but variation in host reaction to infection by *P. weirii* suggests that differences in susceptibility occur. Methods for inoculating test trees have been developed by Wallis and Reynolds (1962) and Nelson (1962). They are, however, time consuming and expensive and do not permit screening of large numbers of trees. Studies now in progress by Li et al. (unpublished) on the biochemical nature of resistance of red alder and susceptibility of Douglas-fir suggest the phenols and phenol oxidases present in red alder but absent in Douglas-fir may be a primary source of alder resistance. Canadian studies (Barton, 1967*a*; 1967*b*) on extractives from diseased and healthy Douglas-fir roots may help explain differing resistance in Douglas-fir and perhaps aid in development of effective chemicals for control.

The role of spores in establishment of new Poria root rot centers is unknown, but infection by *P. weirii* through above-ground injuries on western hemlock (*Tsuga heterophylla*) has been reported (Wright and Isaac, 1956), indicating spore infection. Thus, stumps and injuries accompanying thinning operations now widespread in the region could greatly increase the prevalence of this disease. Dr. E. E. Nelson, U. S. Forest Service, is currently investigating spore infection of fresh Douglas-fir stumps.

Additional studies being conducted in the United States and Canada include the following:

In British Columbia (Dr. G. W. Wallis, personal communication), inoculations have been made to test (1) effect of host vigor and site on fungus development, (2) tree species susceptibility, (3) time required from infection to host death, and (4) food base required to initiate and sustain infection. Control studies have been initiated to determine (1) feasibility and effectiveness of inoculum removal and (2) effectiveness of mixed stands and mixed rotations. Permanent plots also have been established to determine (1) rate and pattern of root rot development, (2) time of symptom expression and tree death, (3) effects of thinning on disease development, and (4) relation of site factors to disease development.

In the United States, Dr. C. H. Driver at the University of Washington (personal communication) is investigating regeneration of *P. weirii* infected sites by studying the effects of land-clearing practices such as stump removal and severe burning to destroy infected logging residues and to encourage antagonistic soil microorganisms. Research by the U. S. Forest Service, in addition to that already noted, includes long-term studies of Poria root rot spread and damage in natural stands of Douglas-fir.

In northwestern United States and Canada, heart rots, dwarf mistletoes, and root rots are the three principal groups of diseases reducing productivity of the forests. As these old-growth forests are replaced by younger ones, heart rots will be of less significance. Increased research and improved forest practices will reduce the impact of dwarf mistletoes. The significance of root rots, especially *P. weirii*, *F. annosus*, and *Armillaria mellea*, will likely increase·as forest management practices are intensified. The *F. annosus* problem now is receiving some needed attention (Shea, 1968) partly by shifting emphasis from other pressing disease problems. Damage from *A. mellea* is becoming conspicuous, yet little research effort can be spared for this root disease. Relatively, more research has been done on *P. weirii*; but much more is needed, especially on soil microbiological relationships and parasite biology, if adequate control measures, including biological ones, are to be developed.

LITERATURE CITED

BARTON, G. M. 1967*a*. Differences in phenolic extracts from healthy Douglas-fir roots and those infected with *Poria weirii*. Can. J. Botany 45:1545–1552.

BARTON, G. M. 1967*b*. A new C-methyl flavanone from diseased (*Poria weirii* Murr.) Douglas fir (*Pseudotsuga menziesii* [Mirb.] Franco.) roots. Can. J. Botany 45:1020–1022.

BOLLEN, WALTER B., CHI-SIN CHEN, KUO C. LU, and ROBERT F. TARRANT. 1967. Influence of red alder on fertility of a forest soil. Microbial and chemical effects. Oregon State Univ. Forest Res. Lab. Res. Bull. 12. 61 p.

BUCKLAND, D. C., and G. W. WALLIS. 1956. The control of yellow laminated root rot of Douglas fir. Forest. Chron. 32:14–16.

CHILDS, T. W. 1963. *Poria weirii* root rot. Phytopathology 53:1124–1127.

CHILDS, T. W., and K. R. SHEA. 1967. Annual losses from disease in Pacific Northwest forests. Pacific Northwest Forest and Range Exp. Sta. U. S. Forest Serv. Resource Bull. PNW–20. 19 p.

LI, C. Y., K. C. LU, J. M. TRAPPE, and W. B. BOLLEN. 1967. Selective nitrogen assimilation by *Poria weirii*. Nature (London) 213:814.

MOUNCE, IRENE, J. E. BIER, and MILDRED K. NOBLES. 1940. A root-rot of Douglas fir caused by *Poria weirii*. Can. J. Res. (C)18:522–533.

NELSON, EARL EDWARD. 1962. *Poria weirii* root rot of Douglas-fir survival of the pathogen and infection of the host. Ph.D. thesis, Oregon State Univ., Corvallis. 71 p.

NELSON, EARL E. 1964. Some probable relationships of soil fungi and zone lines to survival of *Poria weirii* in buried wood blocks. Phytopathology 54:120–121.

RISHBETH, J. 1951. Observations on the biology of *Fomes annosus* with particular reference to East Anglian pine plantations. II. Spore production, stump infection, and saprophytic activity in stumps. Ann. Botany (London) 15:1–21.

SHEA, KEITH R. 1968. *Fomes annosus*: A threat to forest productivity in the Douglas-fir subregion of the Pacific Northwest? IUFRO Third Int. Conf. *Fomes annosus*, Aarhus, Denmark, July 29–August 3, 1968. 7 p.

TARRANT, ROBERT F., and RICHARD E. MILLER. 1963. Accumulation of organic matter and soil nitrogen beneath a plantation of red alder and Douglas-fir. Proc. Soil Sci. Soc. Amer. 27:231–234.

WALLIS, G. W., and D. C. BUCKLAND. 1955. The effect of trenching on the spread of yellow laminated root rot of Douglas fir. Forest. Chron. 31:356–359.

WALLIS, G. W., and G. REYNOLDS. 1962. Inoculation of Douglas fir roots with *Poria weirii*. Can. J. Botany 40:637–645.

WRIGHT, ERNEST, and LEO A. ISAAC. 1956. Decay following logging injury to western hemlock, Sitka spruce, and true firs. U. S. Dept. Agr. Tech. Bull. 1148. 34 p.

Root Rot Induced by Polyporus tomentosus *in Pine and Spruce Plantations in Wisconsin*[1]

R. F. PATTON and D. T. MYREN—*Department of Plant Pathology, University of Wisconsin, Madison.*

Planted forests are particularly liable to damage from root diseases, and as plantations become older they tend to sustain increasing attacks by root disease pathogens. In Wisconsin, with over 1 million acres of planted forests, root diseases are presenting a growing challenge to the skill of forest managers.

Polyporus tomentosus induces root and butt rot in coniferous forests of the world's North Temperate Zone. Much of our recent knowledge of *P. tomentosus* derives from the intensive investigations of Whitney (1962) on this fungus as a cause of the stand-opening disease in white spruce in Canada. There, in severely diseased stands, losses have eliminated the possibility of a future sawtimber operation. Recently, *P. tomentosus* has been found in some of Wisconsin's conifer plantations, arousing concern over the future impact it may have on these stands. Since most information about this fungus pertains to conditions much different from those in this state, our objective is to determine its potential significance in Wisconsin and to clarify further its biology.

SPREAD OF THE FUNGUS.—*Location of infection centers.*—Infection centers were located in the fall by the presence of sporophores emerging from the soil in a stagnating 40-year-old white spruce plantation of 4.8 acres in Menominee County. A dead tree marked the hubs of some plots, but only rarely were other crown symptoms of root rot observed on adjacent trees. In 1963, 13 separate infection centers in this stand were located and individually mapped. During the next 3 years 13 additional pockets of infection appeared for a total of 26 within the stand.

Radial spread.—To assess the rate of spread of this fungus within the stand, 11 of the original infection centers were marked as permanent plots. The remaining 2 were used for other purposes such as root excavations. From 1964 through 1966 the outermost of annually produced sporophores were mapped as an indication of the approximate perimeter of the plots. Plot areas were then computed on the basis of these boundaries. During this 3-year period the total area

[1] Approved for publication by Director, Wisconsin Agricultural Experiment Station.

of the 11 plots increased by 49%, from 1,078 to 1,614 ft². Also, 13 newly detected infection pockets added 717 ft² to the diseased area of this stand. The total affected area of the plantation was estimated to be about 1%.

It is premature at this time to predict how much of the stand might be affected, say, 25 years hence. Nevertheless, it is probable that the fungus will continue to spread and that new infection centers will continue to appear. As decay proceeds in the root systems, trees will be killed or retarded in growth and vigor, openings may occur in the stand from death or windfall, and the entire aspect of the stand may be expected to change significantly, with considerable loss expected in the future.

Effect on trees within plots.—Observations of the tree crowns within the infection centers gave some indication of disease development. Trees in 13 of the original plots were rated by crown class and vigor. In these plots during the 3-year period 1964–1967, 58 of 559 trees died and 5 additional trees evidenced marked symptoms of crown thinning and decline in vigor. The number of living suppressed trees decreased by 43.0%, intermediates by 10.0%, codominants by 2.0%, and dominants by 1.6%. In check plots there was only a 33.0% decrease (from 15 to 10 trees) in the number of living suppressed trees, and there were no changes in crown classification of the remaining 59 trees. In this stand considerable mortality from natural suppression occurred as a result of the 2 × 4 ft spacing of the original planting. Although losses from natural suppression and from root rot were difficult to separate, most mortality to date has been the result of natural suppression. The slightly higher percentage of suppressed trees that died in infection pockets than in check plots could have been the result of accelerated decline because of root rot. There was no firm indication, however, that suppressed trees were more vulnerable to attack by the fungus than others. More important were the changes that occurred in the other crown classes in infection pockets. For example, 19 of the 58 trees that died were not rated as suppressed in 1964; also, 3 of the dead trees and 3 of the trees that developed advanced crown symptoms were dominants.

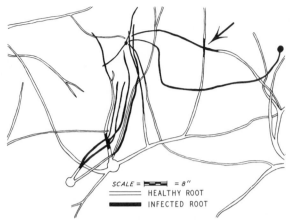

Fig. 2. Map of all the major lateral roots in a small infection center first located in 1966 and washed out in 1967. Healthy roots and roots infected by *Polyporus tomentosus* were intermixed. The arrow marks the site of fungus transfer from one root to another at the point of contact.

Fig. 1. (A) Portion of a root-rot infection center with most of lateral roots exposed by washing. Numerous roots of the two trees in the background were infected by *Polyporus tomentosus*. (B) An example of spread of the fungus by root-to-root contact seen in the exposed root systems in (A). Portions of roots encrusted by resin-infiltrated soil are infected by *Polyporous tomentosus*. The vertical root in the picture was healthy at both ends but was infected by contact with the larger diseased root crossing it diagonally.

Manner of spread and pattern of decay.—The effect of the disease on stand development has been difficult to assess, since the rate of growth of the fungus in the roots and consequent damage to root systems was not known precisely. Hydraulic excavations of the upper 6–8 inches of root systems in representative infection centers indicated the amount and extent of root decay (Fig. 1A). These observations proved the validity of using sporophores as boundary indicators, for root infection usually extended no further than 3 feet beyond the limit of the plot as indicated by their production. Within the plot, infection occurred on scattered individual roots rather than in the entire mass of roots (Fig. 2). At least three examples of spread by root-to-root surface contact were found (Fig. 1B); grafting of roots was not necessary for spread. Usually the fungus grew within the wood, producing a column of decay,

but on some roots it advanced beyond the limit of its extent in the wood by growth in the bark tissues as indicated by thick wefts of mycelium in the inner bark and between bark scales.

The initial points of infection could not be determined in the hydraulic excavations. Isolated infections of segments of lateral roots were occasionally found. Initial infection of such roots must have come from fungus inoculum in the soil either through wounds on the root or by penetration of small rootlets. Thus, it seemed that fungus spread was accomplished primarily by root-to-root contact within infection centers, and possibly also by new infections from fungus inoculum in the soil. The addition of new fungus inoculum to the soil could explain the establishment of the initial root-rot pockets and the continued new appearance of scattered infection pockets in the stand. The abundance of sporophores in the older pockets might result in a concentration of basidiospores in the soil. Infection of pine roots by basidiospores of *Fomes annosus* has been postulated by Hendrix and Kuhlman (1964) and several others. A similar mechanism could operate with this fungus. Whitney (1966) also speculated on the possibility of infection by *P. tomentosus* both by root contacts and inoculation with basidiospores of broken ends of large roots.

OCCURRENCE ON STUMPS OF PLANTED PINES.—In several locations in Wisconsin the fungus was associated with jack pine (*Pinus banksiana*) and red pine (*P. resinosa*) stumps. The observation of sporophores of the fungus on stumps of red pine, left by a thinning operation in the approximately 40-year-old Crandall plantation at Wisconsin Dells, stimulated study of the potential importance of stumps in plantations as infection courts and as sources of inoculum.

Dissection of several of these stumps gave indications of the pattern of entry and movement of *P. tomentosus* within them. In no case did it appear that infection had occurred through the cut surface, but the exact site of infection could not be determined. In at least two stumps decay by *P. tomentosus* was localized in a portion of the root collar and proximal portions of one or more associated lateral roots whose distal portions were healthy. Thus it seemed that the fungus entered the stump at the root collar through a small secondary root or wound, and grew up into the stump, down into the tap root, or out into one or more lateral roots.

In red pine plantations root grafts are fairly common. A root of one of the diseased stumps described above was grafted to a healthy root of an adjacent tree. A column of *P. tomentosus* decay in the root from the infected stump was approaching but had not yet reached the graft. It is likely that the fungus could pass through the graft to the root of the healthy tree. Such infected stumps might serve as sources of inoculum for infection of residual trees either by root contact, as seen in the spruce stand, or through root grafts.

In the Crandall red pine stand the fungus has been observed so far only in association with stumps left after a thinning. If it is present in the roots of living trees, it has produced no indications of its presence. Its rather localized distribution in the excavated stumps and roots suggested that it had not been present in the stumps for many years. It appeared that this plantation, established on a previously cultivated old-field site, was being newly colonized by *P. tomentosus*. Either the stumps or stump roots offered a favorable infection court, or fruiting of the fungus was favored on stumps of previously infected living trees that were cut in the thinning.

BIOLOGY OF THE FUNGUS.—Knowledge of the nature of inoculum and the site of initial infection is critical to our understanding of this disease. We have begun studies on the role of the basidiospore in this regard.

Spore discharge.—Limited studies of the spore-discharge pattern indicated that spore release from sporophores in the field was influenced by relative humidity and temperature. In the laboratory, spore discharge was measured from sporophores held in closed chambers at various relative humidities. Good spore casts were obtained for 2 days at a relative humidity near 0% and for over 5 days at relative humidities between 80 and 100%, and at temperatures between 4° and 24°C, but not at 32°C. The moisture content of the sporophore itself seemed most important, and spore casts were diminished or prevented as the sporophore tissue itself dried out.

Spore germination.—Erratic and inconsistent results have been obtained in trials of germination of spores collected from naturally produced sporophores both by Whitney (1962, 1966) and ourselves. In our pre-liminary trials, germination was less than 1%; in one trial, germination occurred within 48 hours, whereas in another study on the effect of temperature, germination occurred at 12 to 28°C, but only after incubation for 6 weeks. Such spore collections obtained on glass cover slips from naturally produced sporophores were invariably contaminated with microorganisms and were produced under constantly changing environmental influences. These factors undoubtedly affected the results of germination trials. To avoid such influences, a procedure for producing viable basidiospores in pure culture was developed.

Basidiospore production in culture.—Normal sporophores of this fungus have not developed in our cultures, but rudimentary fruiting bodies frequently were formed on petri-dish cultures about 4 weeks old. Whitney (1966) found that addition of cold-water extract of white spruce bark to water agar stimulated basidiospore germination. In some of our exploratory trials with bark extracts, mycelial growth was greater on media to which red-pine-or white-spruce-bark extracts were added than on unamended nutrient media or water agar. In recent trials the addition of pine- or spruce-bark extracts to malt agar favored production of the rudimentary fruiting bodies. Usually only one sporophore formed per plate, beginning always near the colony margin as a localized mound of rather firm mycelial tissue on which characteristic setae developed. Sporophores assumed a variety of forms, from a resupinate layer of shallow pores to a columnar stalk-like structure extending to the top of the petri dish (Fig. 3). When inverted over an agar plate these fruiting bodies cast viable basidiospores onto the agar surface. Casts of viable spores have continued for as long as 12 days, after which the trial was discontinued.

Germination of basidiospores cast from these sporophores onto cornmeal agar plates began within 24 hours. A representative trial of germination based on counts of 1,600 basidiospores in random microscope

Fig. 3. A rudimentary sporophore of *Polyporus tomentosus* that developed on a petri-dish culture.

fields on 5 petri dishes yielded an average germination of 40% after 72 hours incubation in the laboratory at room temperature of approximately 22°C. Our investigations are just beginning, but it seems likely that this method of spore production will provide a reliable source of basidiospores for studies of spore biology and pathogenicity.

ACKNOWLEDGMENTS.—Supported in part by the Division of Conservation, Wisconsin Department of Natural Resources and Wisconsin Agricultural Experiment Station McIntire-Stennis Project 1434. Assistance of E. Herrling and S. Vicen with the illustrations is gratefully acknowledged.

LITERATURE CITED

HENDRIX, F. F., JR., and E. G. KUHLMAN. 1964. Root infection of *Pinus elliottii* by *Fomes annosus*. Nature (London) 201:55–56.

WHITNEY, R. D. 1962. Studies in forest pathology. XXIV. *Polyporus tomentosus* Fr. as a major factor in stand-opening disease of white spruce. Can. J. Botany 40:1631–1658.

WHITNEY, R. D. 1966. Germination and inoculation tests with basidiospores of *Polyporus tomentosus*. Can. J. Botany 44:1333–1343.

The Influence of Soil Bacteria on the Mode of Infection of Pine Roots by Phytophthora cinnamomi

DONALD H. MARX and W. C. BRYAN—*Plant Pathologist and Associate Plant Pathologist, respectively, Southeastern Forest Experiment Station, Forest Service, U.S. Department of Agriculture, Athens, Georgia.*

In recent years, much emphasis has been placed on the biological control of soil-borne plant pathogens. In most instances, successful control has been correlated with an increase of soil organisms antagonistic to a specific root pathogen (Patrick and Toussoun, 1965). An often overlooked segment in the ecology of soil-borne plant pathogens, however, is their possible stimulation by, or dependency on, associative soil organisms for the initiation or completion of certain physiological processes essential for pathogenicity.

An excellent example of this is the apparent dependency of *Phytophthora cinnamomi* on metabolic products of certain soil bacteria for the production of sporangia. Marx and Haasis (1965) isolated two bacterial species from forest soil capable of inducing sporangia in this fungus. After numerous chemical and physical tests, they obtained aseptic sporangia only in a partition-tube assembly in which vegetative mycelium of *P. cinnamomi* was separated from organisms in nonsterile soil leachate by a millipore filter membrane. The membrane functioned as a physical barrier to the biological systems but allowed diffusion of metabolic products which induced aseptic sporangial production. Zoospores obtained from these assemblies were infectious to aseptic feeder roots of shortleaf pine, but sporangia did not form on the roots in the absence of an active leachate (Marx and Davey, in press). Zentmyer (1965) isolated *Chromobacterium violaceum* from soil in Australia which also induced sporangia in this fungus.

This unique dependency on soil bacteria for sporangial production by *P. cinnamomi* provides an excellent experimental tool (1) to determine the influence of soil organisms apparently essential to this fungal pathogen for production of its infectious spore stage and (2) to determine, in the absence of these organisms, the role of vegetative mycelium of *P. cinnamomi* in the infection of aseptic roots of shortleaf *(Pinus echinata)* and loblolly *(P. taeda)* pine seedlings.

Pine feeder-root infection by zoospores occurs only when certain soil bacteria which trigger sporangial production are present in soil solution. Sporangial production is correlated with inhibition of mycelial growth of *P. cinnamomi* caused by the bacteria. How-

ever, in the absence of these soil bacteria and with vigorous mycelial growth, *P. cinnamomi* is highly infectious to pine feeder roots as vegetative hyphae. This suggests that environmental conditions which favor sporangial and zoospore production, such as periodically water-saturated heavy soils (Campbell and Copeland, 1954), are not necessarily the only conditions predisposing to feeder-root damage by *P. cinnamomi*. Any condition favoring mycelial growth, either on the root surface or in the rhizosphere or rhizoplane, may be conducive to root damage from hyphae. The hyphal stage of *P. cinnamomi* is potentially infectious to feeder roots and should not be considered as the nonpathogenic phase (Zentmyer and Mircetich, 1966) in its life cycle. Hyphae of *P. cinnamomi* are also much more tolerant of antibiotics produced by mycorrhizal fungi than are zoospores (Marx, 1969). Hyphae, being more tolerant to antibiotics, should persist longer than zoospores in soil and potentially provide a greater threat to infection of plant roots.

Nonsterile leachates from soils supporting plants with heavy root damage by *P. cinnamomi* are extremely active in inducing sporangia. Conversely, leachates from noninfested soil with healthy plants, or from soil with only mildly infected plants, are only weakly active. The relative numbers of bacteria in all leachates in our tests were similar, but different qualitatively. The activity of leachates from samples of soil profiles in high hazard littleleaf sites revealed a similar correlation. Upper-soil profiles, including the humus layer which contained the majority of feeder roots and recoverable propagules of *P. cinnamomi*, were the most active for sporangial induction. Samples from soil profiles from a similar but healthy loblolly pine site were much less active. Samples from depths of 16 inches or more failed to induce sporangial formation, which suggests rhizosphere effect by the host roots on the sporangial-inducing soil bacteria and accounts for their greater populations in the root zones.

Histological examination reveals that infection of cortex and vascular tissues of roots from zoospore inoculum is similar to that previously reported for these pathogen and host associations (Marx and Da-

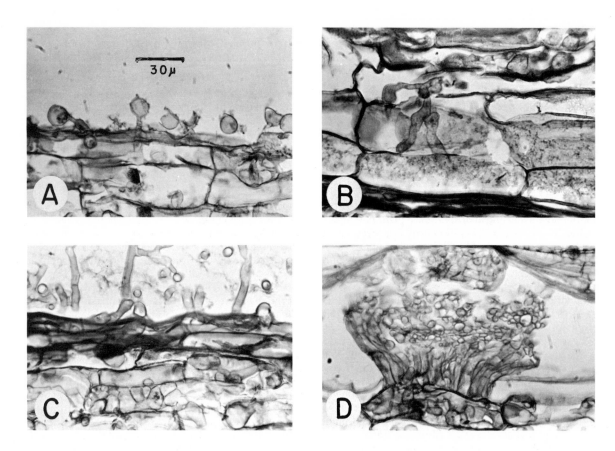

Fig. 1. Shortleaf and loblolly pine-root infection by *Phytophthora cinnamomi* after 12 days. *A*, Direct zoospore penetration of root surface; *B*, mild cortex infection from zoospores; *C*, direct hyphal penetration of root surface; and *D*, mat from hyphal infection exerting physical pressure on adjacent cortex cells.

vey, in press). However, *P. cinnamomi*, in the absence of zoospores, and existing only as vegetative hyphae with vesicles, causes a much heavier infection of cortical and vascular tissues of both shortleaf and loblolly pine roots than do zoospores. An unusual mycelial mat from hyphal infection was observed in all shortleaf and in a few loblolly pine roots. These mats exert a noticeable physical pressure on adjacent cortical and vascular cells and cause some cell displacement (Fig. 1). Infection from hyphae also differs from infection by zoospores in that the cortical and vascular cells are filled with vesicles, which are morphologically identical to those on the root surfaces.

LITERATURE CITED

CAMPBELL, W. A., and O. L. COPELAND, JR. 1954. Littleleaf disease of shortleaf and loblolly pines. USDA Circular No. 940. 41 p.

MARX, D. H., and F. A. HAASIS. 1965. Induction of aseptic sporangial formation in *Phytophthora cinna-*momi by metabolic diffusates of soil micro-organisms. Nature (London) 206:673–674.

MARX, D. H. 1969. The influence of ectotrophic mycorrhizal fungi on the resistance of pine roots to pathogenic infections. II. Production, identification and biological activity of antibiotics produced by *Leucopaxillus cerealis* var. *piceina*. Phytopathology 59:411–417.

MARX, D. H., and C. B. DAVEY. The influence of ectotrophic mycorrhizal fungi on the resistance of pine roots to pathogenic infections. III. Resistance of aseptically formed mycorrhizae to infection by *Phytophthora cinnamomi* Rands. Phytopathology. (In press.)

PATRICK, Z. A., and T. A. TOUSSOUN. 1965. Plant residues and organic amendments in relation to biological control, p. 440–454. *In* K. F. Baker and W. C. Snyder [ed.], Ecology of soil-borne plant pathogens. Univ. of Calif. Press, Berkeley and Los Angeles.

ZENTMYER, G. A. 1965. Bacterial stimulation of sporangium production in *Phytophthora cinnamomi*. Science 150:1178–1179.

ZENTMYER, G. A., and S. M. MIRCETICH. 1966. Saprophytism and persistence in soil of *Phytophthora cinnamomi*. Phytopathology 56:710–712.

Phytophthora cinnamomi *in New Zealand*

F. J. NEWHOOK—*Botany Department, University of Auckland, New Zealand.*

In a large and important genus of plant destroyers *Phytophthora cinnamomi* commands special interest because of its wide host range (over 300 species, Zentmyer, personal communication), its widespread occurrence, and its remarkable uniformity.

Some of its spectacular "successes" are littleleaf disease of *Pinus echinata* in southeastern USA, avocado root rot in California, and pineapple root rot in Hawaii and Queensland. It has been suggested (Crandall, Gravatt, and Ryan, 1945) that it was *P. cinnamomi* that caused the disappearance of chestnuts in southeastern USA. It is now in a position to do the same for jarrah *(Eucalyptus marginata)*, the main timber tree in Western Australia (Podger, 1968).

In New Zealand *P. cinnamomi* regularly causes serious losses of ornamentals in nurseries and gardens and severe losses in coniferous shelterbelts and peach orchards in some seasons. It is associated with littleleaf disease in plantations of exotic pines on clay soils. It is also established in many indigenous plant communities where it could have important ecological significance.

NURSERY LOSSES.—Nurserymen throughout the country regularly suffer heavy losses in a number of species. Amongst ornamental hosts, members of the Proteaceae, Epacridaceae, and Ericaceae are particularly susceptible. The indigenous kauri, *Agathis australis*, and many conifers are severely affected.

Pinus radiata has been killed by *P. cinnamomi* in forest nurseries in many parts of the country. Bassett and Will (1964) reported heavy losses at Thames where they obtained control by soil fumigation with a mixture of methyl bromide and chloropicrin. In most forest nurseries fumigation has not been practised, although losses may be heavy in years of high summer rainfall. Vigorous root regeneration in surface soil frequently ensures recovery of infected seedlings.

FARM PLANTINGS.—*Phytophthora* spp., especially *P. cinnamomi*, cause spectacular mortality in small farm woodlots and in shelterbelts of *P. radiata, Cupressus macrocarpa, Chamaecyparis lawsoniana*, and other conifers (Fig. 1). This problem, which occurs on fertilized pastures throughout the country, has been discussed in detail by Newhook (1959) and Sutherland, Newhook, and Levy (1959). It occurs in years

of unusually heavy autumn rainfall, when soil temperatures are suitable (Chee and Newhook, 1965) for *P. cinnamomi* sporulation.

EXOTIC FORESTS.—The bulk of New Zealand's one and a quarter million acres of exotic forests are planted with species susceptible to *P. cinnamomi*. Despite the apparently widespread distribution of the fungus, there are only a few plantations where *P. cinnamomi* is associated with significant loss. In the Nelson Province, root and collar rot result in sapling losses of *Pseudotsuga menzezii* and *Larix* spp. (Bassett, personal communication). Over much of the area of *Pinus* plantings on heavy clay soils, promising early growth may be seriously checked by a combination of *P. cinnamomi* and poor soil physical conditions. On eroded and poorly aerated ridge tops, decline may commence soon after the pines are planted. At Cornwallis, for example, some 44-year-old *P. radiata* is only 4–6 ft high.

Littleleaf disease.—Hepting and Newhook (1962) reported a decline of *P. echinata* in Waipoua State Forest, with symptoms indistinguishable from those of littleleaf disease in southeastern USA; *P. radiata* showed similar but less severe symptoms; *P. taeda* and *P. palustris* were symptomless except on eroded ridge tops. The seriousness of the disease is determined largely by edaphic factors. At Cornwallis, where *P. cinnamomi* populations are high, Atkinson (1959) has shown that vigour is correlated with internal drainage and aeration of soil on the three main soil types.

Indirect control of littleleaf disease.—It has been shown experimentally (Weston, 1956; Atkinson, 1959) and by aerial application (Conway, 1962) that on the P-deficient clay soils of North Auckland a single topdressing of 5 cwt of superphosphate per acre produces a spectacular improvement in stand health (Fig. 2). This response, which has persisted at Riverhead for 16 years, is not due to destruction of *P. cinnamomi*, since populations of the fungus in both treated and untreated forests have remained high (e.g. 65% positive samples from treated and 60% from untreated forests). Control of the disease seems to be due to a complex series of changes which improve growth conditions from the host and at the same time

173

Fig. 1. Shelterbelt mortality of 35-year-old *P. radiata.* The row of younger trees on the left subsequently died after about 20 years of age. (Photo courtesy J. W. Endt.)

Fig. 2. Indirect control of littleleaf disease of *P. radiata.* Left, untreated plot and, right, plot treated with single application of superphosphate at 1 ton/A 12 years previously.

limit opportunities for infection by zoospores of *P. cinnamomi.* This may be illustrated by an account of the situation in treated and untreated stands of 35- to 42-year-old *P. radiata* at Riverhead.

In much of the untreated forest the pines have suffered a drastic check in growth. The open-canopy stands have trees with sparse crowns and chlorotic foliage and, frequently, dead tops. The understorey is characterised by *Leptospermum* spp. There is almost no litter, F, H, or A horizon. The B horizon is a structureless massive clay which is at or near saturation for much of the year; when dry in summer it is almost rock hard. Fine root development is almost entirely restricted to the top inch. Mycorrhizas are rare and usually restricted to a few Gasteromycete associations. These factors, coupled with *P. cinnamomi* root rot, aggravate an already serious P-deficiency.

By contrast, trees in treated parts of the forest have dense crowns with dark green foliage and vigorous shoot growth. The canopy closes usually within three years of treatment. A deep litter accumulates, with a well-developed F-H layer. Rootlet growth is extensive, with considerable mycorrhizal association both in the F-H layer and to a depth of 2–3 ft or more in the cracks between aggregates. In these cracks there is a copious development of buff or white Basidiomycete mycelium (Fig. 3). Fructifications of *Amanita* spp. and *Boletus* spp. are seasonally plentiful.

The improvement in crown health follows so soon after application of fertilizer that the initial response to the nutrients in superphosphate is obviously direct. This leads over the next few seasons to the marked

Fig. 3. Blocks of soil from beneath *P. radiata*, Riverhead State Forest. Left, treated with superphosphate, showing dense wefts of mycelium surrounding rootlets and mycorrhizas. Center, untreated, almost structureless clay with a thin layer of topsoil and few rootlets. Right, unfertilized and underscrubbed plot. (Photo courtesy S. A. Rumsey.)

improvement in roots and mychorrhizas described above, which ensures better uptake of water and nutrients. Will (1965) has shown that there was an increase in foliar levels of P and, in living bark, improved levels of other elements, notably N, K, and Mn as well as P. As there has been no increase in soil macroporosity (Atkinson, personal communication) it is apparent that the enhanced root and mycorrhizal growth is made possible by improved aeration due to reductions in the frequency and length of periods of waterlogging. This undoubtedly results from the increased transpiration and the reduction in the amount of rain reaching the forest floor that follow the improvement in crown cover. Healthy pine crowns are capable of intercepting large amounts of rain, e.g. up to 32% of the annual rainfall (Rutter, 1963), while litter also retains a large quantity.

The improved soil-moisture situation limits activity by *P. cinnamomi* zoospores and at the same time favours root regeneration and mycorrhizal development, with the nutrient benefits already noted, and possibly mycorrhizal protection (Marx and Davey, 1967) of rootlets against further infection.

The importance of transpiration and rain interception in modifying the site for pine growth is further illustrated by increased waterlogging of the soil and accelerated deterioration of littleleaf stands from which the understorey has been removed.

INDIGENOUS COMMUNITIES.—*P. cinnamomi* is widespread under many indigenous plant communities including: (1) *Leptospermum* scrub at several widely separated places in Auckland and Nelson Provinces; (2) cutover forest dominated by *Coprosma arborea* and *Leptospermum ericoides* at Cornwallis; (3) submontane forests dominated by beech (*Nothofagus*

fusca) and southern rata (*Metrosideros umbellata*) respectively (Bassett, personal communication); (4) dense rimu (*Dacrydium cupressinum*) forest on wet terrace soils in south Westland (Podger, personal communication); and (5) kauri (*A. australis*) at several localities including stands near the northern and southern limits of its range as well as on Little Barrier Island sanctuary.

In all but one of these situations, there was no obvious evidence of plant damage attributable to *P. cinnamomi*. On the Cascade Ridge in the Waitakere Ranges however, Podger and Newhook (unpublished) located a patch of dying pole-sized kauris in regenerating *Agathis* forest. Many kauris were dead, as well as *Phyllocladus trichomanoides* and *Cyathodes fasciculata*, from which *P. cinnamomi* was isolated. Chlorosis and stunting were common in almost every other component of the community.

DISCUSSION.—In New Zealand, *P. cinnamomi* causes major root and collar rot in a few species, but in the majority of susceptible hosts infection is limited to rootlet invasion. Thus severity of host reaction depends not only on the variations in site conditions outlined above but also on the degree to which transpiration reduces plant moisture reserves before rootlet regeneration restores the root-shoot balance.

This is clearly illustrated by the reactions of *P. radiata* to severe rootlet loss in the autumn in shelterbelts, in untreated, and in superphosphate-treated forests.

The deep, exposed crowns of vigorous shelterbelt trees rapidly deplete water reserves, leading to severe defoliation or death. By contrast, mortality is negligible in littleleaf areas where transpiration losses from sparse crowns are low. In treated forests, where crown improvement has occurred following application of superphosphate, depletion of moisture reserves is nevertheless insufficient to cause outward symptoms, even in the worst years for shelterbelt mortality. Transpiration losses from treated forest trees would be intermediate between those of littleleaf and shelterbelt trees because crowns are shallow and afford mutual protection from sun and wind.

Root-shoot balance and moisture stress probably play a similarly important role in stands of *Agathis* and perhaps other indigenous communities infected by *P. cinnamomi*. Deaths are not common amongst mature plants, but crowns of kauri and several understorey species frequently show symptoms analogous to those of littleleaf disease. These host reactions and the high susceptibility of some species to damping-off by *P. cinnamomi* undoubtedly influence ecological succession within these communities.

The lack of extensive mortality in infected native-plant communities might also be interpreted as evidence that *P. cinnamomi* is an indigenous pathogen to which the flora is adapted. Much of the incidence of the fungus, however, can largely be accounted for by inadvertent transport by man and animals. Even the occurrence on the almost inaccessible northern slopes of Little Barrier Island could be accounted for by autonomous dispersal for vast distances from early Maori settlements. The rates of spread required by this explanation are well within those demonstrated by Podger (1968) in the jarrah forests of Western Australia.

LITERATURE CITED

ATKINSON, I. A. E. 1959. Soils and the growth of *Pinus radiata* at Cornwallis, Auckland. N.Z. J. Sci. 2:443–472.

BASSETT, C., and G. M. WILL. 1964. Soil sterilisation trials in two forest nurseries. N.Z. J. Forest. 9:50–58.

CHEE, K. H., and F. J. NEWHOOK. 1965. Variability in *Phytophthora cinnamomi* Rands. N.Z. J. Agr. Res. 8:96–103.

CONWAY, M. J. 1962. Aerial application of phosphate fertilisers to radiata pine forests in New Zealand. Commonw. Forest. Rev. 41:234–245.

CRANDALL, B. S., G. F. GRAVATT, and M. M. RYAN. 1945. Root disease of *Castanea* species and some coniferous and broadleaf nursery stocks, caused by *Phytophthora cinnamomi*. Phytopathology 35:162–180.

HEPTING, G. H., and F. J. NEWHOOK. 1962. A pine disease in New Zealand resembling littleleaf. Plant Disease Reptr. 46:570–571.

MARX, D. H., and C. B. DAVEY. 1967. Ectotrophic mycorrhizae as deterrents to pathogenic root infections. Nature (London) 213:1139.

NEWHOOK, F. J. 1959. The association of *Phytophthora* spp. with mortality of *Pinus radiata* and other conifers. I. Symptoms and epidemiology in shelterbelts. N.Z. J. Agr. Res. 2:808–843.

PODGER, F. D. 1968. Aetiology of jarrah dieback. A disease of dry sclerophyll *Eucalyptus marginata* Sm. forests in Western Austral. M.Sc thesis, Univ. of Melbourne.

RUTTER, A. J. 1963. Studies on the water relations of *Pinus sylvestris* in plantation conditions. I. Measurements of rainfall and interception. J. Ecol. 51:191–203.

SUTHERLAND, C. F., F. J. NEWHOOK, and J. LEVY. 1959. The association of *Phytophthora* spp. with mortality of *Pinus radiata* and other conifers. II. Influence of soil drainage on disease. N.Z. J. Agr. Res. 2:844–858.

WESTON, G. C. 1956. Fertiliser trials in unthrifty pine plantations at Riverhead Forest. N.Z. J. Forest. 7:35–46.

WILL, G. M. 1965. Increased phosphorus uptake by radiata pine in Riverhead Forest following superphosphate applications. N.Z. J. Forest. 10:33–42.

PART VII

◄

ROOT DISEASES OF TROPICAL PLANTATION CROPS

A Comparison of Methods of Dispersal, Survival, and Parasitism in Some Fungi Causing Root Diseases of Tropical Plantation Crops

RONALD A. FOX—*Scottish Horticultural Research Institute, Invergowrie, Dundee, Scotland.*

INTRODUCTION.—This paper is concerned with the behaviour of four pathogens, *Fomes lignosus, Ganoderma pseudoferreum, Fomes noxius,* and *Armillaria mellea,* as found on plantation crops established in lowland rain forest areas of the tropics.

Distribution of the pathogens.—*F. lignosus* is limited to lowland forests; *G. pseudoferreum* and *F. noxius* may be found at somewhat higher elevations; whilst *A. mellea* is better known as a pathogen in temperate or cold temperate regions. The distribution of all four may be affected not only by the range of their potential hosts but by their temperature requirements. Table 1 shows the effect of temperature

TABLE 1. The effect of temperature on growth and development of mycelial strands of *Fomes lignosus* from malt agar food bases (averaged over 4 pH levels) into empty glass tubes, at saturated humidities

Temperature °C	cm/day	mg/day†	mg/cm
32.2	0.8 (1.1)*	1.6	2.0
26.7	0.8 (1.0)	0.8	1.0
21.1	0.6 (0.7)	0.5	0.8
15.6	<0.1	not measurable	

* The first figure is the overall mean, that in brackets is the mean for the first week; the difference indicates that the rate of growth diminished with increasing distance from the food base.
† Oven dry weight.

on the production of mycelial strands by *F. lignosus* from a food base into glass tubes. The linear growth rate is not greatly affected at 21°C, but the quantity of mycelium and the density of the strands are markedly affected. The lowest temperature investigated (15.6°C) in growth experiments was found to be lethal when many isolates were exposed to it for 2 months. The figures in Table 2 indicate a lower optimum for *G. pseudoferreum;* its growth is somewhat inhibited at 32°C and virtually ceases at 16°C. *F.*

TABLE 2. The effect of temperature on the growth rate of *Ganoderma pseudoferreum* on malt extract agar

Temperature °C	32.2	26.7	21.1	15.6
Growth rate, cm/day	0.15	0.8	0.6	0.0

noxius has temperature requirements similar to those of *G. pseudoferreum,* but it is less restricted by lower temperatures.

The distribution of *A. mellea* in the tropics is puzzling and cannot be explained by its host range nor, as yet, by the effects of temperature. In the Far East the record of its occurrence in the lowlands of Malaya on *Hevea brasiliensis,* cited by Raabe (1962), proved to be erroneous when the original publication was examined. Attacks on other crops in the East, so far as I have been able to ascertain from published records, have been at elevations of about 1,000 meters or higher. The few records from the New World tropics are also montane. Wallace (1935a) and Gibson (1962) found a lower limit of about 1,000 m in East Africa. The first record on *H. brasiliensis* was made by Small (1924) in Uganda in plantations 950 km inland at elevations exceeding 1,000 m in climatic conditions unsuited for the commercial growth of *Hevea.* It was reported on plantations of *H. brasiliensis* in the Congo at Yangambi by Depoerk (1946) at about 450 m, where it had evidently been present for some years and eventually proved a serious problem in both rubber and oil palm plantations. In West Africa it was recorded on a number of plants in the Gold Coast and Togoland Territory in the early 1920s. A full account of its occurrence on cacao in this region was published by Dade (1927) who first found it at altitudes of about 850 m but later as low as 130 m. In 1963 I could not find it on oil palms in Eastern Nigeria, but it was becoming a serious root disease of *H. brasiliensis* down to elevations as low as 20 m (Fox, 1964a). It has also been found recently on rubber at similar low elevations in Liberia (J. M. Ross, 1968, personal communication) and in the Federal Republic of Cameroon (E. A. Rosenquist, 1969, personal communication).

Bliss (1946), working with isolates from California, found optima in the range 20°–24°C with growth retarded at 27.4°C. From experiments on several host plants he concluded that the upper critical temperature for root infection was 26°C but the optimum could be either low or high, being inversely related to the optimum for root growth of a given host. Gibson (1961), however, using many isolates from widely varied geographic sources, found optima for mycelial and rhizomorphic growth ranging from 22°–24°C to 27°–29°C—the upper and lower limits which he examined. Further, he found no correlation between optima and climate of origin when he compared isolates from temperate and tropical regions.

Raabe (1966) found much variability in mycelial characteristics and rhizomorph production in the laboratory, even between single-spore isolates derived from the same sporophore. Gibson (1961) found differences in mycelial characteristics and rhizomorphic habit according to the continent of origin of his isolates.

Rhizomorphs are invariably produced in the field in temperate zones. Bliss (1946) regarded them as necessary for infection, but Butler (1928), Wallace (1935b), and Leach (1939) found them to be rare at intermediate altitudes in East Africa, infection invariably occurring by root contact. Gibson (1960) found them to be frequent in forests in the Kenya highlands at altitudes above 2,000 m. In contrast, at low elevations Dade (1927) in the Gold Coast, Pichel (1956) in the Congo, and Fox (1964a) in Nigeria, found that they were never formed, infection always arising by root contact. Raabe (1967) found that variations in pathogenicity or of virulence were not correlated with rhizomorph production *in vitro;* all his isolates would produce rhizomorphs, and he maintained that they are necessary for infection, citing Thomas (1934) in support of this statement, as did M. O. Garraway and A. R. Weinhold in a paper presented to this Congress. In Rhodesia their absence has been attributed, at least in the high-veld savannah soils, to the presence in the soil of a thermolabile partially water-soluble toxin (Boughey et al., 1964) though isolates from this region will produce abundant rhizomorphs *in vitro.* Dade (1927) similarly found that rhizomorphs were readily produced from pieces of infected cacao roots when they were buried in wet moss and kept in the laboratory. It is evident that much needs to be done to resolve some of these apparently contradictory findings, which are vexed by taxonomic problems involving the status of *Clitocybe tabescens* and *Armillariella elegans.*

The unfortunate practice, which has produced some absurd synonyms and taxonomic chaos (Fox, 1961b; Corner, 1968), of assigning names to tropical fungi by comparing them with named herbarium specimens from temperate regions, must throw doubts on the host range of the four pathogens considered in this paper. Undoubtedly all have a very wide range and their occurrence on major plantation crops is well attested. On a given crop their severity may vary with topography and, in particular, with rainfall and soil type, crop loss varying from trivial to catastrophic.

Life Cycles.—In the forest, their life cycles are similar to those of many other root-infecting and timber-rotting fungi of temperate climates, and follow a pattern of sporophore formation, spore dissemination, spore colonisation of exposed wood, its decay, sporophore formation, and so on, but with variations in the extent of root invasion, the importance of epiphytic growth, and the persistence of potential inocula. Although the overall picture is the same, the details differ, an apparently weak point in the infection cycle in a given species being compensated by a strong link elsewhere. Comparisons between them are interesting, as are the changes of importance of different parts of their cycles in the transition from forest to plantation.

Sporophore and spore production.—In temperate regions sporophore production in higher fungi usually occurs in seasonal cycles. *A. mellea* behaves in a like manner in the lowland tropics, its evanescent fruit bodies usually appearing only during or toward the end of the rainy season. Sporophores of *F. lignosus* also tend to be more abundant following periods of high rainfall, though they may appear at any time of the year when especially humid conditions prevail, as on infected stumps or timber overgrown by creeping legumes planted as a ground cover in plantations. On the sides and lower surfaces of infected logs, resupinate hymenia may develop sometimes extending for 1 or even up to 3 m². Masses of imbricate fruit bodies are produced on an infected stump and may sometimes wholly encircle it, extending outwards for 30 cm or more. Although inherently persistent, their effectiveness in producing spores is usually rapidly diminished by fungivorous insects. Sporophores of *G. pseudoferreum* and *F. noxius* are usually initiated during or after wet spells, but like *F. lignosus*, they may be affected by local conditions. Those of the first named are relatively infrequent, usually attaining a radius of some 15 cm, though occasionally twice this size or more, and are durable and perennial, sometimes becoming imbricate. Sporophores of *F. noxius* are comparatively rare, irregular in shape, and often partly resupinate; again, they are persistent. Their rarity may be illusory; being small, dark-coloured, and irregular, they tend to be camouflaged against a background of dark decaying timber.

Spore production by *A. mellea* and *F. lignosus* is prolific; the latter may produce from 10^5 to 10^6 spores per square cm of hymenium per day for many days. *G. pseudoferreum* and *F. noxius* are less productive (Hilton, 1961). All tend to show diurnal cycles of spore production, which may be affected by changes of temperature and relative humidity.

Aerial dissemination.—Success of aerial dissemination (when assessed on the basis of equal numbers of spores released) is apparently accompanied by increasing specialisation. The spores of *A. mellea* and *F. lignosus* will readily germinate in the presence of free water or on plain water agar, the initial percentage being little affected by added nutrients. Most writers are noncommittal or evasive about the role of spores in disseminating *A. mellea* and these are often assumed to have no function. Until fairly recently spore infection of stumps by *F. lignosus* was considered unimportant, though the subject was hotly debated—mostly on the basis of opinions, not evidence (Fox, 1961a).

Many hundreds of stumps of *H. brasiliensis* have been inoculated with spore suspensions of *F. lignosus* without success by R. N. Hilton at the Experiment Station of the Rubber Research Institute of Malaya (Rubber Research Institute of Malaya, Reports, 1959, et seq.). Only John (1965) has had some success, for reasons suggested later. There is, however, indirect evidence that infection occurs where stumps of thinned trees, or of those felled at replanting, have been examined to reveal this fungus growing downwards from the cut surface, and from field experiments where creosoting stumps has reduced the number of centres of infection.

Rishbeth (1964) concluded that stump infection by *A. mellea* was rare but might be important in temperate zones in the establishment of new infection centres; again but few of his experimental inoculations were successful.

The spores of *F. noxius* require nutritional stimuli before germinating, such as wood extracts or expressed sap. Although apparently fragile and thin-walled, like those of *F. lignosus* and *A. mellea*, they may readily initiate new disease centres, and success is obtained with them in stump inoculation experiments. The thick-walled spores of *G. pseudoferreum* may also colonise stumps left during thinning, but inoculation experiments have always given negative results. Until recently only R. P. N. Napper (reported by Sharples, 1936) had succeeded in germinating them. His findings were surprising; germination on agar plates was obtained with spores collected in the field only where they had been exposed to heat and desiccation after deposition by eddy currents on the upper surface of a sporophore. Spores caught on slides suspended beneath fruit bodies kept in moist chambers in the laboratory all failed to germinate. Hilton (1961) treated spores in various ways but failed to obtain germination. Recently T. M. Lim (1968, private communication) obtained low levels of germination with spores from insect frass collected from the upper surfaces of sporophores. Detailed investigation of many fungivores associated with fructifications showed that only two species of millipedes and the larvae of two species of Tupilid flies were able, by ingestion, to break the constitutive dormancy of the spores. After such biological activation the percentage germination, optimal at 30°C, was enhanced by added nutrients including extracts from the wood of rubber stumps, but it was depressed by desiccation or solar radiation.

Although new disease centres are initiated by aerial dissemination, we know little about factors permitting the invasion of exposed wood. All four fungi are similar, requiring comparatively large inocula to establish infection on the roots of living trees; yet the spores, of low inoculum potential, invade exposed wood in competition with other organisms. One may postulate

that after a certain lapse of time a stump may be so moribund that spores of pathogens can invade it, but the dying tissue is still sufficiently viable to inhibit colonisation by saprophytes. R. N. Hilton (Rubber Research Institute of Malaya, 1962) tested this hypothesis by inoculating stumps of *H. brasiliensis* with spores of *F. lignosus*, in both the morning and evening, immediately after felling and at intervals ranging from half-an-hour up to 64 days, without success. Rishbeth (1951) found that infection on pine stumps by basidiospores and conidia of *Fomes annosus*, decreased as the interval between felling and inoculation increased. His finding that the spores could survive in soil and from it infect the surface of covered stumps was unexpected. The few successes obtained by John (1965) with *F. lignosus* occurred only where the stump surfaces were covered with soil after inoculation. The results of Rishbeth (1964) are relevant to John's observations; he found that inoculation of hardwood discs with proteolytic bacteria predisposed them to spore colonisation by *A. mellea*.

The frequency of naturally occurring stump colonisation after thinning, as opposed to its apparent rarity at the time of replanting, may be related to the effects of shading on atmospheric humidity and on the moisture content and surface temperature of the exposed wood; the last, in replantings, doubtless sometimes reaching inhibitory or lethal levels. Until we know more about these effects, of tissue predisposition, synergism, and the role of insects—perhaps in dissemination too—inoculation of stumps with spores of these pathogens cannot be used experimentally. Potential stump protectants will have to be evaluated in large-scale, expensive, and time-consuming experiments relying on the whims of natural infection.

Once the initial invasion of exposed wood has occurred, subsequent events differ. On young roots, up to about the age of 10 years, pathogenesis by *F. lignosus* and *G. pseudoferreum* is dependent upon epiphytic growth; if this is checked, host reaction halts further longitudinal expansion within the infected root. In older trees, or in the large tap roots of some young trees, both fungi may progress slowly in heartwood. In *Hevea* plantations, I have never seen a heart rot in the butt of a tree caused by either fungus extending more than about 50 cm above ground level. The rate of progress in these cases is so slow in comparison with the rate of root invasion (aided by epiphytic growth) that windthrow occurs before the rot has proceeded much above ground level. In contrast, pathogenesis by *F. noxius* is independent of epiphytic growth, and typical butt rots, not confined to heartwood, have been observed extending a meter or more above ground level. Pathogenesis by *A. mellea*, at least in the tropics, is independent of the presence of epiphytic rhizomorphs. A continuous felt of mycelium is produced between the bark and the sapwood of both roots and trunks, and rots originating in roots may extend upwards in the trunk. On one estate in Eastern Nigeria, in 10-year-old plantings of *H. brasiliensis*, the stand had already been reduced to nearly one half by root diseases, and where in some fields 50% of the remaining trees were infected, it was not unusual to find standing trees invaded on one side by *A. mellea* to a height of 2 m or more.

Because pathogenesis is independent of epiphytic growth, it is perhaps not surprising that potential centres of root disease caused by *F. noxius* may arise not only from stumps but from spore infections originating 5 m or more above ground on the stubs of badly pruned branches or where pruning wounds were inadequately treated. In 1960 and again in 1962 I observed extensive infection of pruning wounds on two estates where a clone, particularly susceptible to wind damage, had been partly pollarded to alleviate this problem. Both fields which I examined were adjacent to forests near which the incidence of infection was greatest. In the first, both *F. noxius* and *Ustilina zonata*, were implicated; the latter fungus can also cause butt and root rots in *H. brasiliensis*. In the second, infection was almost wholly due to *F. noxius*, and the number of trees affected was so high, with infection extending from the branches into the trunks, that although the stand was but 12 years old, the only advice which could be given was to destroy it and replant. In one field of the Experiment Station of the Rubber Research Institute of Malaya, 20% of the trees examined in 1965 had pruning wounds infected by *F. noxius* (Rubber Research Institute of Malaya, 1966) and in some of those selected for detailed investigation the pathogen later extended down to the roots (T. M. Lim, 1968, personal communication).

The practice of "brashing" in pine plantations can likewise lead to new root-disease centres when *F. annosus* infects branch stubs. It is possible that some doubts of the importance and frequency of aerial dissemination of *A. mellea* might be resolved if evidence was sought for spore colonisation above eye level, since it, like *F. annosus* and *F. noxius*, can progress in the absence of epiphytic growth. Also, like the last named, it might be found to colonise branch prunings on the ground and from them invade roots near or at the soil surface. In one field of *H. brasiliensis*, about a year after pruning following storm damage in mature trees and where many of the larger branches had not been removed from the field for firewood, as is customary, every piece examined had been colonised by *F. noxius*. From a few, the fungus was growing superficially on fibrous roots and thence downwards to the main lateral roots.

F. lignosus, because of its dependence on epiphytic growth, must be inherently less successful in aerial dissemination than *F. noxius*. Only under exceptionally wet conditions is epiphytic growth formed above ground; further, little heartwood is exposed by pruning because it is usually undertaken as a corrective action when trees are young and only rarely on mature trees following storm damage. On bare stumps of felled trees one must assume, from what is known of its pathogenesis in roots, that its spores can colonise only heartwood where there may be greater competition from saprophytes than on the still-living sapwood.

Teleologically, it might be said to compensate for this disadvantage and for that of its spores not having durability nor the selectivity imposed by need for nutritional stimuli, by producing them in vast numbers. *G. pseudoferreum*, with the same disadvantage of dependence on epiphytic growth, might be said to compensate by again producing large numbers of spores, though fewer than *F. lignosus*, which are thick-walled and resistant and whose transmission and germination is perhaps abetted by insects.

When a potential source of infection has been established by spore colonisation, its danger depends largely on two factors: the volume of wood invaded in relation to the minimum volume needed to establish self-perpetuating infections; and the ability of the fungus to persist in the invaded timber until such times as roots of potential new hosts come in contact with it.

Subterranean spread and persistence.—Once these four fungi have penetrated a stump to ground level, epiphytic growth may enhance their ability to extend over and in the root system. *F. lignosus* has the greatest advantage, for its mycelial strands can grow up to 30 cm a month, whereas *G. pseudoferreum* and *F. noxius* have rates of between 5 and 10 cm, and *A. mellea* totally lacks epiphytic growth. Unfortunately, very little is known about the behaviour of this last fungus in the lowland tropics, about the rate of its extension on roots, and of the role in pathogenesis of the mycelial felt which forms between the bark and wood. Judged by the time of onset of evident infection in plantations, its rate of spread along or within roots is about 1–2 mm per day, the same or somewhat slower than that of *G. pseudoferreum*.

The large volume of wood which *F. lignosus* invades quickly diminishes as a potential source of infection. The rate of decay is high and the potential food reserves are rapidly utilised by prolific production of sporophores and epiphytic mycelia which grow on any continuous surface above or below ground if temperature and moisture conditions are suitable. The use of leguminous creeping covers, which provide suitable conditions for and encourage growth and sporophore production, have been discussed by Fox (1965a) in examining aspects of the biological control of this fungus. *F. noxius* persists longer and in infected wood forms a characteristic honeycomb of zone plates, the rate of decay within each "cell" of the honeycomb sometimes being apparently retarded. This slower rate of decay, and the observation that the inoculum volume for the establishment of a self-perpetuating infection is modest compared with that needed by *F. lignosus*, partly accounts for its persisting for up to 10 rather than the 4 to 5 years generally observed for *F. lignosus*. In addition, invaded roots are less affected by soil environment than are those infected by *F. lignosus*. The epiphytic growth of *F. noxius* develops during pathogenesis as a continuous investment and secretes a mucilage to which soil particles firmly adhere, transforming a root into a pseudosclerotium which remains dormant. It is thus little affected by the favourable physical conditions for growth provided, for example, by planted creeping ground-cover plants. Few sporophores are produced and the profuse development of mycelial strands, so characteristic of *F. lignosus*, is absent. *G. pseudoferreum*, less successful than *F. noxius* in colonising new potential inocula and less successful because of its slow rate of epiphytic growth in exploiting them than *F. lignosus*, compensates by its ability to persist. The tough mycelial investment, sometimes alone or in association with zone plates of other fungi, converts roots or even entire boles of trees into pseudosclerotia which may be effective inocula for upwards of 30 years. Further, many observations on trees where infection had been irreversibly established have shown that the inoculum was far smaller than that required by *F. lignosus*. In addition, it appears that a stimulus by the host root may be necessary before dormancy of the pseudosclerotium is broken, contact between them sometimes lasting for upwards of a year before infection is initiated. The zone plates of *A. mellea* may afford some protection against displacement by saprophytes, but their efficacy in enabling it to persist in infected wood under lowland tropical conditions is unknown.

In a new planting or replanting, infection by these four fungi occurs only by root contact: once initiated, the pattern of infection differs with each fungus. Pathogenesis by *F. lignosus* and *G. pseudoferreum* depends on the extension of epiphytic growth, in the absence of which both *F. noxius* and *A. mellea* can still progress. Thus the first pair should be more affected by the physical, chemical, and biological factors of the soil than the second pair. Within each pair there are differences too. *In vitro*, *F. lignosus* may grow at 1.6 cm or more per day; the maximum rate recorded for *G. pseudoferreum* is only half this figure, 0.8 cm. In contrast, the rate at which the epiphytic growth of *G. pseudoferreum* extends along roots (0.2 cm) is only about one fifth that of *F. lignosus* (1.0 cm). There are two factors which seem to account for this disparity. Although *F. lignosus* is susceptible to antagonism by soil microorganisms, it will nevertheless grow from an inoculum disc over the surface of natural soils in petri dishes, either in the absence of water agar on the soil, or in its presence (Fox, 1965*a*, Fig. 2). *F. noxius* will occasionally grow onto the soil, but even then for only 2–3 mm from the inoculum disc; the rare and sparse growth of *G. pseudoferreum* has to be sought with the microscope. Thus *F. noxius* is markedly and *G. pseudoferreum* exceedingly susceptible to soil antagonists. In addition, *F. lignosus* will extend from a food base for up to 40 cm (a limit imposed by the experimental setup) forming typical mycelial strands over an inert surface such as glass, but *G. pseudoferreum* and *F. noxius* behave differently. Neither is efficient in translocating, never developing more than a thin weft of hpyhae which might slowly extend up to 10 cm from a food base, though they both produce dense *apparent* epiphytic growth when progressing on roots. Growth of isolates of *F. noxius* from culture collections is slow and irregular, but when newly isolated it will grow at least as rapidly as *F. lignosus*. Thus its slow progress on roots may well be limited, like that of *G. pseudoferreum*, by poor translocation and antagonistic effects. It appears that soil conditions can have little direct effect on the pathogenesis of *A. mellea*; its spread from centres of infection at about 1–2 mm per day is roughly the order found for its growth *in vitro*.

It has long been assumed that the epiphytic growth habit of *F. lignosus*, *F. noxius*, and *G. pseudoferreum* confers on them the advantage of being able to bypass or "leapfrog" the host's defences. Comparisons of the *in vivo* and *in vitro* growth rates suggest that this ability is probably of major significance in the rate of pathogenesis only for the first named, the rate of extension within roots of the last two usually being little different from that of *A. mellea*, which lacks the epiphytic habit.

One of the more obvious soil factors affecting these fungi is moisture. *F. lignosus* is most severe in wet soils; its growth rate is significantly depressed by moisture levels greater than 50% water-holding capacity, but it compensates by producing more and larger mycelial strands as moisture levels increase (Table 3).

Although *in vitro* experiments have shown this fungus to translocate efficiently, growing from a food base for long distances on glass or into moist soil (sterilized or not), or into acid-washed, quartz sand—quite unlike the behaviour of *A. mellea* reported by Garrett (1956) —growth is totally suppressed if the relative humidity falls much below saturation point. In the Ivory Coast, 95% of infection of *Hevea* roots is stated to occur in soil near the surface. Biological control is being attempted, with some reported success, by using certain ground-cover plants, especially *Tithonia diversifolia* (Compositae), which dry out the top layers of the soil presumably by high rates of transpiration (Bouychou, 1966).

Whether there is also an effect of soil moisture on host resistance is not known, though this indirect fac-

TABLE 3. The effects of varying the percentage water-holding capacity of soil on the mean rate of growth and on the mean number and size of mycelial strands produced by *Fomes lignosus* in 2 weeks*

	Percentage of water holding capacity							
	91	86	81	66	52	38	24	12
Linear growth, cm	11.4$_a$†	11.4$_a$	11.4$_a$	12.0$_a$	11.6$_a$	13.8$_b$	13.3$_b$	13.1$_b$
Number of strands	33.5$_a$	34.5$_a$	34.5$_a$	24.0	12.5$_b$	12.5$_b$	6.0$_b$	5.5$_b$
Strands wider than 1 mm	14.0$_a$	12.0$_{ab}$	10.0$_{abc}$	7.5$_{abcd}$	5.0$_{bcd}$	4.5$_{cd}$	2.0$_d$	1.5$_d$

* Technique as illustrated, Fox, 1965*a*, p. 357, Fig. 4.

† Values with the same subscript do not differ significantly at the 5% level.

tor may operate with *A. mellea*. There is general agreement that wherever this pathogen occurs in the tropics it is worse in wet situations. On *H. brasiliensis* in Nigeria both distribution and severity were strikingly influenced by clay content and drainage (Fox, 1964a).

The growth *in vitro* of *G. pseudoferreum* is strongly affected by humidity (Table 4). As is the case with several timber-decaying house fungi such as *Serpula lacrymans*, near theoretical water production occurs in the breakdown of wood. The moisture produced may be conserved by its tough continuous skin of epiphytic growth retarding evaporation and enabling it to occur, where commonly found, in relatively dry, well-drained sites. It can, however, be found in wet sandy soils. The reason for its virtual absence in wet coastal or riverine alluvial clay soils might be explained by the suspected greater antagonism in such soils. The typical encrustation of soil bound with mucilage to mycelium which surrounds roots infected by *F. noxius* may also conserve moisture, as may the honeycomb of zone plates formed within infected wood. It too is more common in better-drained soils, but there is no experimental data on its sensitivity to varying moisture levels or humidities, or of differential antagonistic effects in different soils.

TABLE 4. The effects of different conditions of humidity and temperature on the growth of *Ganoderma pseudoferreum* in 9 days

Temperature °C	Humidity	
	Unsaturated	Saturated
26.7	4.0 cm	6.0 cm
32.2	2.5 cm	3.8 cm

Nutrition and pathogenesis.—There are neither experimental data nor adequate field observations to show that the severity or frequency of infection by these fungi is markedly influenced by soil nutritional levels. One aspect of the metabolism of wood-destroying fungi, their nitrogen nutrition, does seem worthy of speculation in relation to the epiphytic growth habit. Both *F. noxius* and *F. lignosus* have the same inherently high growth rates when examined *in vitro*, but their field behaviour is quite different. The ability of *F. noxius* to invade the trunk downwards from pruning wounds, or upwards from root infections, suggests that it recycles its nitrogen supply by a process of autolysis and reuse of the nitrogenous constituents of its older mycelium, or that it may, by physiological adaptation, preferentially allocate the limited supply of nitrogen to metabolic pathways necessary for the efficient exploitation of the woody substrate. These hypotheses, to explain the disparity between the nitrogen requirements of wood-destroying fungi and the nitrogen content of wood, have been expounded by E. B. Cowling and his co-workers in a recent series of papers on the role of nitrogen in wood deterioration. The carbon:nitrogen ratios of mycelia *in vitro* usually range from about 3.5:1 up to 20:1, varying both with the fungus and the medium used. A survey of the literature indicates that average "normal" ratios would lie in the range 10:1 to 15:1, but as Merrill and Cowling (1966) have pointed out, when fungi grow on wood, the substrate may have ratios exceeding 1,000:1, and correspondingly high ratios may be expected in mycelia.

Young *Hevea* wood has a ratio of about 160:1 and Table 5 shows the effect on *F. lignosus* of a range of ratios in the food base from double this value down to 5:1. The rate of growth, and the number of strands produced from the food base, are significantly depressed at the higher ratios.

From these results it seems not unreasonable to postulate that an advantage in pathogenesis conferred on *F. lignosus* by its epiphytic habit, lies not only in an ability to bypass the host's defences, but in the ability of its mycelium to absorb and translocate nitrogen directly from the soil or from organic debris. Indeed, it is difficult to explain its profuse growth on roots and in the litter layer unless this supposition is correct. The extensive analytical data provided by Merrill and Cowling (1966), taking into account not only the nitrogen requirement of mycelium, but that removed from the wood by sporophore growth and spore production, suggest that in at least some instances, where wood decay is rapid, it would be surprising if there were not an extraneous source of nitrogen. These workers recognised that in living trees additional nitrogen might come from the transpiration stream, but that in dead timber it must be extraneous to the decaying wood per se. In a later paper, Levi, Merrill, and Cowling (1968), and Levi and Cowling (1969), produced evidence to support both the hypothesis of an efficient nitrogen recycling system and that of physiological adaptation to substrates low in nitrogen. The slow rate of pathogenesis of *F. noxius* on roots, despite its having an inherent growth rate at least as high as *F. lignosus* (whilst allowing for pathogen-host interactions), could be explained by the observed poor capacity for translocation *in vitro* and the partial inhibition of its epiphytic growth by antagonists. Thus in soil it may partly be denied ex-

TABLE 5. The effects of varying the carbon:nitrogen ratio* in the food base on the mean rate of initial growth (4 days) and later growth (4th–7th day) and on the number of mycelial strands produced by *Fomes lignosus* growing into acid-washed quartz sand

Carbon:nitrogen ratio	5:1	10:1	20:1	40:1	80:1	160:1	320:1
Linear growth, 4 days, cm	5.0†	5.0	5.1	4.6$_a$	3.4$_b$	3.1$_b$	1.6
Linear growth, 4–7 day, cm	3.3$_a$	3.3$_a$	3.0	2.7	2.3	1.9	0.9
Number of strands	54.8$_a$	52.0$_a$	44.0	36.3	23.3	11.3	0.3

* Dextrose, asparagine, thiamine, mineral salts medium; dextrose constant.
† Values with the same subscript do not differ significantly at the 5% level.

traneous nitrogen and largely relies on a recycling mechanism or on physiological adaptation; either process could account for its ability to progress reasonably quickly in branches. Doubtless in both situations some additional nitrogen will be available by host translocation in the immediate vicinity of the dying tissue. The very slow progress of *G. pseudoferreum* might also be explained by antagonism inhibiting its epiphytic growth and limiting its supply of extraneous nitrogen. Carrying these general arguments to what is hoped is not a too illogical conclusion poses one question. Longitudinal progress of both *F. lignosus* and *G. pseudoferreum* in young roots is halted by host resistance if the epiphytic growth is checked; do they then lack the ability to adapt and/or lack or have so inefficient a recycling system that they require extraneous sources of nutrients to maintain sufficient inoculum potential to overcome the resistance of young roots in which only living wood is present?

TABLE 6. Comparative growth rates in cm/day of *Fomes lignosus* and *Ganoderma pseudoferreum*, *in vitro* and *in vivo*

	F. lignosus	G. pseudoferreum	
In vitro	1.6	0.8	on malt extract agar
In vitro	0.7	0.4	on wooden dowels
In vivo	1.0	0.2	on roots in soil

The figures for *F. lignosus* and *G. pseudoferreum*, contrasted in Table 6, emphasise some points of the foregoing arguments. The growth rate of *F. lignosus* on malt agar is more than double that obtained on dowels made from *Hevea* root wood—a reflection of its poor nutrient status. *In vivo*, root decay proceeds at a greater rate than growth on the dowels—a result interpreted as the effect of nutrient uptake by epiphytic growth more than offsetting any effects of host resistance. The low *in vivo* rate for *G. pseudoferreum* compared to the two *in vitro* figures, suggests a combination of the effects of host resistance and the depressing effect of antagonism on epiphytic growth which, in turn, could affect nutrient uptake. It is noteworthy that in an experiment where soil round the boles of several trees had been partially sterilised by organo-mercury drenches, and where the trees were then artificially infected on lateral roots, pathogenesis by *G. pseudoferreum* proceeded at an estimated rate of nearly 1 cm per day.

Rennerfelt and Tamm (1962) observed potassium accumulation in decayed wood of standing spruce infected by *F. annosus*, and the accumulation of phosphorus also, in infected blocks of wood in contact with soil; they suggested that these elements might accumulate by hyphae translocating ions from the soil. In a number of root-infecting fungi with an epiphytic growth habit, such as *Poria weirii*, the mycelium has been observed to extend for a few millimetres into soil from the bark (Wallis and Reynolds,

1965). Garrett (1966), following his comments on the length of time it took to solve the problem of the role of ascospores in the dissemination of *Ophiobolus graminis*, drew attention to a quotation from Gregory (1952): "The hypothesis that a structure is functionless stifles inquiry while the hypothesis that it is probably functional stimulates experiment and observation."

Root pathogens of trees, producing apparently trivial hyphal growth into the soil or with marked ectotrophic growth may, in contrast to their mycorrhizal relatives, use extraneous N, P, and K to the detriment of their hosts.

A series of ad hoc studies over the decade 1955–1965 on *F. lignosus* have shown that, in general, when physical and nutritional factors were optimal for mycelial growth, rhizomorphs (or more correctly mycelial strands) were not formed. They only developed under adverse conditions, e.g., nutritionally unbalanced media, suboptimal pH values, nutrient absence (i.e., growth over an inert surface from a food base), or in the presence of antagonistics. These observations suggest that strand formation is a form of abnormal growth induced by an adverse environment enabling the fungus to extend more successfully in that environment. The energy available for growth on a root surface is probably more efficiently utilized when canalised into a compact strand than if expended in the production of an undifferentiated mycelium with a large surface area which would be more susceptible to antagonistic effects or to adverse physical or chemical factors in the soil. Despite their economic importance, little is known about factors affecting the formation of the mycelial investments of *F. noxius* and *G. pseudoferreum* which are important, though in differing ways, to their pathogenesis and persistence. Apart from the immediate practical value, a comparative study of the pathogenesis of *F. noxius* on roots and on branches (and perhaps of *F. annosus* also), combined with appropriate laboratory studies on nutrition, might yield results of wider interest.

CONTROL PROCEDURES.—Tree or stump poisoning, especially the former, reduces the incidence of *F. lignosus* in replantings, but often gives disappointing results with *G. pseudoferreum*. Observations, on the consequences of cutting a root infected by *G. pseudoferreum* with a knife or axe and the failure of large volumes of sections of naturally infected roots to induce infection when used in inoculation experiments, showed that once its pseudosclerotial skin has been broken, this pathogen is readily displaced by saprophytes. This finding explains the unexpected amount of reduction in incidence in replantings following complete mechanical clearing (Fox, 1965a; Newsam, 1967) and has led to field experiments, already showing promise, to evaluate the use of deep-set disc ploughs or rooting tines in the old stand at replanting. It also suggests investigations on the merits of biological control by stump inoculation, shown to be feasible by Rishbeth (1961), with spore dispersions of

Peniophora gigantea, in order to displace or inhibit *F. annosus*. The concept of competition and displacement is not applicable to *F. lignosus*; it is rarely succeeded by other microorganisms, roots even in an advanced stage of decay often being almost "pure cultures" of this pathogen. In addition, experiments with 60 fungi and actinomycetes, commonly isolated from dead wood of *H. brasiliensis,* showed that their antagonism to *F. lignosus* was mostly indifferent or nil.

The procedures advocated by the Rubber Research Institute of Malaya since 1962 for the control of *F. lignosus, F. noxius,* and *G. pseudoferreum* in plantations of *H. brasiliensis* have previously been summarised (Fox, 1965a), and modifications to control *A. mellea* were included in the tentative scheme put forward by Fox (1964a). The former paper briefly mentioned the potential value of collar-protectant dressings for use on young trees adjacent to known sources of effective inocula to protect the vital root collar. Field observations and histological studies by the author suggested that pathogenesis of *F. lignosus* was dependent on its epiphytic growth, and the experiment reported by John (1958) proved that this hypothesis was correct. Formulations based on quintozene (PCNB), easy to apply, cheap, and of negligible mammalian toxicity—essential in the tropics where they must be used throughout the year by the same labourers—were then developed (Fox, 1965b; 1966). They persist effectively for up to 2 years, a more than adequate period, and commercial products are now available and widely used. No commercial material is yet available for use against *G. pseudoferreum,* but a persistent alkaline formulation might be effective against this fungus and *F. lignosus,* as their growth is inhibited above pH 7.5. The same principle of collar protection does not, of course, apply to *F. noxius* because its pathogenesis is not dependent on its epiphytic growth. Its rate of progress in branches, about one third of that observed in roots of similar size and age, does suggest, however, that its epiphytic growth, always absent on branches, may aid pathogenesis on roots either by "leap-frogging" or by nutritional effects. A collar protectant might therefore be of some practical value by increasing the interval between, and thus reducing, the number of costly collar inspections of trees adjacent to known sources of infection.

The procedures recommended by the Rubber Research Institute of Malaya are applicable, in principle, to other plantation crops. They ensure that most original sources of infection are eradicated, in that their inoculum potential is no longer of practical significance, by not later than 8 to 10 years after planting. A few effective centres of *G. pseudoferreum* will doubtless persist throughout the life of a plantation, but these, if dealt with as they appear, should be of little economic consequence. The initiation of new centres of infection by spore colonisation of wood exposed during routine thinning, branch pruning, or storm damage, is of more importance than had previously been recognised. First, the frequency of branch stub and stump infection by *F. noxius* and of the latter by *G. pseudoferreum* had not been appreciated; second, it usually occurs when the stand has been reduced to its desired final level by thinning. At this stage in the life of a plantation the loss of any individual tree represents an actual loss of capital, whereas loss in the early years may be of no financial significance because the initial planted stand exceeds that which is finally required (Fox, 1964b).

Infection of pruning wounds by *F. noxius* should become less frequent, now that this danger has been recognised, by greater care in pruning operations (notably lacking in the past) and the use of wound dressings. Infection by this fungus of felled branches left on the ground is a matter for elementary estate sanitation; usually they are removed for firewood. The initiation of new disease centres during the life of a stand, by spore colonisation of stumps of thinned or storm-damaged trees, should become of less importance as a result of current investigations on stump inoculation and treatment.

The danger of spore colonisation of stumps at the time of replanting or in a new planting may be less serious than was once thought, earlier results being exaggerated by the procedures used in detecting and eradicating sources of infection. Fox (1964b) pointed out that root infections occur, apart from those detected by collar inspections, which are quite irrelevant to the well-being of the host. Likewise, the numerical incidence of spore colonisation of stumps may be irrelevant in terms of costs and disease incidence in the new stand, depending on the control procedures employed. In two experiments initiated in 1956 creosote was used in replantings to inhibit stump colonisation. In one, the postplanting control procedure was that advocated by Napper (1932; 1938); infection was detected by collar inspection and the sources traced and eradicated. Within only 18 months from planting, the costs for creosote and its application were more than recovered in comparison with expenditure in check plots where creosote had not been used. In the other experiment, where disease was detected by foliage symptoms and eradication confined to the clean-weeded planting rows (the procedure currently advocated, Fox, 1965a), savings were only marginally beneficial whether assessed in terms of expenditure or the potential value of the additional number of trees saved (approximately 1 per acre). In both experiments the pathogen was almost exclusively *F. lignosus.* Further work is needed on the economics of stump protection in new plantings or in replantings where *F. noxius, G. pseudoferreum,* or *A. mellea* are prevalent.

ACKNOWLEDGEMENTS.—The experiments and observations reported in this paper were done at the Rubber Research Institute of Malaya, 1954–1965. I am deeply indebted to Enche Foong Kum Mun for technical assistance.

LITERATURE CITED

BLISS, D. E. 1946. The relation of soil temperature to the development of *Armillaria*. Phytopathology 36:302–318.

BOUGHEY, A. S., J. MEIKLEJOHN, P. E. MUNRO, R. M. STRANG, and M. J. SWIFT. 1964. Antibiotic reactions between African Savannah species. Nature (London) 203:1302–1303.

BOUYCHOU, J. G. 1966. White root disease of *Hevea* in Africa, especially in the Ivory Coast. Rubber Res. Inst. Ceylon Bull. [N.S.] 1:34–35.

BUTLER, E. J. 1928. Report on some diseases of tea and tobacco in Nyasaland. Dept. Agr. Nyasaland. 30 p.

CORNER, E. J. 1968. Mycology in the tropics—*Apologia pro monographia sua secunda*. New Phytol. 67:219–228.

DADE, N. A. 1927. Collar crack of cacao (*Armillaria mellea* [Vahl] Fr.). Gold Coast Dept. Agr. Bull. 5. 21 p.

DEPOERK, R. 1946. Sur un nouveau procédé de lutte contre les pourridiés en hévéa culture. Bull. Inst. Colon. Belge 17:980–986.

FOX, R. A. 1961a. White root disease of *Hevea brasiliensis*: recent developments in control techniques. Commonwealth Mycol. Conf., 6th (1960) Rept., p. 41–48. Commonwealth Mycol. Inst.

FOX, R. A. 1961b. White root disease of *Hevea brasiliensis*: the identity of the pathogen. Proc. Nat. Rubber Res. Conf., Kuala Lumpur, 1960, p. 473–482. Rubber Res. Inst. Malaya.

FOX, R. A. 1964a. A report on a visit to Nigeria undertaken to make a preliminary study of root diseases of rubber. Rubber Res. Inst. Malaya Res. Archive 27, 34 p. (cyclostyled).

FOX, R. A. 1964b. The principles of root disease control. Rubber Res. Inst. Malaya Planters' Bull. 75:210–217.

FOX, R. A. 1965a. The role of biological eradication in root-disease control in replantings of *Hevea brasiliensis*, p. 348–362. *In* K. F. Baker and W. C. Snyder [ed.], Ecology of soil-borne plant pathogens. Univ. of California Press, Berkeley and Los Angeles.

FOX, R. A. 1965b. Formulations of collar protectant dressings for use against *Fomes lignosus*. Rubber Res. Inst. Malaya Res. Archive 50, 11 p.

FOX, R. A. 1966. White root disease of *Hevea brasiliensis*: collar protectant dressings. J. Rubber Res. Inst. Malaya 19:231–241.

GARRETT, S. D. 1956. Rhizomorph behaviour in *Armillaria mellea* (Vahl) Quél. II. Logistics of infection. Ann. Botany (London) [N.S.] 20:193–210.

GARRETT, S. D. 1966. Spores as propagules of disease. Colston Pap. 18:309–319.

GIBSON, I. A. S. 1960. *Armillaria mellea* in Kenya Forests. E. Africa Agr. For. J. 26:142–143.

GIBSON, I. A. S. 1961. A note on variation between isolates of *Armillaria mellea* (Vahl) Kummer. Trans. Brit. Mycol. Soc. 44:123–128.

GIBSON, I. A. S. 1962. A note-book on pathology in Kenya forest plantations. 2nd ed. Government Printer, Kenya. 56 p.

GREGORY, P. H. 1952. Fungus spores. Trans. Brit. Mycol. Soc. 35:1–18.

HILTON, R. N. 1961. Sporulation of *Fomes lignosus*, *Fomes noxius* and *Ganoderma pseudoferreum*. Proc. Nat. Rubber Res. Conf., Kuala Lumpur, 1960, p. 496–502. Rubber Res. Inst. Malaya.

JOHN, K. P. 1958. Inoculation experiments with *Fomes lignosus* Klotzsch. J. Rubber Res. Inst. Malaya 15:223–230.

JOHN, K. P. 1965. Some observations on spore infection of *Hevea* stumps by *Fomes lignosus* (Klotzsch) Bres. J. Rubber Res. Inst. Malaya 19:17–21.

LEACH, R. 1939. Biological control of *Armillaria mellea*. Trans. Brit. Mycol. Soc. 23:320–329.

LEVI, M. P., and E. B. COWLING. 1969. Role of nitrogen in wood deterioration. VII. Physiological adaptation of wood-destroying and other fungi to substrates deficient in nitrogen. Phytopathology 59:460–468.

LEVI, M. P., W. MERRILL, and E. B. COWLING. 1968. Role of nitrogen in wood deterioration. VI. Mycelial fractions and model nitrogen compounds as substrates for growth of *Polyporus versicolor* and other wood-destroying and wood-inhabiting fungi. Phytopathology 58:626–634.

MERRILL, W., and E. B. COWLING. 1966. Role of nitrogen in wood deterioration: amount and distribution of nitrogen in fungi. Phytopathology 56:1083–1090.

NAPPER, R. P. N. 1932. A scheme of treatment for the control of *Fomes lignosus* in young rubber areas. J. Rubber Res. Inst. Malaya 4:34–38.

NAPPER, R. P. N. 1938. Root disease and underground pests in new plantings. Planter (Kuala Lumpur) 19: 453–455.

NEWSAM, A. 1967. Clearing methods and root disease control. Rubber Res. Inst. Malaya Planters' Bull. 92: 176–182.

PICHEL, R. J. 1956. Les pourridiés de l'Hévéa dans la cuvette congolaise. Pub. Inst. nat. Étude agron. Congo belge, (sér.tech), 49. 480 p.

RAABE, R. D. 1962. Host list of the root rot fungus *Armillaria mellea*. Hilgardia 33:25–88.

RAABE, R. D. 1966. Variation of *Armillaria mellea* in culture. Phytopathology 56:1241–1244.

RAABE, R. D. 1967. Variation in pathogenicity and virulence in *Armillaria mellea*. Phytopathology 57:73–75.

RENNERFELT, E., and C. O. TAMM. 1962. The contents of major plant nutrients in spruce and pine attacked by *Fomes annosus* (Fr.) Cke. Phytopathol. Z. 43:371–382.

RISHBETH, J. 1951. Observations on the biology of *Fomes annosus*, with particular reference to East Anglian pine plantations. II. Spore production, stump infection and saprophytic activity in stumps. Ann. Botany (London) [N.S.] 15:1–21.

RISHBETH, J. 1961. Inoculation of pine stumps against infection by *Fomes annosus*. Nature (London) 191: 826–827.

RISHBETH, J. 1964. Stump infection by basidiospores of *Armillaria mellea*. Trans. Brit. Mycol. Soc. 47:460.

RUBBER RESEARCH INSTITUTE OF MALAYA. 1962. Rep. Rubber Res. Inst. Malaya, 1961.

RUBBER RESEARCH INSTITUTE OF MALAYA. 1966. Rep. Rubber Res. Inst. Malaya, 1965.

SHARPLES, A. 1936. Diseases and pests of the rubber tree. Macmillan and Co., London. 480 p.

SMALL, W. 1924. Annual Report of the Government Mycologist. Dept. Agr. Uganda 1922, p. 27–29.

THOMAS, H. E. 1934. Studies on *Armillaria mellea* (Vahl) Quél. Infection, parasitism and host resistance. J. Agr. Res. 48:187–218.

WALLACE, G. B. 1935a. Report of the Mycologist. Rep. Dept. Agr. Tanganyika 1934, p. 90–93.

WALLACE, G. B. 1935b. *Armillaria* root rot in East Africa. E. Africa Agr. J. 1:182–192.

WALLIS, G. W., and G. REYNOLDS. 1965. The initiation and spread of *Poria weirii* root rot of Douglas fir. Can. J. Botany 43:1–9.

Studies on the Parasitism and Control of Tea Root Disease Fungi in Ceylon

N. SHANMUGANATHAN—*Tea Research Institute, Talawakelle, Ceylon.*

Since the classical experiments of Petch in the 1920s, tea root diseases have received little attention in Ceylon until the last six years. Neglect of the problem for nearly 40 years resulted in a surprisingly large area of tea plantings becoming infested with root pathogens. A survey by Mulder and Redlich in 1962 revealed that one root disease alone had destroyed more than a thousand acres of tea in the high-country plantations, while damage by other, less serious diseases was also appreciable. These observations prompted a revival of interest in root disease investigations at the Tea Research Institute, and this review summarizes the results of recent studies.

There are four root rots of tea that are economically important in Ceylon, but only Poria root disease (*Poria hypolateritia*) occurs widely (Fig. 1). It is also the most difficult one to control. For years tea root diseases were controlled by the laborious and expensive method of digging and removing infected roots and replanting the cleaned area. The efficacy of this practice was poor and this accounts for the present widespread occurrence of Poria root disease on many plantations. This method of control failed because of incomplete root extraction. Also, the high cost of the operation caused many estates to abandon it altogether. Recent studies were directed, therefore, towards devising a cheaper and more effective method of control using soil fungicides.

Control by Soil Fumigation with D-D.—The first material tested was D-D, which is used widely in tea nurseries as a nematocide. In view of its fungicidal properties, its efficacy in controlling *Poria* was investigated in field experiments. Results showed that the best control was obtained when D-D was applied at the rate of 2,000 lb per acre at 6 in depth. This treatment, however, killed *Poria* effectively only down to 18 in. As the depth of infested soil is about 30 in, D-D was provisionally recommended with the reservation that Guatemala grass, a nonhost for *Poria*, be grown on treated land for two years before tea was replanted, to enable the inoculum in the 18 to 30-in layer to perish on its own after exhaustion of root reserves. In instances where replanting could not be delayed, it was suggested that after fumigation *Tephrosia vogelii* be planted for one year as an indicator of remaining infection before replanting (Shanmuganathan, 1964).

Control by Soil Fumigation with Methyl Bromide.—The success of American workers in controlling *Armillaria mellea* in citrus orchards by methyl bromide fumigation prompted experiments with it in 1964. It was found that fumigation with methyl bromide under a polythene cover at the rate of ½ lb/100 ft^2 of soil controlled *Poria* effectively down to 3 ft. This treatment appeared more effective, economical, and less laborious than fumigating with D-D and was, therefore, recommended to plantations for *Poria* control (Shanmuganathan and Redlich, 1965). Methyl bromide is effective at a relatively low dosage presumably because tea soils are well drained and soil temperatures seldom fall below 60°F. Control has been obtained in soils with moisture content up to 40%. At present the only drawback with this method of control is that the polythene cover degenerates rapidly under tropical conditions. Certain brands of reinforced polyvinyl chloride are more durable but expensive.

While penetration of methyl bromide occurs in the soil directly under the cover, the fumigant does not seem to diffuse laterally outside this area even at twice the standard dosage. When large patches are treated, no unfumigated alleys should, therefore, be left behind. There is evidence that control is superior if the infested patch is divided into units of 200 ft^2 and each unit fumigated separately (Shanmuganathan and Fernando, 1967). Fumigation is usually carried out under polythene tarps 24 ft by 14 ft; the methyl bromide is applied from one-pound cans containing 2% chloropicrin as a warning agent.

The mode of action of methyl bromide on *P. hypolateritia* appears to be direct as well as indirect. *Trichoderma viride* was consistently found on test inocula recovered after fumigation, while a *Penicillium* sp. appeared frequently. On many occasions, the buildup of *T. viride* could be seen with the naked eye. Both these fungi are antagonistic to *P. hypolateritia*, with *T. viride* showing several types of antagonism. We have observed that nearly a week is required to kill the bulk of the deep-seated inoculum, whereas the surface inoculum is killed within 48 hr. It is possible

Fig. 1.   A sheet of tea showing damage by Poria root disease.

that the action of the fumigant in the upper layers is direct, while in the lower layers, where the fumigant may not reach sufficient concentration for direct killing, the action is indirect.

SOME SIDE EFFECTS OF FUMIGATION.—There is an increased growth response of tea planted on treated land. This has been shown to be a direct result of fumigation and is not due to the destruction of root pathogens. A similar effect has also been observed in tea nurseries when nursery soil was fumigated with methyl bromide, but not with D-D (Kerr and Vytilingam, 1966). Soil nitrogen determinations indicate clearly that this increased growth is related to a shift in nitrogen nutrition following the accumulation of ammonia nitrogen after fumigation. We have observed that for several weeks the fumigated soils contained a greater amount of ammonia nitrogen than the untreated controls, probably because of inhibition of nitrification. Further, when three different rates of methyl bromide were used for fumigation, the increase in ammonia nitrogen was linearly related to the dose, and this relationship was significant. These observations provide some indirect evidence that tea can utilize ammonia nitrogen more effectively than nitrate nitrogen.

Methyl bromide has so far shown no adverse effects such as stimulation of other root pathogens. It effectively controls *Rosellinia arcuata* and *Ustulina deusta*, and is also an excellent nematocide. It also suppresses weed growth on treated land for over two months.

EPIDEMIOLOGY AND SAPROPHYTIC SURVIVAL.—Examination of the root system of infected plants shows that *P. hypolateritia* colonizes only the top 30 in of the root system. The fungus invades practically all lateral roots and the main stump and causes a soft rot. Sterile fructifications are frequently formed on the collar of dead bushes. Infection occurs both by contact between healthy and infected roots and by growth of mycelium through the soil. As the majority of infections appear to arise by way of lateral roots and only a few from the main stump, the former is probably the more common method.

In the absence of air-borne spores, the only known source of inoculum for new outbreaks is infected roots of tea or shade trees. As the planting of shade trees susceptible to *P. hypolateritia* has long been abandoned, it appears unlikely that shade-tree roots can still serve as sources of inoculum. New infections can arise, therefore, only in two ways: in young plantations through infected roots left over during the uprooting of old tea, and in both young and old plantings by accidental spread of infected material by man. While the former can be prevented by fumigating all infested patches in the old tea before uprooting, the dispersal of infected roots within plantations is more difficult to prevent.

We now recommend that all diseased plants be grubbed by winching before the affected land is fumigated, so that the fumigant has to deal with only the residual, not easily recoverable inoculum. Much inoculum is dispersed during the initial digging-up operation, and some suggest that such dispersal would be unimportant if the uprooting is carried out after the fumigation, for the pathogen would be dead in roots taken out after treatment. Experiments indicate that this is feasible if the dosage of methyl bromide applied is increased to 1–2 lb/100 ft².

CHARCOAL STUMP ROT *(Ustulina deusta)*.—Unlike *P. hypolateritia, U. deusta* spreads mainly by airborne spores discharged from fructifications formed on infected shade-tree or tea stumps. The fungus invades exposed stump surfaces of *Grevillea robusta* and then spreads to the adjoining tea. Field observations have established that if shade trees are ring-barked in full leaf before felling, the incidence of charcoal stump rot is greatly reduced. Ring-barking is effective only if trees are felled after complete defoliation. This may take two years for fully grown *G. robusta*, because depletion of root reserves is slow. Killing trees with arboricides like 2, 4, 5–T is not as effective as ring-barking because defoliation occurs before root reserves are sufficiently depleted.

SUMMARY AND OUTLOOK.—Because of its versatility, its effectiveness at low dosage, and its desirable side effects, soil fumigation with methyl bromide has now established itself as a routine measure for root disease control on tea plantations in Ceylon. Methyl bromide is easy to apply and the cost is less than that for manual cleaning. Using methyl bromide, large areas of infested land can be cleaned rapidly and with good measure of success. The complete eradication of the insidious Poria root disease seems now possible, if the current interest shown by estate superintendents continues.

There is no satisfactory alternative at present to ring-barking shade trees in order to eliminate infections by *Ustulina*. The simpler method of painting the stump surfaces with a bituminous paint immediately after felling is not adequate, for this will not prevent the extension of any dormant lesions already present. Though standing trees are not known to be killed by *Ustulina*, inoculation experiments indicate that such trees can become infected, but further development is probably restricted by host resistance.

LITERATURE CITED

KERR, A., and M. K. VYTILINGAM. 1966. Fumigation of nursery soil with methyl bromide. Tea Quart. 37:162–163.

MULDER, D., and W. REDLICH. 1962. Results of a survey of red root disease (*Poria hypolateritia* Berk). Tea Quart. 33:141–145.

SHANMUGANATHAN, N. 1964. Recent developments in the control of *Poria* root disease. Tea Quart. 35:22–31.

SHANMUGANATHAN, N., and S. R. A. FERNANDO. 1967. Some observations on *Poria* control by soil fumigation with methyl bromide. Tea Quart. 38:311–319.

SHANMUGANATHAN, N., and W. W. REDLICH. 1965. Control of *Poria* root disease with methyl bromide. Tea Quart. 36:144–150.

Economics of Control of the White Root Disease (Fomes lignosus) of Hevea brasiliensis in Ceylon

O. S. PERIES—*Vidyodaya University, Nugegoda, Ceylon.*

▶

Economically, rubber is a marginal crop in Ceylon at present. Therefore, the industry is vitally interested in economizing on all costs of production, including disease control. Fox (1964) and Newsam (1964, 1967) have discussed the economics of White Root disease (*Fomes*) control in Malaysia. However, the data they provide are not directly applicable to Ceylon conditions, as rubber in Malaysia is planted in relatively flat land with an average annual rainfall of about 80 in, as compared to the distinctly hilly and boulder-strewn areas, with a precipitation of approximately 150 in per year, in Ceylon. Labour wages too are higher in Malaysia than in Ceylon. These factors influence the choice of control methods adopted in the two countries.

This paper discusses the results of a series of experiments conducted specifically to assess the costs of controlling the disease.

EXPERIMENTAL.—Experiments were conducted on cooperating estates. All were designed with a view to assessing the costs and efficacy with reference to *Fomes* control of different methods of clearing the old stand and detecting and treating the disease. A standard randomized block layout was used for all experiments, with four treatments replicated seven times, the plot size being 1 acre.

Use of fungicides.—The results of a number of experiments (Peries, 1965) have shown that the removal of the food base and excision of infected roots, so as to eradicate a self-propagating infection on the root system, is sufficient to control the disease. The use of a fungicide in the treatment of *Fomes*-infected trees has been proved to be uneconomical (Peries, Fernando, and Samaraweera, 1963, 1965). Except where otherwise stated, eradication of the food base and excision of the infected roots was adopted as the standard method of treatment in all subsequent experiments.

Eradication prior to replanting.—Four treatments were tested to establish the most economical method of clearing: (1) complete manual clearing of roots of all standing trees, followed by stacking and burning of all timber; (2) uprooting all trees, but root clearing confined to trees found to be infected when uprooted; (3) standard method, i.e. all trees uprooted, stacked, and burnt, no root clearing; (4) as in (3), but timber removed from site. The results of the experiment are presented in Table 1.

TABLE 1. Average cumulative losses in 5 years, through *Fomes* infection, after different clearing methods

Treatment index (as in text)	1	2	3	4
Av. no. of infections per acre	1.0	1.6	9.6	10.8

S.E. of Means 0.81   L.S.D. 5%: ± 1.71; 1% ± 2.35

The incidence of *Fomes* was negligible in plots where the root system of the old stand was completely cleared. This confirms the theoretical expectation that uprooting and burning the bulk of the old timber, which acts as a food base for the fungus, would virtually eradicate the disease. This method, however, is not economical, as clearing costs are excessive. The eradication of the roots of old trees found to be infected at the time of uprooting (treatment 2) is economically sound.

Method of detection.—There are two standard methods of detecting *Fomes*-infected trees: (1) root inspection (RI), i.e. periodical examination of the upper part of the root system, to a depth of about 9 in, and (2) foliar inspection (FI). The disease can be detected at an early stage, before the collar area is infected, by the former method. However, this may lead to the treatment of trees which may not succumb to the disease if left untreated (Fox, 1961). It also invariably injures the root system to some extent, facilitating penetration by *F. lignosus* (Peries, 1963; Newsam, 1964), thus increasing percentage of infection. Foliar symptoms are seen at a later stage of infection, and only on trees that are likely to succumb to the disease. Treatment is likely to be more successful in the former than in the latter method of detection.

Table 2 shows the results of an experiment conducted to compare the efficacy and costs of these two methods. The area was heavily infected with *Fomes*, and planted at a slightly higher (10%) density

TABLE 2. Efficacy of two methods of disease detection. Average number of trees/acre infected, and recovering after treatment

Detection and treatment	RI+treat.	RI+up-root	FI+treat.	FI+up-root
No. infected	26	22	15	12
No. recovered	20	—	6	—

than normal in order to allow for removal of infected trees.

Two labourers would take 1 day for root inspection of 1 acre, whereas one labourer can conveniently carry out foliar inspection of 5 acres in a day. There would be less trees for treatment in the latter method; therefore, its cost would be lower.

Collar protectant dressings (CPD).—John (1958) has established that infection by *F. lignosus* can be prevented by checking the epiphytic growth of the fungus on rubber roots. Pentachloronitrobenzene (PCNB), in a suitable greasy carrier, has been used successfully in Malaysia as a prophylactic fungicide (Fox, 1966) for this purpose.

The economics of the use of CPDs were assessed in a repetition of the above experiment, where the normal treatment was supplemented by the application of a CPD to the roots of treated trees and those adjacent to them in half the treatments. The results of the experiment are summarized in Table 3.

TABLE 3. Average number of infected trees recovering after treatment with and without a CPD—two methods of detection

Method	RI+CPD	RI−CPD	FI+CPD	FI−CPD
Treated	20	25	12	16
Recovered	16	20	8	7

A comparison of the results presented in Tables 2 and 3 show that the overall percentage of infection is lower, and recovery higher, in areas where CPDs are used. It was observed that the incidence of the disease was lower mainly because the occurrence of diseased trees in runs was almost totally prevented by the use of the CPD on trees adjacent to those infected; treatment was more successful as the epiphytic growth of the pathogen on the roots was prevented by the CPD. Unlike Malaysia, where the disease occurred in runs of two or more adjacent trees in less than 45% of cases (Newsam, 1967), in Ceylon the disease tended to occur in groups in over 60% of cases.

Biological Control.—Peries (1965) referred to the possibility of reducing the incidence of *Fomes* by mixing small quantities (¼ lb) of sulphur with the soil in the planting hole. The results of an experiment to investigate the effect, on the incidence of disease, of introducing ¼ lb. of sulphur into the planting hole are presented in Table 4.

TABLE 4. Average incidence of *Fomes* per acre in areas treated with ¼ lb sulphur (S) as compared to controls

Method	S+RI	No S+RI	S+FI	No S+FI
Infected	6	14	4	12
Recovered	5	12	1	5

The addition of sulphur reduced the pH of the soil. This acid medium favoured the growth of *Trichoderma viride*, which parasitises other fungi through the medium of the antibiotics gliotoxin and viridin, both of which are only stable in acid medium (Garrett, 1965). The extra cost of the treatment is the cost of sulphur. However, the use of sulphur in all replantings is not advocated, as it would be economically unsound. Sulphur should be used only in areas known to be heavily infected with *Fomes*, at the time of replanting.

CONCLUSIONS.—The most economic procedure for *Fomes* control in Ceylon is to eradicate all roots of infected trees at the time of uprooting, and to plant at a slightly higher density than normal, to enable uprooting of infected trees as a part of routine thinning out. Detection of infected trees is best done by the method of foliar inspection. Collar protection of trees adjacent to those infected is economical. Sulphur can be added to the soil profitably at planting time, in areas known to be heavily infected by the disease. This will reduce clearing and treatment costs.

Basic differences in methods of root disease control in different countries can be caused by differences in environmental, topographical, and soil factors. The writer is of the opinion that the spread of *Fomes* from tree to tree is more rapid in Ceylon than in Malaysia because of the higher rainfall recorded in Ceylon. This is the basic factor which has lead to differences in control methods between the two countries.

LITERATURE CITED

Fox, R. A. 1961. White Root disease of *Hevea brasiliensis*: recent developments in control techniques. Rep. 6th Common. Mycol. Conf. 1960, p. 41–48.

Fox, R. A. 1964. The principles of root disease control. R.R.I.M. Planters' Bull. No. 75, p. 210–217.

Fox, R. A. 1966. White Root disease of *Hevea brasiliensis*: collar protectant dressings. J. Rubber Res. Inst. Malaya 19:231–241.

Garrett, S. D. 1965. Toward biological control of soil-borne plant pathogens, p. 4–17. *In* K. F. Baker and W. C. Snyder [ed.], Ecology of soil-borne plant pathogens. Univ. of California Press, Berkeley and Los Angeles.

John, K. P. 1958. Inoculation experiments with *Fomes lignosus* Klotzsch. J. Rubber Res. Inst. Malaya 15:223–233.

Newsam, A. 1964. Effects of clearing methods on root disease incidence. R.R.I.M. Planters' Bull. 75:225–232.

Newsam, A. 1967. Clearing methods for root disease control. R.R.I.M. Planters' Bull. 92:176–182.

Peries, O. S. 1963. Review of the Plant Pathology Department. Ann. Rev. Rubber Res. Inst. Ceylon. 1962, p. 57–70.

Peries, O. S. 1965. Recent developments in the control of the diseases of the *Hevea* rubber tree. Rubber Res. Inst. Ceylon Quart. J. 41:33–43.

PERIES, O. S., T. M. FERNANDO, and S. K. SAMARAWEERA. 1963. Field evaluations of methods for control of White Root disease (*Fomes lignosus*) of *Hevea*. Rubber Res. Inst. Ceylon Quart. J. 39:9–15.

PERIES, O. S., T. M. FERNANDO, and S. K. SAMARAWEERA. 1965. Control of White Root disease (*Fomes lignosus*) of *Hevea brasiliensis*. Rubber Res. Inst. Ceylon Quart. J. 41:81–89.

Some Factors in the Control of Root Diseases of Oil Palm

P. D. TURNER—*Harrisons & Crosfield (Malaysia) Sendirian Berhad, Oil Palm Research Station, Banting, Selangor, Malaya.*

Compared with a number of other crops, the oil palm (*Elaeis guineensis*) is susceptible to comparatively few diseases, although disease incidence varies from country to country (Turner and Bull, 1967). The aim, however, of agricultural practice is to establish those conditions which maximise yield in the shortest possible time, and in this respect disease control becomes of increasing importance. Also, with the production of higher yielding oil-palm varieties, the levels of disease incidence at which control measures become economical will fall.

Root diseases of oil palms fall into two distinct groups. In the first group are those infections which are dependent on the presence of a massive source of inoculum, from which the pathogen spreads to the palm either by direct root contact with the source or spreads from the source to the palm by rhizomorphs. In this group, disease etiology is fairly well understood, but much remains to be learned of the etiology of the diseases in the second group, where the causal pathogens are common soil saprophytes and infection appears to follow predisposition of the palms in some way or other. As the differences between the two groups in disease etiology are so great, it is not surprising that the approach to control is also very different.

DISEASES REQUIRING LARGE INOCULA.—The most important root diseases which require the presence of large infection foci are basal stem rot, caused by *Ganoderma* spp., and Armillaria root and stem rot. There are also a few other minor root diseases of similar etiology, such as charcoal base rot associated with *Ustulina zonata*, but these are of little significance. In all instances the fungi are primarily saprophytes occurring naturally on various decaying jungle timbers.

The incidence of basal stem rot in palms of up to thirty years old is greatly influenced by the composition of the former stand at the time of replanting (Turner, 1965). When jungle is felled and partially cleared, the decaying timber and stumps are colonised by a wide range of fungi, of which the species of *Ganoderma* pathogenic to oil palm form only a small fraction. Hence, the subsequent levels of disease are very low. Similarly, *Ganoderma* is found in very small quantities as a coloniser of rubber stumps, and here again the incidence of basal stem rot in former rubber areas is very low. Where the area was previously occupied by coconuts, however, serious outbreaks of disease are common since coconut stumps and trunks appear to have a selective medium effect for colonisation by *Ganoderma* which largely restricts invasion by other decaying microorganisms. Contact by oil-palm roots with these large inocula usually ensues and it is not uncommon for 50% of a palm stand to have been destroyed when it is only half-way through its economic life-span of 30 years. Old oil palms also present a potential disease hazard after replanting, but since numerous other fungi which are nonpathogenic to oil palms also colonise stumps, subsequent disease incidence is usually lower.

Old jungle stumps and trunks form sources of infection by *Armillaria mellea*. Rhizomorphs either penetrate the oil-palm roots and grow through the inside of the root into the palm or spread around the base of the palm and cause destruction of the stem tissue there (Wardlaw, 1950).

SOIL-BORNE DISEASES.—Under this heading are (1) the important nursery disease, blast, caused by species of *Pythium* and *Rhizoctonia*; (2) vascular wilt, caused by a strain of *Fusarium oxysporum*; and (3) dry basal rot, associated with *Ceratocystis paradoxa*. Some years ago in Malaya an undefined disease, nursery root rot, was the cause of some concern since losses were considerable, but the disease has since virtually disappeared. Control in this instance relied upon not using an affected area again as a nursery site.

Almost wherever oil palm is grown, blast disease has been found. This disease, which is brought about by co-infection of the fleshy feeding roots by species of *Pythium* (the primary invader) and *Rhizoctonia* (Robertson, 1959), has been responsible for extremely serious losses of up to 80% of the nursery. From the aspect of disease control, the critical factor is that control must be entirely by prevention since the development of even slightly affected palms is severely retarded. The disease occurs when soil conditions change so that the status of the normally saprophytic fungi is changed to that of pathogenic, i.e. when the soil moisture content drops to about 10% of its moisture-holding capacity and when soil temperature rises.

Outbreaks of Fusarium wilt occur especially in countries where prolonged dry periods are a regular seasonal feature and where palms are grown in poorer soils. It is believed that the feeding roots produced close to the soil surface are killed by drought, with the resulting dead tissues providing courts of entry for the pathogen. Although the possible absence of the pathogen cannot be discounted, it is this close association with periods of drought and poor nutrient status which offers the likely explanation for the absence of the disease from countries with a more evenly distributed rainfall and better growing conditions, such as Malaysia.

An atypical and comparatively rare combination of climatic factors has been suggested as an underlying cause for dry basal rot, a disease which has given rise to highly significant economic losses in West Africa (Robertson, 1962). Whether these conditions predispose palms to infection, or whether they so favour the development of the causal organism to permit it to become pathogenic, is unknown. The pathogen is one of the commonest fungi found throughout the tropics and can be isolated from virtually every part of healthy oil palms where it is, presumably, normally a casual inhabitant. Any change in the status of the fungus, therefore, brought about by the causative factors takes place in the soil population.

CONTROL OF ROOT DISEASES.—*Removal or treatment of infection foci.*—For those diseases where infection is dependent upon the presence of a massive inoculum, the best control method is through preventing infection by removing potential inocula. This is essential with basal stem rot since by the time external symptoms of disease appear, the lesion has come to occupy about one-half of the cross-sectional area of the palm base. The case for the removal of coconut tissue is thus clear, although costly. Fortunately, the fact that it requires a minimum inoculum size of 45 cu in to initiate and establish infection (Navaratnam and Chee, 1965) makes it necessary to remove only the bole and the thick root crust immediately surrounding it; the mass of small roots may be left to decay since they contain insufficient inoculum potential to become infective. The necessity for removal of oil-palm stumps is not so clear, but a decision is made easier by the desirability of their removal through their becoming breeding sites for the very harmful rhinoceros beetle.

With *Armillaria* infection, however, the practicability of removing jungle stumps as a preventative measure is limited, since the cost of removing stumps other than the smallest would be prohibitive. In this instance, ring-barking prior to felling to exhaust carbohydrate reserves beneath the point of ringing would appear worthwhile where a high incidence of *Armillaria* infection might be expected. More economic control of *Armillaria* would seem to be by regular inspection and surgical treatment where necessary to excise affected tissues and remove sources of infection.

Stump treatments, which result either in exclusion of *Ganoderma* or more rapid breakdown through encouraging extensive invasion by saprophytes, have been found to have no value in former coconut areas where stumps have been left in the ground. Inducing more rapid breakdown could be of some practical value where old oil-palm stumps remain.

In all instances of root diseases, sanitation is important. Whilst there is no evidence of infectious spread of blast, wilt, or dry basal rot, removal of dead tissues prevents their colonisation by potential pathogens or injurious pests. Infectious spread does occur in *Ganoderma* attacks, and this is also highly likely where *Armillaria* is present.

Fungicides.—There appears to be little potential value in the use of fungicides against oil-palm root diseases, although the development of a suitable systemic fungicide could well be of value against wilt. Numerous soil treatments in nurseries have been carried out for the control of blast, but with very little success, and more satisfactory control measures are available. Soil treatment by fungicides on an estate scale is quite impracticable for both vascular wilt and dry basal rot. With *Ganoderma* infections, successful fungicidal treatment would depend upon detection of the lesions long before external symptoms appear. Fungicidal protectants are of value following surgery in both this instance and *Armillaria* infections.

Agronomic practices.—With oil-palm diseases generally, those palms which are poorly developed appear to be most susceptible to disease. Vigour has been suggested to influence blast and wilt, and this in turn is connected with nutrition. Wilt in particular has been closely associated with nutritional status, with disease levels being much lower in palms where the potassium levels were high (Prendergast, 1957). Application of magnesium limestone has been reported to reduce the incidence of *Armillaria* attacks, but there is no obvious relationship between palm nutrition and basal stem rot or dry basal rot. Nutrition often reflects soil conditions, and light sandy soils favour wilt, blast, and *Armillaria* infection.

Nursery practice is of primary importance in controlling blast, and is aimed at preventing those conditions which permit the causal fungi to become pathogenic. The main factor controlling blast is water, and provision of adequate water, especially by an overhead sprinkler system, is essential. This, in its turn, requires siting the nursery close to an adequate and continuous water supply. It should be noted, however, that sufficient water is in any event a prerequisite for good nursery growth, and if blast occurs, this indicates that markedly suboptimal conditions are being provided. The necessity for watering is greater where the polybag technique is used for raising seedlings, especially where the soil used for filling the bags is of a rapidly draining sandy type, since water loss is accentuated.

Where the water supply is limited, and this will occur more in those countries with a pronounced dry season than in the more equitable climate of Malaysia, other methods to protect the seedlings must be used. Water loss can be reduced by adding a mulch to the soil surface, although care should be taken in selecting the type of mulch used so as to avoid any nutritional disturbances through the excessive addition of any element. Where the polybag system is used, lining the rows of bags in an east-west direction reduces insolation. Erection of lateral shade may be worthwhile; overhead shade, although reported to be effective, is expensive.

Breeding for resistance.—With neither *Armillaria* nor *Ganoderma* infections does there appear to be any point in attempting to prevent disease by developing resistance in palms, since the volume of inoculum involved makes it unlikely that such resistance could ever be achieved. With Fusarium wilt, however, the situation is very different. In this instance there exists an unchanging pattern of events which predetermines the onset of disease. In the absence of irrigation, control can be satisfactorily achieved only by the development of resistant varieties. This method of control would also seem to be the only satisfactory method for avoiding dry basal rot. In water-short areas, blast is likely to be most effectively controlled by the development of resistant varieties, towards which end some progress has been made in West Africa (Prendergast, 1963).

LITERATURE CITED

NAVARATNAM, S. J., and K. L. CHEE. 1965. Root inoculation of oil palm seedlings with *Ganoderma* sp. Plant Disease Reptr. 49:1011–1012.

PRENDERGAST, A. G. 1957. Observations on the epidemiology of vascular wilt disease of the oil palm. J. W. Afr. Inst. Oil Palm Res. 2(6):148–175.

PRENDERGAST, A. G. 1963. A method of testing oil palm progenies at the nursery stage for resistance to vascular wilt disease caused by *Fusarium oxysporum* Schl. J. W. Afr. Inst. Oil Palm Res. 4(14):156–175.

ROBERTSON, J. S. 1959. Blast disease of the oil palm. Its cause, incidence and control in Nigeria. J. W. Afr. Inst. Oil Palm Res. 2(8):310–330.

ROBERTSON, J. S. 1962. Dry basal rot, a new disease of oil palms caused by *Ceratocystis paradoxa* (Dade) Moreau. Trans. Brit. Mycol. Soc. 45:475–478.

TURNER, P. D. 1965. The incidence of *Ganoderma* disease of oil palms in Malaya and its relation to previous crop. Ann. Appl. Biol. 55:417–423.

TURNER, P. D., and R. A. BULL. 1967. Diseases and disorders of the oil palm in Malaysia. Incorporated Society of Planters, Kuala Lumpur.

WARDLAW, C. W. 1950. *Armillaria* root and trunk rot of oil palms in the Belgian Congo. Trop. Agr. (Trinidad) 27:95–97.

▶

Banana Root Diseases Caused by Fusarium oxysporum f. sp. cubense, Pseudomonas solanacearum, and Radopholus similis: A Comparative Study of Life Cycles in Relation to Control

R. H. STOVER—*Tela Railroad Company (a subsidiary of United Fruit Co.), La Lima, Honduras.*

▶

A healthy banana plant with its attached suckers sends out adventitious main roots in a radius of about 8 ft and to a depth of 4 ft in well-drained loam. These 600 cu ft of soil are explored by up to 500 main roots and innumerable smaller secondary and tertiary roots. In the process of invading the soil, the roots and rhizome are invaded by a large number of fungi (Stover, 1966), nematodes (Stover and Fielding, 1958; Wehunt and Holdeman, 1959), and two species of bacteria (Stover, 1959; Buddenhagen, 1961). However, only a single fungus species *(Fusarium oxysporum* f. sp. *cubense)*, one species of nematode *(Radopholus similis,* Cobb, 1893; Thorne, 1949), and one species of bacteria *(Pseudomonas solanacearum)* are strong pathogens and cause heavy losses of production in the tropics. A comparative study of these three major banana pathogens will indicate at which point the life cycle of each is most vulnerable. The degree of vulnerability will determine how easily and in what form the pathogen can be suppressed or eradicated. A comparison of the three pathogens will also bring out the diversity of techniques required to maintain losses at a minimum.

LIFE CYCLE OF FUSARIAL WILT.—*F. oxysporum* f. sp. *cubense* remains dormant and immobile as chlamydospores in the remains of decayed host tissue until stimulated to germinate by host roots, root excretions from nonhost roots, or contact with pieces of fresh noncolonized plant remains (Stover, 1962a, b). Following germination, a thallus is produced from which conidia form in 6–8 hours and chlamydospores in 2–3 days if conditions are favorable. Invasion of roots of a susceptible banana plant is followed by the development of a systemic vascular disease. In the advanced stages of disease, the fungus grows out of the vascular system into adjacent parenchyma, producing vast quantities of conidia and chlamydospores. The conidia can also develop into chlamydospores. The chlamydospores are returned to the soil when the dead plant or plant part decays, where they can remain dormant and viable for several years. The cycle is repeated when chlamydospores germinate and again grow

either saprophytically or by invading the host. There is no natural means of dispersal away from the diseased plant.

VULNERABILITY OF THE FUSARIAL WILT PATHOGEN.—Because of the ability of chlamydospores to survive long periods in the absence of the banana host, and saprophytic growth on non-banana-plant substrates, it is impossible to eradicate *Fusarium* from soil. Once the soil is widely infested it is economically impossible even to reduce the population to a level that will permit banana growth for more than a few years. The only vulnerable point in the life cycle of *Fusarium* is its lack of a natural dispersal mechanism. Unless man or floodwaters intervene, *Fusarium* moves at a tortoise pace. This makes exclusion and early quarantine measures the only effective methods of control apart from the growing of resistant varieties. This is illustrated by the recent history of wilt disease in Colombia.

In the Santa Marta area of Colombia fusarial wilt of banana was excluded until 50 years after the introduction of the susceptible Gros Michel variety (Stover, 1962a). Even 5 years after its first appearance less than 100 acres out of 30,000 had been destroyed. The main dispersal agent was water from extensive surface-irrigation canals along which diseased bananas were growing. It was not until 10 years after the first wilt cases were recorded that large abandonments appeared and several hundred acres of bananas had been destroyed. More recently, spread of wilt from a single primary source of infection and in a nonirrigated area has been observed in the Turbo area of Colombia. Wilt first appeared about 3 years after bananas were planted in 1960 and involved around 25 plants in one location. These were destroyed along with adjacent healthy plants, the roots of which may have contacted diseased roots. Wilt has not yet spread from this location to 30,000 acres of surrounding bananas, and the area affected amounts to less than 2 acres. The origin of the disease was traced to a small former planting made with rhizomes imported in 1945 from the Acandí area, where wilt was known to have been present. The remainder of the Turbo area was planted with rhizomes imported from Santa Marta around 1914, an area then free of wilt. The history of fusarial wilt in Colombia shows that where there is no dispersal in planting stock or flood and irrigation waters, exclusion and early quarantine of the first disease outbreaks is effective. However, the Santa Marta and Turbo situations are extremely unique because man, usually unknowingly, moved *Fusarium* in planting stock to wherever Gros Michel bananas were grown commercially on a large scale. From these primary foci of infection, introduced in the early stages of commercial development, the disease has spread until wilt-susceptible varieties can no longer be grown successfully.

LIFE CYCLE OF PSEUDOMONAS SOLANACEARUM.—The bacteria ooze from diseased roots, fruit, flower-bract scars, and man- or animal-made wounds. Depending on the strain and in the absence of mechanical transmission on pruning tools, bacteria are dispersed primarily by insects to healthy flower-bract scars or fresh wounds usually made by pruning tools, or are dependent on root-to-root spread (Buddenhagen, 1965). Root spread results from the growth of healthy roots of adjacent plants into the vicinity of diseased roots. Once a banana plant is invaded through a natural or man-made opening, a systemic vascular disease results. In the advanced stages of disease, bacteria ooze from flower bracts, wounds, and roots, and the cycle is repeated with insect dispersal or root spread to healthy plants. The fate of bacteria that remain in the plant when it dies, and the bacteria that are released into the soil, is not clear. It is known that survival in soil varies from a few months up to 18 months depending on strain and location (Sequeira, 1958, 1962). No form of the bacterium resistant to adverse conditions outside the host is known, nor has saprophytic growth apart from the host or on root exteriors been reported. It seems unlikely, however, that *Pseudomonas* would survive more than a year without some form of growth and regeneration. In addition to banana, *P. solanacearum* (race 2) is found in a few other hosts (Buddenhagen, 1960; Sequeira and Averre, 1961) but the host range reported thus far is much less extensive than that of race 1 (Kelman, 1953; Buddenhagen and Kelman, 1964). Some of the wild hosts show few or no symptoms and it is not clear how they are infected.

VULNERABILITY OF THE BACTERIAL WILT PATHOGEN.—In the absence of a resistant spore form, *P. solanacearum* must disperse or die. Thus, the most vulnerable point in the life cycle of this pathogen is dispersal. In the case of the soil-borne B strain, there is no effective means of dispersal in the soil, and the disease spreads slowly by root contact at the rate of 30 to 50 ft per year, except where mechanical spread on contaminated tools occurs. There is a small amount of insect spread from banana to banana but this is not common (Buddenhagen, 1965). As with fusarial wilt, widespread dispersal of soil-borne bacterial wilt has occurred primarily as a result of movement in planting

stock (Buddenhagen, 1961). Since *P. solanacearum* spreads slowly and does not persist long in soil, it should be possible to eradicate bacterial wilt caused by soil-borne strains. In fact, eradication has been obtained in certain areas of Honduras and the Bocas del Toro area of Panama by the prompt destruction of diseased and adjacent healthy mats followed by a fallow period before replanting. However, following absence of the disease for 2 or 3 years, the B strain will mysteriously reappear in a few plants. To explain the reappearance of wilt after several years' absence, it must be assumed that there is a reservoir of inoculum in wild hosts in woodland and small farms surrounding the banana zone. Undoubtedly, apart from *Heliconia* spp., many of these hosts remain to be identified. Also, it remains to be determined how the vectors acquire inoculum and bring the B strain back into bananas. If the sources of nearby inoculum could be found and eliminated, then the B strain could be excluded permanently from banana plantations instead of for a few years only.

The SFR strain of *P. solanacearum* is much less vulnerable than the B strain because of the ease with which the SFR strain can be acquired and transmitted by insects. The only feasible method of preventing dispersal is to eliminate the source of inoculum. Since insect dispersal occurs so readily, new sources of inoculum are greatly increased not only within banana plantations but in wild and domestic *Musa* spp. outside the plantations. Since all of these cannot be eliminated, a reservoir of inoculum is always available somewhere to keep the chain of dispersal and infection unbroken. Therefore, it is unlikely that the SFR pathogen can be eradicated. Banana growers will have to live with insect-transmitted bacterial wilt and strive to maintain a low level of disease by constantly seeking out sources of inoculum and destroying them. It remains to be determined how much of this inoculum may be coming from weed and forest plants within and without the plantation.

LIFE CYCLE OF RADOPHOLUS SIMILIS.—The adult worm burrows into young cortical tissue of the main roots and into the rhizome. A maximum of 76 eggs are laid at the rate of 2 to 3 per day and hatch in 8–10 days. The larvae mature and begin egg-laying in 12 to 15 days (Loos, 1962). The larvae may continue to extend the lesions made by adults or may migrate to new areas along the root or rhizome. *Radopholus* cannot survive more than 5 months in the absence of a host (Loos, 1961; Tarjan, 1961). There is no effective means of dispersal in the soil except by movement of the worms along roots from infected plants to adjacent healthy plants. In this way, disease spreads from 10 to 40 ft per year (Christie, 1959). Widespread dispersal has resulted from movement in planting stock. However, Loos (1961) found *Radopholus* in river water and suggests that irrigation water may be one mode of dispersal. In contrast to the citrus strain of *Radopholus*, the banana strain has a small wild-host range (Blake, 1961; Ayala and Roman, 1963).

TABLE 1. Survival time in soil and vulnerability of banana root pathogens in relation to methods and ease of control

Pathogen	Survival time in soil	Most vulnerable character	Least vulnerable character	Method of control	Ease of control
F. oxysporum f. sp. *cubense*	Years	Lack of dispersal mechanism	Chlamydospore stage and saprophytic growth	Exclusion and quarantine	Very difficult
P. solanacearum Strain B	12–18 months	Lack of resistant spore form; slow dispersal	Wild hosts	Eradication	Not difficult
P. solanacearum Strain SFR	2–3 months	Lack of resistant spore form	Insect dispersal; wild and domestic hosts	Reduction of inoculum sources*	Difficult
R. similis	5 months	Lack of resistant cyst stage	Wild hosts	Eradication	Not difficult

* By early detection and destruction of infected *Musa* and wild hosts; early removal of male flower buds of *Musa* varieties susceptible to insect-disseminated strains.

VULNERABILITY OF THE BURROWING NEMATODE.—There are two vulnerable points in the life cycle of *R. similis*: the absence of a stage that can survive long periods outside the host, such as occurs in the cyst-forming species of nematodes, and an extremely inefficient form of natural dispersal. *R. similis* is easily eradicated from the soil by destruction of all hosts and fallowing for 5 to 6 months (Loos, 1961). Rhizomes can be easily freed of nematodes by hot water treatment (Blake, 1961; Small and Bomers, 1962). Thus, by a simple combination of fallow and planting nematode-free rhizomes, *Radopholus* can be eradicated.

CONCLUSION.—The most vulnerable points in the life cycle of all three banana pathogens are dispersal and/or mode of survival in the absence of bananas (resistant spore forms or wild hosts). The degree of vulnerability determines the method and ease of control (Table 1). The dispersal link in the life-cycle chain can be readily broken if the source of inoculum is easily eliminated. With *R. similis*, inoculum sources are easily eliminated because there is no cyst stage and few wild hosts. It is impossible to eliminate *Fusarium* because of the chlamydospore stage and saprophytic growth on plant substrates other than banana. With *P. solanacearum*, breaking the dispersal link is possible with some strains, but with others it is complicated by the presence of insect vectors and wild hosts. The role of wild hosts in maintaining a reservoir of inoculum for infection of banana plantations is not clearly defined, and much more research is needed in this area. In the evolution of *P. solanacearum* from a soil-borne to an air-borne pathogen, the ability to survive more than a few months in soil has been lost. However, survival in soil is no longer important or necessary. A great increase in the efficiency of dispersal ensures more numerous sources of inoculum.

The more numerous these sources, the more difficult it is to break the cycle of dispersal–infection–dispersal. However, insect dispersal in general is much less extensive spacially than the dispersal of wind-borne fungi. The majority of insects move and feed within a radius of a few miles, usually less than 10 miles (Andrewartha and Birch, 1954). Nevertheless, a small percentage of individuals may fly distances of 40 miles or more. Since these is a steeply declining frequency of infection as distances increase from point of origin (Gregory and Read, 1949), this greatly reduces the chances of *P. solanacearum* being dispersed hundreds of miles and infecting a host. Thus, bacterial wilt has been excluded from major banana-growing areas on the Atlantic coasts of Costa Rica and Ecuador, which are isolated from sources of inoculum by a few hundred miles of mountains and forest. When these natural-barrier areas are colonized by farmers who plant *Musa*, then the stage will be set for short-range insect dispersal from farm to farm until eventually *P. solanacearum* will be carried into those areas at present free of bacterial wilt.

LITERATURE CITED

ANDREWARTHA, H. G., and L. C. BIRCH. 1954. The distribution and abundance of animals. Univ. Chicago Press, Chicago, Illinois.

AYALA, A., and J. ROMAN. 1963. Distribution and host range of the burrowing nematode in Puerto Rican soils. J. Agr. Univ. Puerto Rico 47:28–37.

BLAKE, C. D. 1961. Root rot of bananas caused by *Radopholus similis* (Cobb) and its control in New South Wales. Nematologica 6:295–310.

BUDDENHAGEN, I. W. 1960. Strains of *Pseudomonas solanacearum* in indigenous hosts in banana plantations of Costa Rica, and their relationship to bacterial wilt of bananas. Phytopathology 50:660–664.

BUDDENHAGEN, I. W. 1961. Bacterial wilt of bananas: history and known distribution. Trop. Agr. (Trinidad) 38:107–121.

BUDDENHAGEN, I. W. 1965. The relation of plant-pathogenic bacteria to the soil, p. 269–284. *In* K. F. Baker and W. C. Snyder [ed.], Ecology of soil-borne plant pathogens. Univ. of California Press, Berkeley and Los Angeles.

BUDDENHAGEN, I. W., and A. KELMAN. 1964. Biological and physiological aspects of bacterial wilt caused by *Pseudomonas solanacearum*. Ann. Rev. Phytopathol. 2:203–230.

CHRISTIE, J. R. 1959. Plant nematodes: their bionomics and control. Agr. Expt. Sta. Univ. Florida.

GREGORY, P. H., and D. R. READ. 1949. The spatial distribution of insect-borne plant virus diseases. Ann. Appl. Biol. 36:475–482.

KELMAN, A. 1953. The bacterial wilt caused by *Pseudomonas solanacearum*. North Carolina Agr. Expt. Sta. Tech. Bull. 99, 194 p.

LOOS, C. A. 1961. Eradication of the burrowing nematode, *Radopholus similis*, from bananas. Plant Disease Reptr. 45:457–461.

LOOS, C. A. 1962. Studies on the life-history and habits of the burrowing nematode, *Radopholus similis*, the cause of black-head disease of bananas. Proc. Helminthol. Soc. Wash. 29:43–52.

SEQUEIRA, L. 1958. Bacterial wilt of bananas: dissemination of the pathogen and control of the disease. Phytopathology 48:64–69.

SEQUEIRA, L. 1962. Control of bacterial wilt of bananas by crop rotation and fallowing. Trop. Agr. (Trinidad) 39:211–216.

SEQUEIRA, L., and C. W. AVERRE, III. 1961. Distribution and pathogenicity of strains of *Pseudomonas solanacearum* from virgin soils in Costa Rica. Plant Disease Reptr. 45:435–440.

SMALL, C. V. J., and H. B. O. BOMERS. 1962. Hot water treatment of banana planting material against nematodes. Surinaamse Landbouw 10:109–117.

STOVER, R. H. 1959. Bacterial rhizome rot of bananas. Phytopathology 49:290–292.

STOVER, R. H. 1962a. Fusarial wilt (Panama disease) of bananas and other *Musa* species. Phytopath. Paper No. 4, Commonwealth Mycol. Inst., Kew, Surrey.

STOVER, R. H. 1962b. Studies on *Fusarium* wilt of bananas. IX. Competitive saprophytic ability of *F. oxysporum* f. *cubense*. Can. J. Botany 40:1473–1481.

STOVER, R. H. 1966. Fungi associated with nematode and non-nematode lesions on banana roots. Can. J. Botany 44:1703–1710.

STOVER, R. H., and M. N. FIELDING. 1958. Nematodes associated with root injury of *Musa* spp. in Honduran banana soils. Plant Disease Reptr. 42:938–940.

TARJAN, A. C. 1961. Longevity of *Radopholus similis* (Cobb) in host-free soil. Nematologica 6:170–175.

WEHUNT, E. J., and A. L. HOLDEMAN. 1959. Nematode problems of the banana plant. Proc. Soil and Crop Sci. Soc. Florida 19:436–442.

◄

PART VIII

CROP GROWTH RESPONSES TO SOIL FUMIGATION

One Centennium of Soil Fumigation: Its First Years

HELGA TIETZ—*Department of Plant Pathology, University of California, Berkeley.*

▶

The idea of treating soils with volatile, fumigant chemicals to rid them of organisms damaging to crops is now one centennium old. Developed under desperate circumstances of agricultural disaster, the practice of soil fumigation, through its unforeseen benefits, has provided, more than any other discovery in agriculture, stimuli to research into the natures and activities of soil-borne organisms and their influence on plant growth (Wilhelm, 1966).

In 1864 the rootlet-sucking aphid, *Phylloxera vastatrix*, had begun to devastate vineyards of Europe. First noticed in the Rhône valley near Avignon and in 1866 near Bordeaux, the pest spread rapidly and the resulting damage to vineyards became a national emergency due to the eminent position of viticulture in the French economy (Sagnier, 1892; Anonymous, 1881; Fig. 1). France, foremost in wine production among the countries of the world, then derived an estimated 10 milliard (billion) francs annually from its agriculture, of which 1.5 milliard or 15% came from vineyards grown on but 6% of the arable soil (Borde de Tempest, 1873). As a result of the new disease, wine yields dropped to one-tenth of the previous annual production and even to nil in certain areas heavily attacked. Everyone's welfare was threatened (Céris, 1869). Severe losses occurred often within only 3 years after the first detection of the pest (Faucon, 1870).

The cause of the disease was identified (Planchon, 1868; Cornu, 1879), and the search for effective control measures was begun. The Société des Agriculteurs de France, in 1868, organized a special Phylloxera commission headed by la Loyère to study and report on the new disease of the vine. In their report reference was made to the first trials of soil fumigation undertaken by one member of the commission, Baron Paul Thenard (Vialla, 1869). Thenard (Fig. 2) was a chemist and agriculturist. Envisioning the possibility of a chemical control for Phylloxera by soil fumigation without rendering the soil infertile, Thenard conceived the idea of employing carbon disulfide. The volatile compound had long been used as an insecticide in granaries (Barral, 1882) and in zoological laboratories (Dumas, 1874a). In 1869 Thenard conducted the first tests with CS_2 as a soil fumigant.

Thenard described his novel experiments in a letter written in 1869. Holes about 20 cm deep and 35–40 cm apart on a grid were made in freshly cultivated soil with a cabbage planter. A small glass of CS_2 was poured into each hole, and the opening quickly closed. Relating dosage of fumigant to the soil area rather than to the individual plant, Thenard calculated the rate at 1500 kg/ha (1350 lb/acre). After two days, Thenard and members of the Société des Agriculteurs of Bordeaux examined the vine roots and failed to find a single live Phylloxera. Unfortunately, some of the vines showed evidence of chemical injury, and within two weeks many had shed their leaves. Yet Thenard, who was apprehensive about the outcome of the experiments from the start, stressed the fact that Phylloxera had been destroyed by a chemical which left the soil still capable of giving further crops. The primary goal had been achieved, and Thenard, in view of this result, regarded the loss of some of the treated vines as negligible. In another vineyard, Thenard reduced the rate of CS_2 to 300 kg/ha (270 lb/acre) and metered it into an open plow furrow, added peanut-oil cake, and plowed the furrow shut. Vines were still unharmed after four days, but Phylloxera control was incomplete (Thenard, 1869, 1874, 1879; Rommier, 1876).

These experiments with CS_2 showed that eradication of Phylloxera from the soil was possible through the use of a fumigant chemical. But the results, seemingly of little practical value, aroused no excitement. Thenard was disappointed and turned his attention to coal-tar derivatives which had been indicated as promising insecticides by his friend and student A. Rommier.

Jean-Baptiste Dumas was president of the Phylloxera commission named by the Académie des Sciences in 1871. He figured prominently in the battle against Phylloxera and, unlike Thenard, set a high value upon a control treatment that would not injure the vines. Familiar with the physical properties of CS_2, Dumas reasoned that injury to the vines in the experiments of Thenard and others who subsequently had used CS_2 was caused not only by excessive dosages, but also by too rapid volatilization of the chemical. He felt also that the action of CS_2 was not sufficiently lasting to prevent reinfestation from nonkilled eggs. In 1874, Dumas announced that sulfocarbonate salts of sodium and particularly of potassium ideally met the requirements of low volatility and long action. Made by reacting potassium or sodium sulfide with

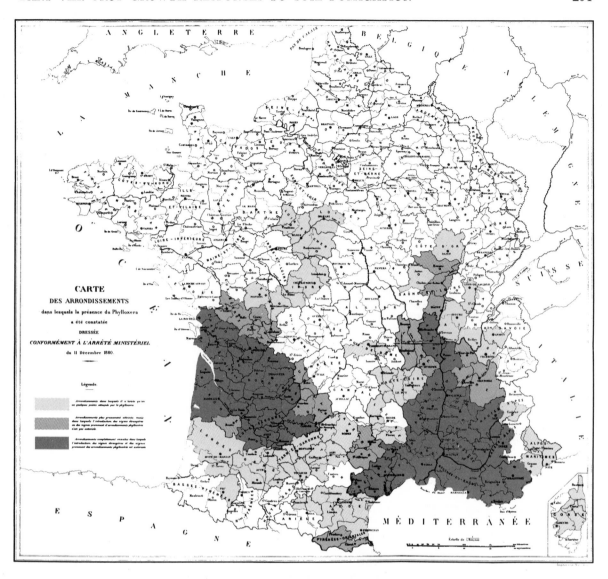

Fig. 1. Map of France showing areas infested by Phylloxera in 1880 (Anonymous, 1881). Shaded areas indicate infested regions with the darkest shading corresponding to devastated areas.

CS_2, the sulfocarbonates, when applied in an aqueous solution, released CS_2 and H_2S vapors upon contact with acids of the soil. Carbon disulfide thus became available in a form easily manageable and transportable; sulfocarbonates were solid, nonflammable, and barely volatile (Dumas, 1873, 1874b). Dumas devised economical procedures for commercial manufacturing of potassium sulfocarbonate, and Mouillefert (1877) did the research to bring it into large-scale application. The optimism, however, soon was dampened. Not only was the effect on Phylloxera of insufficient duration to achieve full control, but the fumigation cost was greater than with liquid CS_2. Large amounts of water were required for application

of the sulfocarbonates and consequently the labor involved was considerable. Thus, even a single treatment with sulfocarbonate was economically justified only in the most remunerative vineyards (Marion, 1876; Mouillefert, 1883; Rommier, 1876). Despite the shortcomings, fumigation with sulfocarbonates was practiced, during the years 1878 to 1884, on about 10% of the area treated against Phylloxera (Tisserand, 1885).

The problem of economically achieving an even distribution of an optimal dosage of CS_2 at the desired depth in the soil, and the hazards of handling and shipping the highly flammable compound, had hindered the development of a practicable control for

Phylloxera on the living vine. Many other fuming chemicals, such as petroleum, benzene, tar, calcium phosphide, even potassium cyanide, were investigated, but gave generally poor results or were unsuitable as field insecticides (Mouillefert, 1874).

Monestier, Lautaud, and d'Ortoman, in 1873, experimented with amounts of 150 to 400 g of CS_2 per vine applied about 80 cm deep, thus placing the CS_2 below the root zone. After 8 days all aphids were destroyed and the vines regained health within a few weeks. The method, though not applicable under all soil conditions, was hailed enthusiastically by the scientific community (Bazille, 1873; Dumas, 1873; Monestier, Lautaud, and d'Ortoman, 1873), and gave a new impulse to the idea of soil fumigation.

Based on Dumas' findings on the effects of low doses (Dumas, 1874a), Alliès in 1876 constructed an apparatus to deliver small amounts of the fumigant and injected 30 g of CS_2 per vine in 4 holes at 40 cm from the trunk. Beginning in May 1875, he treated the same vines five times at monthly intervals and again three times during the following summer. Vines were rendered free from Phylloxera; they also responded with an actively regenerating root system and abundant new growth (Fig. 3). A lasting control had been achieved economically through repeated applications of low doses made possible by employment of a new tool (Alliès, 1876a; 1876b; Marion, 1876). Soon there were several models of soil fumigation guns; the one most used was the injector perfected by Gastine. In designing his "pal distributeur" he realized that the requirements of precise dosage, technical reliability, and simplicity of handling were indispensable for the success of soil fumigation (Fig. 4). The injector basically consisted of a hydraulic pump and a retention valve, both contained in a tube which was pushed into the ground, and a reservoir for 4 kg of fumigant. The operator was able to deliver the quantities of 4 and 8 g of CS_2 at a soil depth of about 35 cm

Fig. 2. Baron Arnould Paul Edmond Thenard (1819–1884), the "father of soil fumigation" (Wilhelm, 1966). Son of the famous chemist Louis Jacques Thenard, he was a distinguished chemist in his own right. In a laboratory on the premises of the château and domaine of Talmay (Cote-d'Or) which he owned, Thenard conducted research on the chemistry, structure, and fertility of soils. In 1864 he was elected to the Académie des Sciences; he was also a member of the Phylloxera Commission of the Académie and vice-president of the national Commission Supérieure du Phylloxera. Engraving reproduced from *L'Illustration*, August 23, 1884, courtesy of Bibliothèque Municipale, Dijon.

Fig. 3. The salutary effect of soil fumigation with carbon disulfide on a Phylloxera-infested vineyard at left, in contrast to the devastations by the pest in an adjacent area not treated, as depicted in 1876 (Rohart, 1876).

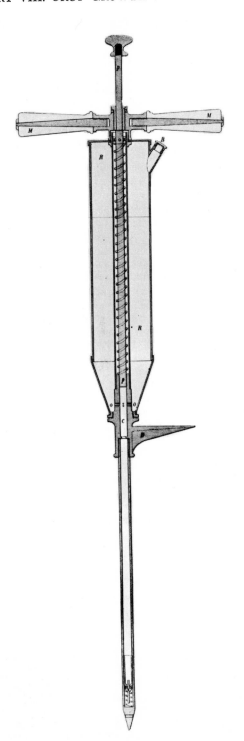

Fig. 5.   A CS$_2$ fumigation team, probably advisors from the Paris-Lyon-Mediterranean Railway Company, at work. Each operator of a Gastine injector was followed by another worker who, with a rod, immediately closed the hole left by the injector (Barral, 1882).

Paris à Lyon et à la Mediterranée became instrumental in the development of soil fumigation with CS$_2$. The railway company made available its large resources for the intensified battle against the Phylloxera and received wide recognition for this move. In 1877, with the cooperation of A. F. Marion, carbon disulfide manufacture was begun on a large scale. A 100-kg barrel was designed for the safe transport of the fumigant. Carbon disulfide sold for 40–50 francs per barrel and was shipped free to the subscribers of the rail company service (Barral, 1884). Before long, thousands of barrels were in circulation. The Gastine gun also went into commercial production and was offered at a reasonable price. By 1882, the Gastine injector was in demand by Phylloxera-infested countries such as Russia, Italy, Spain, and even America, the home of the Phylloxera (Barral, 1882). The P.-L.-M. Railway Company also supplied teams of experienced advisors to give instructions in the proper fumigation technique (Fig. 5; Anonymous, 1883). In 1884, of the total vineyard area in France treated against Phylloxera, over 33,000 ha (8,100 acres) or about 53% were fumigated with CS$_2$ (Tisserand, 1885).

There were reports of damage to vines and other accidents allegedly caused by CS$_2$, but these did not deter the use of soil fumigation. Many of the mishaps reported were due to failure to observe the recommended conditions for handling and applying the chemical (Jaussan, 1882). Vine growers who had become hostile to CS$_2$ were reconverted to its use by the evident general success of fumigation (Anonymous, 1883). It is ironical that Thenard's own vineyard master had to be visually confronted both with the disastrous potential of the Phylloxera and with the benefits derived from CS$_2$ soil fumigation before he was convinced of the necessity of fumigation in the vineyards of the Thenard estate. There are no records that Thenard himself expanded his first fumigation experiments,

Fig. 4.   *Injecteur à sulfure de carbone,* the fumigation gun designed by G. Gastine in 1876.

with a push on the protruding shaft of the piston (Gastine, 1876).

From 1876 the Compagnie des Chemins de Fer de

yet when challenged as to the value of CS_2 soil fumigation for Phylloxera control before an Academy session in 1879, he defended CS_2 and proceeded then to outline the details of the fumigation procedure, giving optimal numbers of injection holes per hectar, the cost of application, labor involved—details which only a man fully informed in the practice of soil fumigation could know. For his own vineyards he had placed an order with the P.-L.-M. Railway Company for 10,000 kg CS_2 for the year 1880 (Thenard, 1879).

The unexpected vigorous growth which had been noticed time after time on vines following soil fumigation with CS_2 was also observed in other crops. Oberlin was first to describe this surprising phenomenon. He conducted experiments on soils fumigated with CS_2, which he planted to various crops such as vetch and beans, oats, alfalfa, and clover. His results proved beyond doubt that the increased growth response observed was, in fact, due solely to the action of CS_2 in the soil (Oberlin, 1894). As the fight against the Phylloxera continued its course, the attention of scientists was soon focused upon this new puzzle. Numerous investigations into the biology of soil-borne organisms and their effects on plants followed. The phenomenon of soil fumigation growth response to this day has not been fully explained, yet the research it inaugurated has enriched our knowledge of soil microorganism-plant root interactions.

Initiated by a single idea and aimed at a single cause, soil fumigation now embraces fields of plant pathology, entomology, nematology, and weed control. Neither Thenard nor any of the men who cooperated to develop the practice of soil fumigation anticipated the many beneficial effects it brought, over and above the conquering of Phylloxera.

LITERATURE CITED

ALLIÈS, F. 1876a. Sur un procédé d'application directe du sulfure de carbone dans le traitement des vignes phylloxérées. Compt. Rend. 82:612–615.

ALLIÈS, F. 1876b. Nouvelle note concernant les résultats obtenus par le traitement des vignes phylloxérées, au moyen de sulfure de carbone; emploi du nouveau pal distributeur. Compt. Rend. 83:1222–1224.

ANONYMOUS. 1881. Commission Supérieure du Phylloxera. Rapport sur les travaux administratifs entrepris contre le Phylloxera et sur la situation du vignoble français pendant l'année 1880. Paris. 23 p.

ANONYMOUS. 1883. Chemins de Fer de Paris à Lyon et à la Méditerranée. Service spécial pour combattre le Phylloxera. Rapport sur les travaux effectués par ce service pendant la campagne de 1882. Marseille. 105 p.

BARRAL, J. A. 1882. Conférence sur le Phylloxera. IV. J. Agr. 17:300–303.

BARRAL, J. A. 1884. Chronique agricole. VI. Le phylloxera. J. Agr. 19:9–10.

BAZILLE, G. 1873. Destruction du Phylloxera. J. Agr. Prat. 37(2):261–263.

BORDE DE TEMPEST, E. 1873. Le Phylloxera. J. Agr. Prat. 37(1):43–48.

CÉRIS, A. DE. 1869. Chronique agricole. J. Agr. Prat. 33(2):315–319.

CORNU, M. 1879. Études sur le *Phylloxera vastatrix*. Mém. Acad. Sci. 26:1–357.

DUMAS, J. B. 1873. (A letter to the Academy answering a note from Mr. Lichtenstein). Compt. Rend. 77:520–522.

DUMAS, J. B. 1874a. Mémoire sur les moyens de combattre l'invasion du Phylloxera. Compt. Rend. 78:1609–1618.

DUMAS, J. B. 1874b. Note sur les sulfocarbonates, p. 40–45. *In* Institut de France, Académie des Sciences, Commission du Phylloxera. Séance du 3 Décembre 1874. Paris. 45 p.

FAUCON, L. 1870. Maladie de la vigne. J. Agr. Prat. 34(2):131–134.

GASTINE, G. 1876. Note sur l'injecteur à sulfure de carbone, p. 32–35. *In* A. F. Marion, Expériences faites pour combattre le Phylloxera. Marseille. 37 p.

JAUSSAN, L. 1882. Rehabilitation du sulfure de carbone. J. Agr. Prat. 46(2):772–778.

MARION, A. F. 1876. Expériences faites pour combattre le Phylloxera. Rapport du Comité Regional institué à Marseille par la Compagnie des Chemins de Fer de Paris à Lyon et à la Méditerranée. Marseille. 37 p.

MONESTIER, C., LAUTAUD, and D'ORTOMAN. 1873. Exposé des mesures que doivent prendre les viticulteurs pour détruire le Phylloxera d'après le système imaginé par M. Monestier et mis en pratique par MM. Lautaud, d'Ortoman et Monestier. J. Agr. Prat. 37(2):263–264.

MOUILLEFERT, P. 1874. Résumé des expériences faites à la Station Viticole de Cognac dans le but de combattre le Phylloxera, p. 35–39. *In* Institut de France, Académie des Sciences, Commission du Phylloxera. Séance du 3 Décembre 1874. Paris. 45 p.

MOUILLEFERT, P. 1877. Le Phylloxera. Expériences du Comité de Cognac. Solution pratique de la guérison des vignes phylloxérées par les sulfocarbonates alcalins, système P. Mouillefert et Felix Hembert. Paris. 80 p.

MOUILLEFERT, P. 1883. Traitement des vignes phylloxérées, par le sulfocarbonate de potassium, en 1882. Compt. Rend. 96:180–181.

OBERLIN, C. 1894. Bodenmüdigkeit und Schwefelkohlenstoff mit besonderer Berücksichtigung der Rebenverjüngung ohne Brache oder ohne Zwischenkultur. Mainz. 19 p.

PLANCHON, J. E. 1868. Nouvelles observations sur le puceron de la vigne (*Phylloxera vastatrix* [nuper *Rhizaphis*, Planch.]). Compt. Rend. 67:588–594.

ROHART, F. 1876. Où en est la question Phylloxera. J. Agr. Prat. 40(2):721–726.

ROMMIER, A. 1876. Les divers procédés essayés jusqu'à présent pour combattre le Phylloxera. J. Agr. Prat. 40(2):762–766, 856–857.

SAGNIER, H. 1892. Phylloxera, p. 159–169. *In* J. A. Barral and H. Sagnier. Dictionnaire d'Agriculture. Paris.

THENARD, P. 1869. A Monsieur le Vicomte de la Loyère. Paris. 7 p.

THENARD, P. 1874. (Observations presented on the subject of a communication of Mr. Dumas regarding means against the Phylloxera). Compt. Rend. 78:1619–1620.

THENARD, P. 1879. Réponse aux questions de M. Fremy relatives à l'emploi du sulfure de carbone appliqué à la destruction du Phylloxera. Compt. Rend. 89:926–931.

TISSERAND, E. 1885. Situation des vignobles phylloxérés. J. Agr. Prat. 49(1):640–642, 663–668.

VIALLA, L. 1869. Rapport de la commission nommée par la Société des Agriculteurs de France pour étudier la nouvelle maladie de la vigne. J. Agr. Prat. 33(2):598–602, 627–636.

WILHELM, S. 1966. Chemical treatments and inoculum potential of soil. Ann. Rev. Phytopathol. 4:53–78.

A Concept of Rootlet Health of Strawberries in Pathogen-free Field Soil Achieved by Fumigation

STEPHEN WILHELM and PAUL E. NELSON—*Professors, Departments of Plant Pathology, University of California, Berkeley, and The Pennsylvania State Uuniversity, University Park.*

▶

INTRODUCTION.—Criteria by which health and disease are distinguished in rootlet systems of perennial plants, exemplified by the strawberry, are poorly defined, and few studies on root diseases have taken into account the phenomenon of rootlet exchange, which we see as having general significance. Recognized first by Jones (1943) in studies on alfalfa, Rogers (1939) on apple, Nelson and Wilhelm (1957) on strawberry, and Newhook (1960) on conifers, rootlet exchange means that, after growth to maturity and function for a relatively short time, rootlets become senescent and die. Under favorable conditions, they are replaced. Replacement occurs at the site where the dead rootlet was attached to the supporting root. The vigor of the shoot system, amount of crop produced, and the soil environment affect the exchange, of which there are qualitative and quantitative aspects. Qualitative aspects include details of rootlet cell-wall composition and structure; quantitative aspects include the numbers of rootlets, their longevity, and their total absorbing area.

Soil fumigation, now used widely in California by strawberry growers for the control of root-infecting fungus pathogens (Wilhelm, 1962; Wilhelm, Storkan, and Sagen, 1961) has provided opportunities to study strawberry-root systems of healthy plants growing under field conditions in soil essentially free from pathogens. Soil fumigation is custom-applied by specialists and, when fungus pathogens are known to be present, consists of the application of a mixture of equal parts by weight of chloropicrin and methyl bromide at rates from 275 to 400 lb/acre. Thirteen chisels equally spaced on a tractor tool-bar net an application width of twelve feet (Figs. 1, 2). Simultaneous with injection of the chemicals, soil is covered with sheets of thin polyethylene glued together to form an uninterrupted cover (Figs. 2, 3, 4, 5). The polyethylene cover remains in place two days before removal.

Strawberries develop extensively branched, fibrous root systems and yield heavily in fumigated soils where prior to fumigation cultivation was unprofitable because of root diseases. These salutary effects of fumigation, observed in numerous fields of many soil types, convinced the authors that, in general, living pathogens and not soil chemical or physical factors determined the suitability of soils for strawberries. They also made it evident that development of a true concept of rootlet health was not possible if it was based only on studies conducted in field soils that were not treated to kill pathogens. We have come to view productiveness of strawberries and fruit size and quality as being determined in large measure by the state of health of the feeder or absorbing rootlet system, by its longevity and capacity to grow, and by the ability of the plant to rapidly replace the rootlets once they have died.

LITERATURE REVIEW.—Early investigators did not neglect roots. For instance, the Royal Horticultural Society of London, in 1886, awarded a prize to Sewell (1886) for an excellent essay "Roots and their work" and Freidenfelt (1902) cited 504 papers on roots published before 1900. Many of the papers dealt with description and classification of root-system types. There were Ernahrungswurzeln (feeder rootlets), Befestigungswurzeln (supporting roots), Zugwurzeln (anchoring roots), Speicherwurzeln (storage roots), and others, and often disagreement among investigators as to which was which (Rimback, 1899; Tschirch, 1905). In general a dimorphism in the gross structural aspects of roots was recognized in perennial plants. Dark, hard, woody roots were more obviously supporting in function, and white, softer roots absorbing. However, there were no references to rootlets as being noncambial or deciduous in nature, nor to their replacement or exchange during the growing season. This knowledge brings us well into present times.

Mann and Ball (1927) reported that second- and third-order branch rootlets of strawberries frequently died back to the supporting root, which was perennial. Later Mann reported that fine lateral rootlets were short-lived and that the entire fibrous root system, formed from a branch root of the first order, may decay to the point of union with the branch root (Mann, 1930). No attempt was made in Mann's work to assess the role of pathogens in this decay. Nelson and Wilhelm (1957) reported that the fine lateral rootlets

Figs. 1–5. Fig. 1. Chisels on a tractor tool bar used in applying soil fumigants. Note the roll of polyethylene used to cover the soil after the fumigant is applied. Fig. 2. Front view of the tractor tool-bar showing the chisels inserted into the soil and the applicator (a) used to apply glue to the polyethelene covering previously fumigated soil. Fig. 3. Detail of the glue applicator. Fig. 4. Rear view of the fumigation rig showing the polyethylene being unrolled over the fumigated soil and a rubber wheel pressing the glue joint together. Fig. 5. View of fumigation rig and field showing the extent of the area covered by polyethylene. Courtesy of R. C. Storkan, Pres. Tri-Cal Inc.

of strawberry plants growing in soil under as nearly ideal conditions as could be provided in the field, were transient, though the main root system was perennial. The transient rootlets were composed almost entirely of primary tissues, and new rootlets were produced from sites on the perennial roots where the previous rootlet died. Thus, they presented evidence that the strawberry rootlet, after a variable but comparatively short existence, dies and is replaced by a new one, initiated at the same site. During a year several crops of new rootlets may have formed and died.

Rootlets of fruit trees were reported to die within a few weeks after their formation and to be replaced. Prior to their death suberization occurred (Childers and White, 1942; Kinman, 1932; Kolesnikov, 1930). Newhook (1960) reported that rootlet replacement balanced rootlet death in the root systems of conifer trees infected by *Phytophthora* sp.

NATURE OF THE STUDY AND METHODS.—In the present study, anatomical changes accompanying rootlet dying, here called nonpathological dying, were studied in an effort to learn the pattern, if any, by which tissues became senescent. An attempt was also made to correlate rootlet dying with seasonal phases of vegetative growth of the strawberry. This, however, could not be done in detail. During the study a root pathogen—*Ceratobasidium* sp.— was occasionally encountered which hastened and intensified rootlet dying, but rootlet dying reported herein occurred apart from all recognized forms of pathogenesis. The pathological effects of *Ceratobasidium* have been studied and are being published elsewhere.

The investigation was conducted mainly on the Shasta and Goldsmith cultivars of the garden strawberry, *Fragaria chiloensis* × *Fragaria virginiana*. The plants studied were from commercial fields located in the Santa Clara and Pajaro valleys of California. Previous to planting the crop, the soil had been fumigated with 375 lb/acre of 57–43% mixture of chloropicrin and methyl bromide and covered for two days with thin polyethylene sheeting. The planting was uniform in growth, vigorous, and highly productive. Collections were made by carefully digging entire root systems of plants and washing root fascicles free of soil with a fine spray of water.

Root material collected was fixed in either Rawlins (1933) alcohol-formalin-acetic acid fixing solution no. 1 or alcohol-formalin-propionic acid fixing solution (Johansen, 1940) for 48 hr or longer. The fixed material was dehydrated by the tertiary butyl alcohol schedule of Johansen (1940) and imbedded in Tissuemat held at 55°–57°C. When necessary, root material was softened prior to sectioning in a solution consisting of 90 ml of a 1% aqueous solution of Dreft (sodium lauryl sulfate) and 10 ml of glycerol (Alcorn and Ark, 1953). Material was sectioned on a rotary microtome at 10 μ, and sections were mounted on slides with Haupts' adhesive (Johansen, 1940). Sections were stained with Johansen's quadruple stain (Johansen, 1940).

Microchemical tests for suberin and fats used Sudan IV dissolved in 95% alcohol and glycerine as described by Rawlins and Takahashi (1952) and triethyl phosphate as described by Jensen (1962). Root material for these tests was fixed in both the formal-calcium fixative and in Lewitsky's fluid (Jensen, 1962) and dehydrated, imbedded, and sectioned as above.

MACRO- AND MICRO-ANATOMY OF THE STRAWBERRY ROOT SYSTEM.—Root growth during the first year of a strawberry plant set in the fall consisted of extensive branching and rebranching from roots existing when the plant was set, in production of new adventitious roots from the crown and of extensive rootlet production from these. The terminal branches are referred to as the feeder rootlets. Secondary tissues derived from vascular and cork cambiums develop in the older adventitious roots and their main branches to form the perennial portion of the root system, called herein the structural system. The anatomical details of this development have already been described (Nelson and Wilhelm, 1957).

Main supporting roots develop vascular and cork cambiums, and though perennial, usually do not live more than three years under field conditions. During the third year, the oldest ray cells of the wood lose their ability to store starch (Figs. 6, 7), and saprophytes from the soil have invaded the vessel elements. Cultures made from root heartwood have revealed the presence of a number of different kinds of fungi such as *Fusarium oxysporum*, *F. roseum*, *Cylindrocarpon* sp., *Phoma* sp., Basidiomycetes of the *Rhizoctonia-Ceratobasidium* complex, *Streptomyces*, and bacterial species. Most of these organisms are saprophytes which gained entrance to the wood of the supporting roots through vascular connections of dead or senescent feeder rootlets (Fig. 8). Once within the heartwood, though occupying dead tissues, saprophytes by toxin, or by enzyme production and tissue disintegration, may be pathogenic and may contribute to the eventual death of the supporting root. Whatever the cause, ulimately the supporting root dies. These conclusions are based on several years of field observations and laboratory studies.

In California, rootlet growth is most extensive during the winter and spring. A well-established rootlet system supports the plant during the period of spring growth and fruit formation. Within a fascicle of rootlets on plants established in the field for one year or more, rootlet branching may occur to the fourth or fifth tier (Fig. 9). During the period of spring crop production and into the summer, extensive non-pathogen-induced rootlet dying occurs. By fall, the rootlets which supported the plant in spring and early summer have been replaced by a new but less extensively developed system.

In the process of rootlet exchange occurring within a single fascicle of rootlets, a sequence of anatomical changes occurs. The first evidence of rootlet senescence is the collapse of the single outermost tier of cells that meets the soil (Fig. 10). This collapse occurs shortly after the metaxylem has differentiated. The

Figs. 6–9. Fig. 6. Portion of a cross section of a two-year-old strawberry root stained with I-KI solution showing starch storage in the polyderm and xylem ray cells (× 134). Fig. 7. Portion of a cross section of a three-year-old strawberry root stained with I-KI solution showing limited starch storage in the xylem ray cells (× 145). Fig. 8. Lateral rootlet from the root system of a strawberry plant taken directly from soil, showing fungus hyphae growing out from areas where branch rootlets have died. Fig. 9. A fascicle of rootlets formed on one of the main strawberry roots. Note the extensive branching of rootlets in the fascicle.

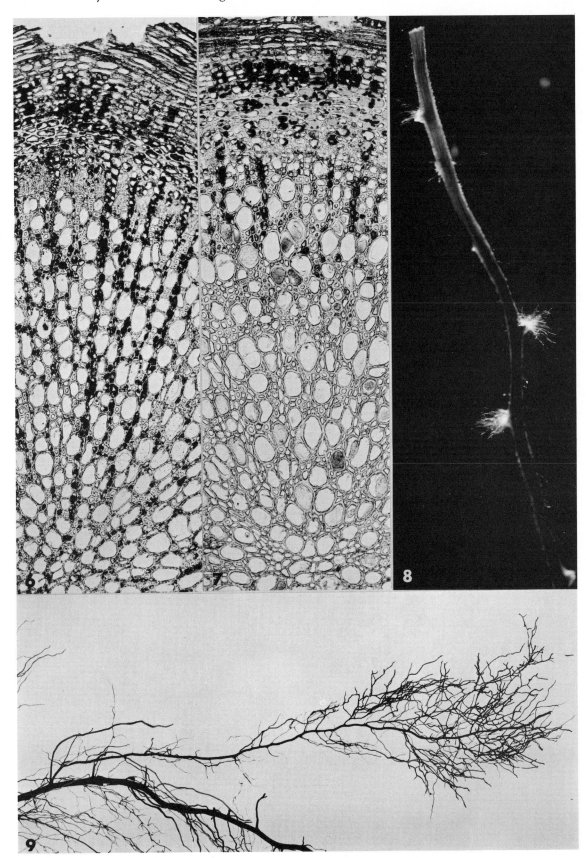

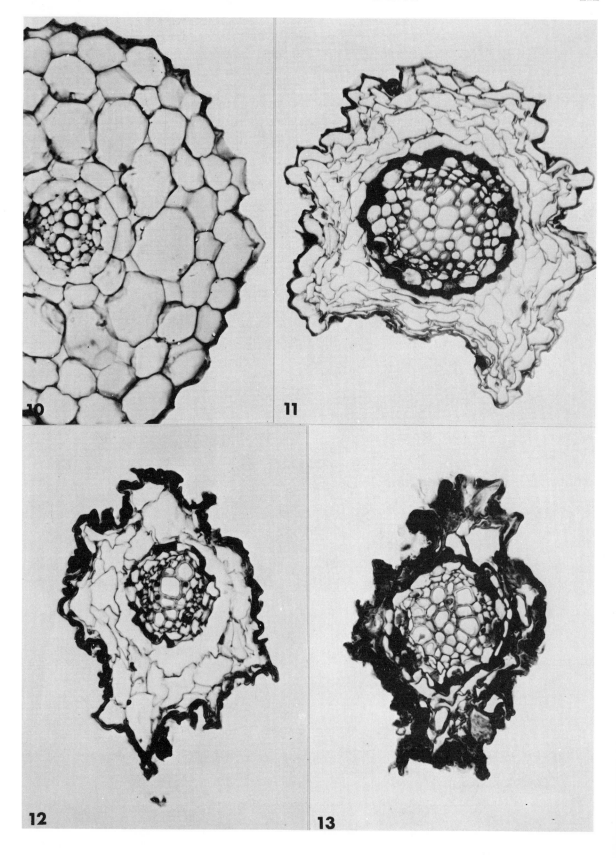

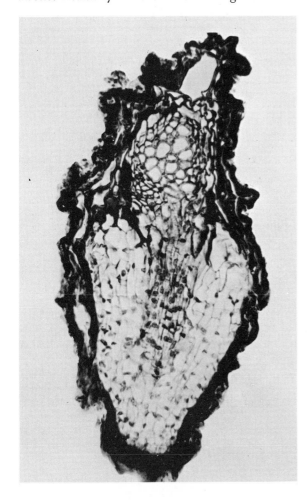

Fig. 14. Cross section of a senescent strawberry rootlet showing the development of a branch rootlet (× 285).

collapsed cell-tier contains suberin and forms a protective cuticle-layer around the rootlets. Next, cortical cells, except the endodermal tier collapse, become suberized and form a bark-like layer which encloses the stele (Figs. 11, 12, 13). As if to insure further protection, cells of the pericycle may divide to form a thin-layered polyderm. A vascular cambium does not differentiate in these rootlets. Collapsed rootlets, devoid of living cortex, are found in greatest abundance during midsummer and may be quiescent in the soil for periods of various durations. During this time of summer slump in new rootlet formation, fruit production and plant vigor decline. In late summer and extending into the fall, the quiescent rootlet may resume growth either by extension of length or by branching (Fig. 14). Ultimately, however, the rootlet

dies, as do all rootlets of the fascicle, and a new fascicle is or may be initiated (Fig. 15).

DISCUSSION AND SUMMARY.—Rootlets of the strawberry plant after maturation pass into senescence and death through a sequence of changes in form. The changes also undoubtedly affect their function. Health and productiveness of the plant depend upon the ability of the root system to maintain a favorable equilibrium between rates of death and rates of replacement of the rootlets. Many factors internal to the plant such as foods available from the shoot, and of the external environment such as soil conditions of moisture and aeration, affect the equilibrium. Root-infecting fungus pathogens may drastically upset the equilibrium, but they were not the subject of this study and had been controlled by preplant soil fumigation, though saprophytes were found to occupy the heartwood of the oldest supporting roots. Invasion of old xylem tissues by saprophytes probably limits the life-span of the supporting perennial root.

The fact that the epidermal tier of cells of strawberry rootlets collapses to form a suberized layer around the mature portion of the rootlet raises questions about the function of absorption. Absorption of water through suberized tree roots (Addoms, 1946; Hayward, Blair, and Skaling, 1942) and through dead roots has been demonstrated (Kramer, 1933; Kramer, 1946), but mineral absorption is probably a different matter. In only immature portions of rootlets, such as their tips or portions just mature, do cells not covered by the collapsed suberized tier contact the soil. If nonsuberized cells only are able to absorb nutrients from the soil, then it follows that the rootlet must be actively growing to absorb nutrients.

Prior to the present practice of soil fumigation with mixtures of chloropicrin and methyl bromide in California, strawberries responded little or not at all to inorganic fertilizer applications, and growth except on "new land" was often poor. Often on the richest soils productivity was most uncertain, and cultivation of the everbearing types, i.e. first-year bearers, was unsuccessful. Perhaps water, but little mineral, absorption seemed possible through the dark, senescent, and dead rootlets.

In soils rendered free of pathogens by fumigation, strawberries develop vast and extensively branched rootlet systems and respond promptly and noticeably to select fertilizer applications, but for maximum productivity they must be kept heavily irrigated. Collapse of the epidermal-cell tier of rootlets, followed by suberization, undoubtedly reduces the capacity of the rootlet to absorb water. Taken in the aggregate, considering the entire rootlet system and its organization into fascicles, the process of rootlet exchange is exceedingly complex, and rootlet death and replacement

Figs. 10–13. Cross sections of strawberry rootlets showing various stages in rootlet senescence. Fig. 10. Early stage as indicated by collapse of some cells in the outermost cell layer (× 545). Fig. 11. Collapse of many cortical cells and some suberization of the outer cell layer (× 370). Fig. 12. Collapse of most cortical cells and heavy suberization of outer cortical cells (× 500). Fig. 13. Collapsed suberized cortex surrounding the intact stele (× 500).

are the rule of health. Likewise, pathological necrosis of rootlets must be distinguished from the non-pathological.

LITERATURE CITED

ADDOMS, RUTH M. 1946. Entrance of water into suberized roots of trees. Plant Physiol. 21:109–111.

ALCORN, S. M., and P. A. ARK. 1953. Softening paraffin-embedded plant tissues. Stain Technol. 28:55–56.

CHILDERS, N. F., and D. G. WHITE. 1942. Influence of submersion of the roots on transpiration, apparent photosynthesis, and respiration of young apple trees. Plant Physiol. 17:603–618.

FREIDENFELT, T. 1902. Studien über die Wurzeln krautiger Pflanzen. Flora 91:115–208.

HAYWARD, H. E., WINIFRED M. BLAIR, and P. E. SKALING. 1942. Device for measuring entry of water into roots. Bot. Gaz. 104:152–160.

JENSEN, W. A. 1962. Botanical histochemistry. W. H. Freeman and Co., San Francisco. 408 p.

JOHANSEN, D. A. 1940. Plant microtechnique. McGraw-Hill, New York. 523 p.

JONES, F. R. 1943. Growth and decay of the transient (noncambial) roots of alfalfa. J. Am. Soc. Agron. 35:625–634.

KINMAN, C. F. 1932. A preliminary report on root growth studies with some orchard trees. Proc. Am. Soc. Hort. Sci. 29:220–224.

KOLESNIKOV, V. A. 1930. The dying off of rootlets of fruit trees. J. Pomol. Hort. Sci. 8:204–209.

KRAMER, P. J. 1933. The intake of water through dead root systems and its relation to the problem of absorption by transpiring plants. Am. J. Botany 20:481–492.

KRAMER, P. J. 1946. Absorption of water through suberized roots of trees. Plant Physiol. 21:37–41.

MANN, C. E. T. 1930. Studies in the root and shoot growth of the strawberry. V. The origin, development, and function of the roots of the cultivated strawberry (*Fragaria virginiana* x *chiloensis*). Ann. Botany 44:55–86.

MANN, C. E. T., and E. BALL. 1927. Studies in the root and shoot development of the strawberry. II. Normal development in the second year. J. Pomol. Hort. Sci. 6:81–86.

NELSON, P. E., and S. WILHELM. 1957. Some anatomic aspects of the strawberry root. Hilgardia 26:631–642.

NEWHOOK, F. J. 1960. Some comments on the relationship of climate to *Phytophthora* infection of *Pinus radiata*. N. Z. J. Forest. 8:261–265.

RAWLINS, T. E. 1933. Phytopathological and botanical research methods. John Wiley and Sons, New York. 156 p.

RAWLINS, T. E., and W. N. TAKAHASHI. 1952. Technics of plant histochemistry and virology. The National Press, Millbrae, California. 125 p.

RIMBACK, A. 1899. Beiträge zur Physiologie der Wurzeln. Berichte der Deutschen Botan. Gesellsch. 17:18–35.

ROGERS, W. S. 1939. Root studies. VIII. Apple root growth in relation to rootstock, soil, seasonal and climatic factors. J. Pomol. Hort. Sci. 17:99–130.

SEWELL, P. 1886. Roots and their work. Gardener's Chron. 25(II):197–199; 235–237.

TSCHIRCH, A. 1905. Uber die Heterorhizie bei Dikotylen. Flora 94:68–78.

WILHELM, S. 1962. The control of Verticillium wilt of strawberry and of weeds by preplant soil fumigation with chloropicrin and chloropicrin-methyl bromide mixtures. 16th Int. Hort. Cong. 3:263–269.

WILHELM, S., R. C. STORKAN, and J. E. SAGEN. 1961. Verticillium wilt of strawberry controlled by fumigation of soil with chloropicrin and chloropicrin-methyl bromide mixtures. Phytopathology 51:744–748.

Fig. 15. Portion of the root system of a strawberry plant showing a dead rootlet fascicle (left) and a living rootlet fascicle (right). Note the initiation of a new rootlet just above the living rootlet fascicle (right).

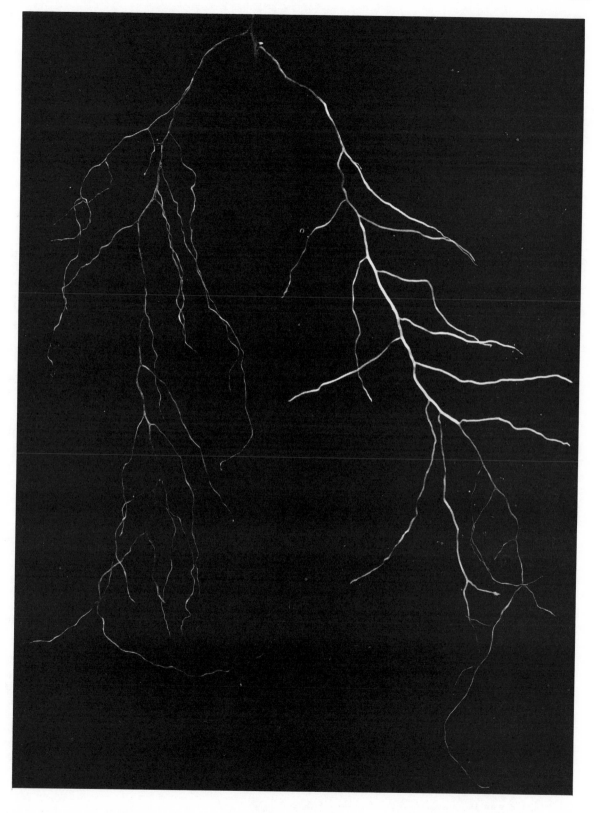

Increased and Decreased Plant Growth Responses Resulting from Soil Fumigation[1]

JACK ALTMAN—*Department of Botany and Plant Pathology, Colorado State University, Fort Collins.*

In recent years, investigators have been examining the nature and fate of chemicals applied to soils, plants, and animals. Studies have been limited to the degradation of soil chemicals by microorganisms, the physical-chemical binding processes of various types of soils, and direct chemical or photo-oxidation of soil chemicals. Wilhelm (1966) and Kreutzer (1965) have reviewed chemical treatments of soil in regards to inoculum potential and ecology.

In most of the earlier research, studies were restricted to herbicides or insecticides. Lorenzoni (1962) reported the stimulatory effect of dilute Simazine on plant growth and seed germination, whereas Bollen, Morrison, and Crowell (1954) reported inhibitory effects due to insecticides. Information regarding nematocides and soil fumigants has not been as prevalent.

Fumigants like carbon disulphide, formaldehyde, chloropicrin, or methyl bromide can partially sterilize soil; they reduce numbers of microorganisms temporarily, and the new population of microorganisms sometimes differs from the original, although a normal one is eventually restored. The study of interactions between microorganisms and these substances that result in stimulation or inhibition of soil microbiological processes is therefore more complicated. Fungi are generally more susceptible to fumigants than bacteria, although there are resistant species of both. In most instances, beneficial responses occur (Altman, 1965, 1966; Wilhelm, 1966). However, in some isolated instances, the alteration and subsequent establishment of a new population results in detrimental interactions (Tam, 1945; Kreutzer, 1965). This paper is an attempt to compare some of the growth responses that result from soil fumigation.

INCREASED GROWTH RESPONSE.—Hiltner and Störmer (1903) and Störmer (1907) showed that many toxic compounds such as carbolic acid, cresol, toluol, and xylol, deadly to plants on direct contact, stimulated plant growth if applied to soils before planting. The action of these chemicals in giving yield increases did not depend on bacterial metabolism, because other compounds such as carbon tetrachloride, chloroform, benzol, and CS_2, which bacteria could not metabolize, also gave salutary effects.

Moritz and Sherpe (1905) worked on the hypothesis that oxidation of CS_2 in soils to yield carbonic and sulfuric acids augmented availability of minerals. It is questionable whether their extensive data confirmed the hypothesis, but the growth response of potato and oats following fumigation of field soil was undisputed. They rejected the oligodynamic principle that traces of chemicals poisonous in large quantities were stimulatory, and they suggested that research be directed to the biology of soil microorganisms.

Starkey (1955), Waksman (1932), and ZoBell (1964a) reported that organic pesticides may influence the activities of soil microbes by serving as carbon and energy sources for the selective growth and multiplication of specific organisms. Starkey (1955) and ZoBell (1946a,b) concluded that any organic compound could be decomposed by one or more soil microorganisms. Waksman (1932) had earlier suggested that compounds, such as lignin, containing aromatic ring structures are more resistant and hence are decomposed at a lower rate.

Martin (1963) reported that the growth of microorganisms and the disappearance of compounds in enrichment-culture studies demonstrated the ability of the microorganisms to utilize these compounds. ZoBell (1946a,b) stated that nonspecific organic compounds such as crude oils, illuminating gases, petroleum ethers, gasolines, fuel oils, paraffin oil, asphalts, waxes, and natural and synthetic rubbers were readily degraded and in many instances resulted in increased microbial growth.

Altman and Lawlor (1966) reported the effects of chlorinated hydrocarbons, including D-D (a chlorinated propane-propylene mixture) and fractions of D-D on the growth of certain indigenous soil-borne bacteria. These compounds, in low concentrations, could also be utilized by the indigenous soil microflora. Waksman (1932) and Alexander (1961) made similar observations. They proposed that the indigenous microflora could then exert a direct or an indirect stimulatory effect on plants grown in the treated soil.

Effects of soil fumigation.—Nitrogenous Responses. —Nitrification processes in soil were stressed in the

[1] Published with the approval of the Director of the Colorado Experiment Station as Scientific Journal Series Paper No. 1335.

216

late nineteenth century with the assumption that total plant growth was a reliable indication of increased microbial activity. Warington (1878) showed that soil possessed the capability of converting ammonia-N to nitrate-N and that CS_2 and other chemical fumigants stopped the process. That soil fumigation for the control of fungi and nematodes results in a temporary disruption of nitrification and an increase of ammonia-N has also been demonstrated (Waksman and Starkey, 1923; Newhall, 1955; Tam, 1945; Warcup, 1957).

Tam (1945) made one of the initial observations regarding nitrification inhibition following soil chemical treatment with halogenated hydrocarbons. This inhibition was shown to persist for 8–23 weeks, depending upon the concentration of chemical applied to the soil. Bollen, Morrison, and Crowell (1954) have reported similar effects with soil insecticides.

Apart from inhibition of nitrification, most researchers have noted a stimulation of plant growth following soil treatment. Lorenzoni (1962) reported the effect of dilute Simazine in stimulating plant growth and increasing seed germination. The author's experiments (Altman, 1964; Altman and Tsue, 1965) indicate that there is an increase in ammonium, nitrate, and some amino acids following soil fumigation. Sugar beets grown in fumigated soil showed increases in glutamic and aspartic acid in their roots (Altman, 1963). These nitrogenous compounds could account, in part, for stimulated plant growth. Increased growth response may also result from increased nitrogen fixation due to partial sterilization. Partial sterilization of soil is believed to render a greater amount of energy available to the nitrogen-fixing bacteria. Koch (1899, 1911) stated that nitrogen fixation by bacteria is decreased by partial sterilization. The author has found that although this may occur, the ammonia level increased after partial sterilization, and this may be the cause for the increased growth response. McKee (1962) pointed out that plants do, in fact, utilize ammonia more readily than nitrates. These beneficial results also occurred with pineapple (Tam and Clark, 1943). The increased growth response may have been related to the greater availability of nitrogen or some nitrogenous component for plant growth.

In studies on increased plant vigor following soil fumigation with D-D (Altman, 1964; Altman and Tsue, 1965), there was an increase in ammonia for 7 weeks following soil treatment, and this was maintained at a level 25% greater than the control for the remaining 18 weeks. Comparisons with other physical and chemical soil treatments indicated that increased nitrogen could account for the stimulatory effect. The stimulatory effect was observed in the absence of plant pathogens and in nonamended fumigated soil. Chromatographic analyses of extracts from raw, treated soil showed an increase in glutamic and aspartic acids. This increase was detectable 2 weeks after treatment. In addition, a saprophytic bacterium was isolated from this treated soil which produced a stimulatory growth effect on sugar beets in laboratory tests.

In field tests and in greenhouse tests, there was a consistent slight depression of sucrose in sugar beets grown in fumigated soil, but beet tonnage and, therefore, net sucrose was increased (Altman and Fitzgerald, 1960).

Schmehl, Finkner, and Swink (1963) and Gardner and Robertson (1942) showed that increased nitrogen fertilization resulted in a depression of sucrose. They concluded that when nitrogenous fertilizers were made available late in the season, plants continued to grow and carbohydrates were utilized for growth rather than for storage, with the net effect of a lower sucrose percentage in the beet root at harvest. This implied that although nitrification was inhibited, some form of nitrogen was made more available to the plant.

Since fumigation with D-D produced a similar effect, Altman (1963, 1965) and Altman and Tsue (1965) suggested that one mechanism for stimulated seedling germination and increased growth response was biological and that the surviving soil microbial population may have had the ability to make nitrogen more available for plant growth. The possibility also exists that the fumigant was broken down, absorbed into the host, thus resulting in the growth that was observed. Another possible explanation is that the fumigant or one of its by-products may have acted as an auxin or kinin-like material and thus stimulated seed germination.

Microbial-ecological changes.—Altman and Tsue (1965) observed a stimulation of growth of plants grown in D-D treated soil which was analogous to a stimulation resulting from a side dressing of fertilizer. Laboratory tests showed a buildup of nitrogenous compounds in the soil following fumigation. An increase in growth of indigenous, saprophytic microorganisms (*Pseudomonas* spp. and *Arthrobacter* spp.) was also observed. This effect of plant stimulation was maintained for 3 years when fumigants were placed in the row under conventional agricultural practices. Similar results have been obtained with Telone (Altman, unpublished).

Altman (1966) suggested that it was unlikely that any chlorinated hydrocarbon with a low boiling point could last longer than 48 hours at $10°–22°C$. He felt that stimulation of the growth effect was a result of an ecological change in the soil microbiological population, and suggested that initially, many of the soil parasites and many of the non-spore-forming soil saprophytes were killed or injured by the soil fumigants, and in their place the surviving spore-forming microflora developed in such great numbers that they prevented the regrowth and reentry of the detrimental microorganisms into the soil biosphere. He felt that these indigenous saprophytic organisms, once established, became the "new" soil microflora population, and these organisms contributed to the stimulatory effects of soil treatment that could be observed for one year or longer.

Furthermore, the microbial population of the soil, if undisturbed, maintained a dynamic equilibrium once it had reached its climax state. A scheme modified

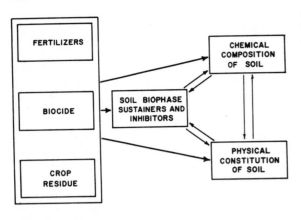

Fig. 1. Effects of various amendments on the soil complex.

after Kreutzer (1960) illustrating microbial changes in the soil resulting from various soil treatments is presented in Fig. 1.

This scheme implies that any treatment applied to the soil, including biocides, may affect microbial sustainers or inhibitors. For example, a fumigant introduced into the soil selectively destroys the most vulnerable saprophytic fungi or bacteria. Survivors include *Trichoderma*, which in the partial vacuum of the biosphere can grow uninhibited and become directly antagonistic to a pathogen such as *Rhizoctonia*. Altman and Lawlor (1966) have shown that *Arthrobacter globiformis* grows readily in the presence of various concentrations of fumigants and can provide a nitrogen component available for better plant growth. Furthermore, Altman and Tsue (1965) have isolated a *Pseudomonas* from fumigated soil that was capable of stimulating plant growth *in vivo*. It is quite possible that any augmentation, be it by fertilizers, biocides, or crop residues, can trigger such a biological mechanism.

Garrett (1965), in a formal definition of biological control, proposed both naturally occurring and artificially contrived biological control. He stated: "Biological control of plant disease may be defined as any condition under which, or practice whereby, survival or activity of a pathogen is reduced through the agency of any other living organism (except man himself), with the result that there is a reduction in incidence of the disease caused by the pathogen. Biological control can be brought about either by introduction or by augmentation in numbers of one or more species of controlling organisms, or by a change in environmental conditions designed to favor the multiplication and activity of such organisms, or by a combination of both procedures." The use of fumigants, resulting in partial sterilization, causes microbial-ecological changes, and the subsequently established soil saprophytic microflora can function in biological control.

Oberlin (1894) and Girard (1894) discussed in-creased plant growth following soil fumigation with CS_2. Oberlin reported yield increase ranging from 50–100% in lettuce, cabbage, beets, cereal, alfalfa, clover, and beans grown in soil fumigated with carbon disulfide. Girard's papers followed four years of field experimentation on control of the sugar beet nematode by soil fumigation as practiced against *Phylloxera*. Wheat, potatoes, beets, and clover gave significantly greater yields in the soils fumigated with 300 g CS_2/m^2. In addition to the immediate favorable growth response, he noted that the salutary effects carried over into the second year, and suggested the possibility of direct CS_2 plant stimulation through the oligodynamic principle, viz. that traces of chemicals, poisonous in large quantities, were stimulatory. This idea originated from investigations by von Nageli (1893).

Experiments with D-D in Colorado have confirmed the observations of Oberlin and Girard. Altman (1965, 1966) observed that salutary effects carried over for three years. The oligodynamic principle did not apply in this research since the fumigant readily dissipated in 72 hours at 10°C. in soil. Altman suggested that a microbial-ecological response occurred and that the saprophytes belonging to the Corynebacteriaceae (Altman and Lawlor, 1966) and Pseudomoniaceae (Altman and Tsue, 1965) predominated. This research was carried out in the field and in the laboratory. *In vitro* studies using such bacteria isolated from fumigated soil showed that germination and growth of sugar beet seeds and seedlings were stimulated when they were inoculated with these organisms. Clark (personal communication) has pointed out that the dominant soil bacteria are genera that belong to the Corynebacteriaceae. This group of bacteria, according to Clark, is isolated 80% of the time from soil. *Arthrobacter globiformis* is an example of one such genus. Altman and Lawlor (1966) were able to show that *A. globiformis* could, in fact, grow profusely in the presence of chlorinated hydrocarbon fumigants.

Russell (1920) synthesized his long experience into three principles basic to an understanding of soil-plant relationships: (a) All of the complex and numerous populations of soil microorganisms are, in the final analysis, dependent upon the green plant; (b) partial sterilization of the soil by heat or by chemicals tends to preferentially reduce harmful components of the soil microflora; (c) simplification of the soil population enhances soil fertility. Russell's conclusion can also be interpreted by examining Fig. 1.

DECREASED GROWTH RESPONSE.—*Direct toxicity.*— Some of the changes in the biophase following fumigation are not beneficial. Soil fungicides such as allyl alcohol, formaldehyde, methyl bromide, and sodium N-methyl dithiocarbamate apparently have a deleterious effect on mycorrhizal fungi (Wilde and Peridsky, 1956; Hacskaylo and Palmer, 1957).

Martin, Baines, and Page (1963) showed that citrus plants were highly sensitive to a temporary toxic condition which sometimes follows partial sterilization of the soil and that the method of treating the soil, whether by dry heat, steam, or volatile fumigants or other types of chemicals, was not of primary significance. They concluded that any treatment which kills soil organisms could produce the temporary toxic condition. However, their illustration implied a varietal response, since of three varieties planted in methyl-bromide-treated soil, two were resistant and one was sensitive to the toxic condition.

Disease increase.—Increases in incidence of crown gall following preplanting fumigation treatments have been rather common, and most evidence suggests that a competition factor favoring the pathogen is involved (Deep and Young, 1965). Dickey (1962) showed that while some fumigants decreased the number of crown gall bacteria in the soil, all treatments left a sufficiently high number of these bacteria to allow high incidences of crown gall.

Soil fumigation with ethylene dibromide (EDB) may cause an increase of Verticillium wilt (McClellan, Wilhelm, and George, 1955).

The vacuum or boomerang effect.—Other unfavorable results of fumigation are those of disease accentuation and disease exchange. Collectively, this has been called the boomerang phenomenon (Kreutzer, 1960). Accentuation of disease can occur as a result of soil treatment for the control of soil inhabitants. He observed rapid reinvasion of soils treated with chloropicrin and chlorobromopropene by *Rhizoctonia solani* and species of *Pythium*. Gibson (1956) stated that disease accentuation from damping-off organisms resulted following the use of organic mercurials. Such effects are undoubtedly the result of the killing or inhibition of antagonists, permitting the reinvading pathogen to grow speedily through the soil without biological opposition. Smith (1939) found that *Trichoderma viride* increased in soil fumigated with chloropicrin. If however, *Pythium ultimum* was introduced into this treated soil, it gave greater damage than in untreated soils. Smith concluded that a likely reason for this effect was not only the direct removal of *Pythium* antagonists by the chemical, but the inhibitory effect of *Trichoderma viride* on surviving antagonists such as *Rhizoctonia*. Butler (1957) has shown that *R. solani* parasitizes species of *Pythium*. When both fungi were added simultaneously to steamed soil, *Rhizoctonia solani* had a definite antagonistic effect on *P. ultimum*.

In experiments with herbicides, Altman and Ross (1967) have shown that an effect analogous to the boomerang effect can occur. Sugar beets grown in preplant Tillam and Pyramin-treated fields were predisposed to more disease than comparable beets grown in nontreated soil. These authors suggested that the herbicide affected the host plant by maintaining it in a juvenescent state for a prolonged period (7–14 days) after emergence. These beets were then more susceptible to extremely low levels of *Rhizoctonia* inoculum present in the field. Laboratory and greenhouse studies confirm this hypothesis. The interactions discussed above are summarized in Fig. 2.

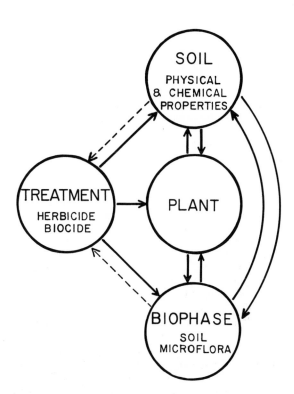

Fig. 2. The influence of biocides on soil-plant interactions.

Disease trading.—There are numerous cases of disease trading following soil treatment. Soil treatments with chlorobromopropene controlled bulb rot of iris, caused by *Sclerotium rolfsii*, but increased infection induced by bulb-borne *Fusarium* (Haasis, 1952). Fumigation of the soil apparently removed the natural antagonists of *Fusarium*.

Kreutzer (1960) stated that disease trading should be more marked when specific and non volatile fungicides are used. He frequently observed increased severity in attack of sugar beet seedlings by *Pythium ultimum* and *P. aphanidermatum* following soil treatment with pentachloronitrobenzene for the control of *Rhizoctonia*. Fulton, Waddle, and Thomas (1956) observed that although chloronitrobenzene eliminated *Rhizoctonia solani* and *Macrophomina phaseoli* in soils, it increased the incidence of disease in cotton seedlings, caused by *Fusarium moniliforme* and *Colletotrichum gossypii*. Organic mercurials can also produce a disease exchange. Gibson (1953) observed that the use of these fungicides decreased preemergence losses in beds of groundnut seedlings, but increased subsequent damage from crown rot, caused by *Aspergillus niger*. In this case, *A. niger* was apparently less susceptible to mercury poisoning than its antagonists.

Huber, Watson, and Steiner (1965) showed that Telone, a compound which inhibits nitrification in the soil, provided excellent control of Verticillium wilt of potatoes, but increased the severity of *Rhizoctonia*. The population of *Verticillium* remained the same or increased in Telone-fumigated soil (unpublished data), so fungitoxicity apparently was not the determining factor. Huber suggested that since the host plant was favored nutritionally in fumigated soil, it was more susceptible to *Rhizoctonia*. Furthermore, the increased level of ammonium-N favored the development of the virulence of the indigenous level of *Rhizoctonia* (Huber, 1968, personal communication).

SUMMARY.—Soil fumigation is one of the oldest and newest ways of successfully fighting soil pests. It reached a high state of development in Europe near the close of the last century where carbon disulfide was used on hundreds of thousands of acres in the fight against the *Phylloxera* of grape. After the First World War, the merits of the war gas chloropicrin revived interest in this method. Other volatile fumigants began pouring from the organic chemical laboratories, and it has now become profitable to treat two or three hundred thousand acres of farmland devoted to tobacco, vegetables, pineapple, and ornamentals in the United States.

The five most popular soil fumigants are ethylene dibromide, the dichloro-propene-propane mixture known as D-D, the 1–3 dichloropropene and related chlorinated hydrocarbon mixture known as Telone, chloropicrin, and methyl bromide. Each has its place, but none is ideal. The least expensive are good chiefly against nematodes and insects, while those with fungicidal properties in addition are more expensive.

The advantages from soil fumigation have been made possible through the development of inexpensive equipment for accurately dispensing metered amounts of the fumigants, often as little as three gallons per acre. Tractor-drawn equipment capable of doing many acres a day is now available.

With the manufacture of over 1,000 new organic chemicals a year by each of several manufacturers, there is every chance that someday better soil fumigants will be available. Several new ones are undergoing tests in various field laboratories all the time.

Repeated annual fumigations have no deleterious cumulative effects over prolonged periods (3–5 years), although temporary suppression of nitrification occurs for a few weeks as it does after steaming, with its attendent accumulation of ammonia.

Tests designed to screen chemicals as volatile soil fumigants must be specific for the pathogen in question and must take into account the type of survival propagule, its nature and distribution in field soils. Interactions with other pesticides (insecticides and herbicides) must also be considered. These requirements pose severe problems for the chemical industries and university researchers and imply that the research man in development must have a working knowledge of the totality of the soil-borne disease problems affecting the crops being grown, or which may be grown, in the fumigated soil.

ACKNOWLEDGMENTS.—The author expresses thanks to Dr. William Kreutzer for his suggestions and to Mrs. Paula Edwards and Miss Jan Kinzie for their competent assistance in the preparation of the manuscript.

LITERATURE CITED

ALEXANDER, M. 1961. Introduction to soil microbiology. John Wiley & Sons, Inc. New York. 436 p.

ALTMAN, J. 1963. Increase in nitrogenous compounds in soil following soil fumigation. Phytopathology 53:870. (Abstr.)

ALTMAN, J. 1964. The effect of chlorinated C_3 hydrocarbons on growth and amino acid production of indigenous soil bacteria. Phytopathology 54:886.

ALTMAN, J. 1965. Plant growth stimulation studies, p. 35–40. 11th Conference on Control of Soil Fungi, Reno, Nevada.

ALTMAN, J. 1966. Side effects associated with the use of nematocides, p. 39–44. Symposium on Nematode Control, University of Hawaii.

ALTMAN, J., and B. J. FITZGERALD. 1960. Late fall application of fumigants for the control of sugar beet nematodes, certain soil fungi, and weeds. Plant Disease Reptr. 44:868–871.

ALTMAN, J., and S. LAWLOR. 1966. The effects of some chlorinated hydrocarbons on certain soil bacteria. J. Appl. Bacteriol. 29(2):260–265.

ALTMAN, J., and M. ROSS. 1967. Plant pathogens as a possible factor in unexpected preplant herbicide damage in sugar beets. Plant Disease Reptr. 51:86–88.

ALTMAN, J., and K. M. TSUE. 1965. Changes in plant growth with chemicals used as soil fumigants. Plant Disease Reptr. 49:600–602.

BOLLEN, W. G., H. E. MORRISON, and H. H. CROWELL. 1954. Effect of field and laboratory treatments with BHC and DDT on nitrogen transformations and soil respiration. J. Econ. Entomol. 47:307–312.

BUTLER, E. E. 1957. *Rhizoctonia solani* as a parasite of fungi. Mycologia 49:354–373.

DEEP, I. W., and R. A. YOUNG. 1965. The role of preplanting treatments with chemicals in increasing the incidence of crown gall. Phytopathology 55:212–216.

DICKEY, R. S. 1962. Efficacy of five fumigants for the control of *Agrobacterium tumefaciens* at various depths in the soil. Plant Disease Reptr. 46:73–76.

FULTON, N. D., B. A. WADDLE, and R. O. THOMAS. 1956. Influence of planting dates on fungi isolated from diseased cotton seedlings. Plant Disease Reptr. 40:556–558.

GARDNER, R., and D. W. ROBERTSON. 1942. The nitrogen requirements of sugar beets. Colorado Agr. Exp. Sta. Tech. Bull. 28, Ft. Collins, 3 p.

GARRETT, S. D. 1965. Toward biological control of soilborne plant pathogens, p. 4–16. In K. F. Baker and W. C. Snyder [ed.], Ecology of soil-borne plant pathogens. Univ. of California Press, Berkeley and Los Angeles.

GIBSON, I. A. S. 1953. Crown rot, a seedling disease of groundnuts caused by *Aspergillus niger*. II. An anomalous effect of organo-mercurial seed dressings. Trans. Brit. Mycol. Soc. 36:324–334.

GIBSON, I. A. S. 1956. An anomalous effect of soil treatment with ethyl mercury phosphate on the incidence of damping-off in pine seedlings. Phytopathology 46:181–182.

GIRARD, A. 1894. Recherches sur l'augmentation des récoltes par l'injection dans le sol du sulfure de carbone a doses massives. Bull. Soc. Nat. Agr. 54:356–363.

HAASIS, F. A. 1952. Soil fumigation with chlorobromopropene for control of *Sclerotium rolfsii* in Dutch iris. Plant Disease Reptr. 36:475–478.

HACSKAYLO, E., and J. G. PALMER. 1957. Effects of several biocides on growth of seedling pines and incidence of mycorrhizae in field plots. Plant Disease Reptr. 41: 354–357.

HILTNER, L., and K. STÖRMER. 1903. Studien über die Bakterienflora des Ackerbodens, mit besonderer Berücksichtigung ihres Verhaltens nach einer Behandlung mit Schwefelkohlenstoff und nach Brache. Arb. Biol. Abt. Land-u Forstwirtsch 3:445–527.

HUBER, D. M., R. D. WATSON, and G. W. STEINER. 1965. Crop residues, nitrogen and plant disease. Soil Sci. 100: 302–308.

KOCH, A. 1899. Untersuchungen über die Ursachen der Rebenmüdigkeit mit besonderer Berücksichtigung der Schwefelkohlenstoffbehandlung. Arb. Deut. Landw. Bes. 40:7–44.

KOCH, A. 1911. Über die Wirkung von Äther und Schwefelkohoenstoff auf höhere und niedere Pflanzen. Antr. Bakteriol. Parasitenk., 2, Abt. 31:175–185.

KREUTZER, W. 1960. Soil treatment, p. 431–476. In J. G. Horsfall and A. E. Dimond [ed.], Plant Pathology. Vol. III. Academic Press, New York and London.

KREUTZER, W. 1965. The reinfestation of treated soil, p. 495–507. In K. F. Baker and W. C. Snyder [ed.], Ecology of soil-borne plant pathogens. Univ. of California Press, Berkeley and Los Angeles.

LORENZONI, G. F. 1962. Stimulant effects of highly dilute simazine. Estratto da Maydica 7:115–124.

MARTIN, J. P. 1963. Influence of pesticide residues on soil microbiological and chemical properties. Residue Rev. 4:96.

MARTIN, J. P., R. C. BAINES, and A. L. PAGE. 1963. Observations on the occasional temporary growth inhibition of citrus seedlings following heat or fumigation treatment of soil. Soil Sci. 95:175–184.

McCLELLAN, W. D., S. WILHELM, and A. GEORGE. 1955. Incidence of Verticillium wilt in cotton not affected by root-knot nematodes. Plant Disease Reptr. 39:226–227.

McKEE, H. S. 1962. Nitrogen metabolism in plants. The Clarendon Press, Oxford. 728 p.

MORITZ, J., and R. SCHERPE. 1905. Über die Bodenbdhandlung mit Schwefelkohlenstoff und ihre Einwirkung auf das Pflanzenwachstum. Arb. Biol. Abt. Land-u Forstwirtsch 4:123–156.

NEWHALL, A. G. 1955. Disinfestation of soil by heat, flooding and fumigation. Botan. Rev. 21 (4):189–250.

OBERLIN, C. 1894. Bodenmüdigkeit und Schwefelkohl-enstoff mit besonderer Berucksichtigung der Rebenverjüngung ohne Brache oder ohne Zwischenkultur. Phillip von Zabern, Mainz. Germany 19 p.

RUSSELL, E. J. 1920. The partial sterilization of soils. J. Roy. Hort. Soc. 45:237–246.

SCHMEHL, W. R., R. FINKNER, and J. SWINK. 1963. Effects of nitrogen fertilization in yield and quality of sugar beets. J. Am. Soc. Sugar Beet Tech. 12:538–544.

SMITH, N. R. 1938. The partial sterilization of soil by chloropicrin. Soil Sci. Am. Proc. (1938) 3:188.

STARKEY, R. L. 1955. Perspectives and horizons in microbiology, p. 179–195. In S. A. Waksman [ed.], Microorganisms and plant life. Rutgers Univ. Press, New Brunswick, New Jersey.

STÖRMER, K. 1907. Über die Wirkung des Schwefelkohlenstoffs und ähnlicher Stoffe auf den Boden. Jahresber Berlin. Agnew. Botan. 5:113–131.

TAM, R. K. 1945. The comparative effects of a 50–50 mixture of 1,3-dichloropropene and 1,2-dichloropropane (DD mixture) and of chloropicrin on nitrification in soil and on the growth of the pineapple plant. Soil Sci. 56:245–261.

TAM, R. K., and H. E. CLARK. 1943. Effect of chloropicrin and other soil disinfectants on the nitrogen nutrition of the pineapple plant. Soil Sci. 56:245–261.

VON NÄGELI, C. 1893. Über oligodynamische Erscheinungen in lebenden Zellen. Denkschr. Schweis. Naturf. Ges. 33:1–51.

WAKSMAN, S. A. 1932. Principles of soil microbiology. Williams and Wilkins Co., Baltimore, 476 p.

WAKSMAN, S. A., and R. L. STARKEY. 1923. Partial sterilization of soil, microbiological activities and soil fertility. Soil Sci. 16:137–156.

WARCUP, J. H. 1957. Chemical and biological aspects of soil sterilization. Soils Fertilizers 20:1–5.

WARINGTON, R. 1878. On nitrification. J. Chem. Soc. 33:44–51.

WILDE, S. A., and D. J. PERIDSKY. 1956. Effect of biocides on the development of autotrophic mycorrhizae in Monterey pine seedlings. Soil Sci. 2:107–110.

WILHELM, S. 1966. Chemical treatments and inoculum potential of soil. Ann. Rev. Phytopathol. 4:53–78.

ZOBELL, C. E. 1946a. Functions of bacteria in the formation and accumulation of petroleum. Oil Wkly. 120:30.

ZOBELL, C. E. 1946b. Action of microorganisms on hydrocarbons. Bact. Rev. 10:1–5.

Nutrition of Young Conifers and Soil Fumigation

BLANCHE BENZIAN—*Chemistry Department, Rothamsted Experimental Station, Harpenden, Herts, England.*

▶

In some of the British Forestry Commission's nurseries (mostly those started on farming land between 1920 and 1940) certain conifers, such as Sitka spruce, (*Picea sitchensis*), Western hemlock (*Tsuga heterophylla*), and Lodgepole pine (*Pinus contorta*), often remained small and stunted, even with ample plant nutrients applied as either compost or fertilizer (Benzian, 1965). This stunted growth occurred mainly on neutral or near-neutral soils, though there was at least one exception, namely the failure on the acid soil of Ringwood Nursery (Hampshire), later found to be associated with nematode damage. Stunting of increasing severity was successfully induced artificially by adding graded dressings of calcium carbonate to seedbed plots on naturally acid soils in two nurseries of contrasted histories: one cleared from heathland, the other established on farming land. Results for the heathland site, Wareham, Dorset, are shown in Fig. 1. Three of the species (Sitka spruce, Lodgepole pine, Western hemlock) have similar well-defined optima at about pH 4.5 (measured in 0.01M $CaCl_2$), and Western red cedar at about pH 5.5. For all four species growth was severely depressed at high pH values.

In nurseries with stunted growth (on neutral or near-neutral soils) improvements resulted from treating the soil with such acidifying agents as sulphur and sulphuric acid, by topdressing the seedlings with ammonium sulphate, or most successfully of all, by applying to the soil partial sterilants, such as steam, formalin, chloropicrin, or chlorobromopropene. Fig. 2 shows the poor growth without partial sterilants and the responses to formalin and chloropicrin with four conifer species at Ringwood Nursery (on an acid soil, where nematode damage is likely to have been the main cause of trouble). The differences were equally spectacular in the nurseries on neutral or near-neutral soils on which stunting normally occurs.

To isolate the factor or factors responsible for stunting and for the dramatic improvement in growth brought about by "partial sterilization," workers in several disciplines (in collaboration with the author) studied different aspects of the problem. The most detailed work has been concerned with the role of pathogenic fungi. Griffin (1965) found a correlation between the amount of root damage (based on the percentage of dead root tips) and loss in height of Sitka spruce seedlings. In the majority of plots examined, even on those producing very poor seedlings, the amount of root damage was small and he concluded that root damage does not provide a general explanation for poor growth, although it may do so sometimes. Ram Reddy, Salt, and Last (1964) reached similar conclusions. In their experiments, fungicidal soil treatments (Maneb and quintozene) applied to nursery soils had little effect on growth of Sitka spruce seedlings, and fungicidal seed dressings had none. They argued that if growth after partial sterilization was related causally to the suppression of pathogens, responses would tend to differ with variations in the activity of the pathogens, which are usually subject to seasonal fluctuations. There was no evidence to support this hypothesis. More recently Salt (1968) tested a wide range of chemicals, particularly those with fungicidal properties, applied to seed and soil, but at best they improved only germination, never growth, thus adding further circumstantial evidence to the lack of injury from fungal pathogens. Ingsted and Molin (1960) and Will (1962), too, have reported growth responses with conifer seedlings to soil fumigation in nurseries where soil pathogens seem not to have been the cause of poor growth.

Brind (1965), reporting the results of studies by different workers on the effect of partial sterilization on the soil micropopulation (bacteria by Crump, amoebae by Singh, and fungi by Mollison), showed some striking and persistent changes after partial sterilization in the balance of the three major groups of soil organisms. She did not, however, attempt to explain how these changes might have affected the growth of young conifers.

Levisohn (1965), studying mycorrhiza-formation, reported that the improvement in growth after partial sterilization bore no relationship to root infection.

To find out whether benefits from formalin depended on it being a source of energy for microorganisms, formalin drench was compared with flour, sugar, and alcohol—all energy-rich materials without toxic properties. None of the materials had any effect on seedling growth (Benzian, 1965). However, Ingestad and Nilsson (1964) did obtain improvements from sucrose in the growth of spruce and pine seedlings in Swedish forest-nursery experiments.

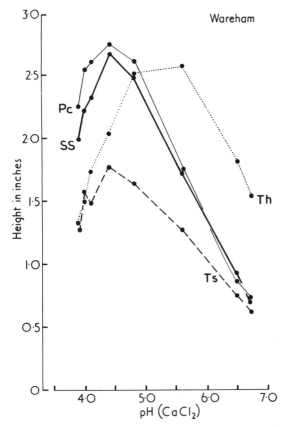

Fig. 1. pH values of soils and heights of four conifer species in Wareham pH trials 1957–1959. SS = Sitka spruce *Picea sitchensis*, Pc = Lodgepole pine *Pinus contorta*, Ts = Western hemlock *Tsuga heterophylla*, Th = Western red cedar *Thuja plicata*.

Soil samples from experiments in many nurseries were examined for nematode infestation by Goodey (1965). Numerous specimens of *Hoplolaimus uniformis (Rotylenchus robustus)* were found at Ringwood Nursery—the only nursery on acid soils where stunting was acute—and almost none anywhere else. Recently A. G. Whitehead (personal communication) found that more specimens of a species of *Trichodorus* occurred on the high-pH than on the acid plots of one of the two pH-range experiments mentioned above. In the same nursery, specimens of *Trichodorus* were more numerous in a naturally high-pH part of a seedbed than in the adjoining acid parts.

At some nurseries poor soil structure may have contributed to stunted growth. Soil from Ringwood Nursery was examined as part of a series of samples taken from agricultural and forest nursery experiments in a laboratory study of pore sizes, aggregate stability and permeability, and of clod strengths (Williams and Cooke, 1961). The Ringwood soil proved to be one of the most unstable, and over half of its total pore space was lost on slaking.

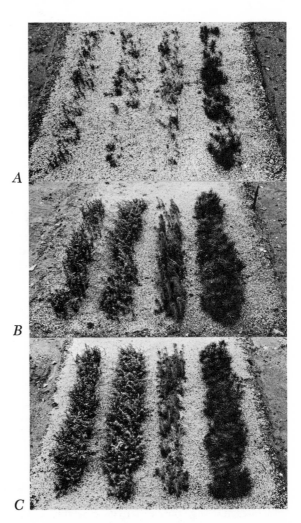

Fig. 2. Partial sterilization. One-year seedlings at end of growing season, Ringwood Nursery. Manured with inorganic fertilizer supplying N, P, K, Mg. Species from left to right: Western red cedar *Thuja plicata*; Western hemlock *Tsuga heterophylla*; Sitka spruce *Picea sitchensis*; Lodgepole pine *Pinus contorta*. A, stunted seedlings: no partial sterilization. B, good seedlings: formalin drench applied to soil during winter. C, good seedlings: chloropicrin injected into soil during winter.

Nitrogen nutrition was suspected to be one of the key factors in the problems under discussion. In high-pH nurseries, where calcium nitrate and "Nitro-Chalk" (a granular mixture of ammonium nitrate and calcium carbonate) had little effect, ammonium sulphate consistently brought about improvements (Fig. 3). Ammonium sulphate may owe its advantage to its acidifying action, or to the fact that it is less easily lost by leaching, or to both. To find out to what extent partial sterilants benefited plant growth by inhibiting nitrification, it was necessary to find a material that acted as a nitrification inhibitor whilst having little or no effect

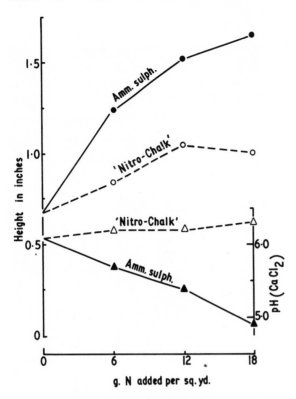

Fig. 3. Comparison between graded additions of "Nitro-Chalk" and ammonium sulphate applied annually (on four occasions) to Sitka spruce *Picea sitchensis* seedbeds. Upper pair of curves shows height responses (circles) for the mean of 3 years; lower pair of curves shows pH values (triangles) at conclusion of experiment.

TABLE 1. Effects of nitrogen fertilizer with and without nitrification inhibitor (at 2% of weight of N) on the growth of Sitka spruce seedlings. Mean of 4 years

Treatment	Height in inches
no nitrogen	0.60
"Nitro-Chalk" alone	1.28
"Nitro-Chalk" with inhibitor	1.30
ammonium sulphate alone	1.69
ammonium sulphate with inhibitor	2.04

soil or foliage with manganese sulphate or manganese chelate have failed, but in a recent nursery experiment on a near-neutral soil, the author obtained a height increase of 13% with powdered wattle bark mixed with manganese sulphate. In the same experiment a wattle-Fe preparation gave no improvement.

When it first became obvious, at the end of the Second World War, that in the Forestry Commission's older nurseries (usually established on old farming land) growth of young conifers was very poor compared with that in the then newly opened heathland nurseries, many different hypotheses were advanced to explain the failure in the so-called Sitka-sick nurseries. Among the factors suspected to be associated with the stunted growth were lack of suitable mycorrhiza in soils previously used to grow agricultural crops; long-continued uninterrupted growing of conifers, leading to a build-up of pests and diseases; and deficiencies of major and minor nutrients. However, it could quickly be shown that one of the most important conditions for satisfactory growth was suitable soil reaction, which for most of the conifers meant a very acid soil. Provided the soil was maintained near optimum pH for the species in question and adequately manured, not only could good seedlings be grown as the first crop after an arable rotation (without deliberate mycorrhizal inoculation), but productivity could be maintained on the same site for a period of 15 years without any signs of deterioration (Benzian, 1966). Although stunting of young conifers is no longer a problem of practical importance in the Forestry Commission's nurseries, because the affected nurseries either have been closed, have changed over to the growing of hardwoods, or (in a few instances) receive regular applications of formalin, some experimental work on high-pH soils has continued in the hope that a better understanding of the underlying causes might help to maintain productivity in the newer nurseries.

Up to now, intensive work by several mycologists has given little indication that fungal pathogens are implicated, though the possibility still remains that new techniques may lead to the discovery of a causal agent. So little work has been done on other microorganisms that one must not ignore the possibility that bacteria, for example, may be involved, especially in view of the strong association between stunting and soil reaction. Nematodes, which for a long time were thought to have caused damage in only one of the

on other soil processes. Potassium dichromate (chosen because chromates were known to inhibit nitrification processes in sewage purification) gave height increases between 10 and 15% in two high-pH nurseries and about 40% at Ringwood Nursery, but even this material has some fungicidal properties, so that it was impossible to attribute its benefit solely to its effect on nitrogen nutrition (Benzian, 1965). In a recent series of experiments on a near-neutral soil, ammonium sulphate and "Nitro-Chalk" were applied as such and also treated with the nitrification inhibitor 2-chloro-6-(trichloromethyl) pyridine (Turner and Goring, 1966). As in earlier experiments, ammonium sulphate was superior to "Nitro-Chalk," though "Nitro-Chalk" gave some improvement. When treated with the nitrification inhibitor, ammonium sulphate improved growth still further in three years out of four with an average increase over the four years of about 20% (Table 1).

Another suspected cause of stunting was manganese or iron deficiency. Both elements tend to be less available on high-pH soils, and manganese concentrations in the plants are increased by partial sterilization. Many attempts to improve seedling growth by treating nurseries, have aroused fresh interest since recent work on other genera has shown an association between large numbers of nematodes and high pH.

Among nutritional factors, nitrogen is likely to be one of the most important, but some paradoxical results are difficult to interpret. Thus, Sitka spruce seedlings grown with formalin have the bluish tinge and bloom characteristic of plants that have received lavish amounts of nitrogen, but the stunted seedlings grown without sterilants on high-pH soils do not exhibit the characteristic symptoms of acute N deficiency, such as the pale green colour of the whole foliage and the pinkish tips of the youngest needles. Although some extra benefit was obtained from ammonium sulphate treated with a nitrification inhibitor, this was small compared with the spectacular improvements resulting from partial sterilants. Equally, the slight increase in growth with a manganese wattle preparation makes it unlikely that manganese deficiency is one of the main causes of stunting.

The most promising approach to elucidating the success of partial sterilization in improving growth still appears to be the testing, in factorial combination, of all the treatments that alter the specific soil conditions likely to be involved. Up to now, few such treatments have been available, though a useful beginning has been made with the nitrification inhibitor mentioned above. The new chemicals now being developed with the aim of controlling specific groups of organisms may afford better opportunities for judging whether, under the conditions examined in our experiments, pests or pathogens are responsible for the poor growth, or whether it is mainly associated with nutritional disorders.

LITERATURE CITED

BENZIAN, BLANCHE. 1965. Experiments on nutrition problems in forest nurseries. Bull. For. Comm. Lond. No. 37. Vol. 1. H.M.S.O. 251 p.

BENZIAN, BLANCHE. 1966. Manuring young conifers: Experiments in some English nurseries. Proc. Fertiliser Soc. 94:5–37.

BRIND, JANET E. 1965. Some studies on the effect of partial sterilization on the soil micropopulation, p. 206–209. *In* Bull. For. Comm. Lond. No. 37. Vol. 1. H.M.S.O.

GOODEY, J. B. 1965. The relationships between the nematode *Hoplolaimus uniformis* and Sitka spruce, p. 210–211. *In* Bull. For. Comm. Lond. No. 37. Vol. 1. H.M.S.O.

GRIFFIN, D. M. 1965. A study of damping-off, root damage and related phenomena in coniferous seedlings in British forest nurseries, p. 212–227. *In* Bull. For. Comm. Lond. No. 37. Vol. 1. H.M.S.O.

INGESTAD, T., and N. MOLIN. 1960. Soil disinfection and nutrient status of spruce seedlings. Physiol. Plantarum 13:90–103.

INGESTAD, T., and H. NILSSON. 1964. The effects of soil fumigation, sucrose application and inoculation of sugar fungi on the growth of forest-tree seedlings. Plant Soil 20:74–84.

LEVISOHN, I. 1965. Mycorrhizal investigations, p. 228–235. *In* Bull. For. Comm. Lond. No. 37. Vol. 1. H.M.S.O.

RAM REDDY, M. A., G. A. SALT, and F. T. LAST. 1964. Growth of *Picea sitchensis* in old forest nurseries. Ann. Appl. Biol. 54:397–414.

SALT, G. A. 1968. Germination losses and stunting of Sitka spruce, *Picea sitchensis*, in nursery seedbeds. First International Congress of Plant Pathology Lond. (Abstr.):170.

TURNER, G. O., and C. A. I. GORING. 1966. N-Serve . . . a status report. Down to Earth 22:19–25.

WILL, G. M. 1962. The uptake of nutrients from sterilised forest-nursery soils. New Zealand J. Agr. Res. 5:425–432.

WILLIAMS, R. J. B., and G. W. COOKE. 1961. Some effects of farmyard manure and of grass residues on soil structure. Soil Sci. 92:30–39.

Growth Response of Rice to Soil Fumigation

KISABU IYATOMI and TSUTOMU NISHIZAWA—*Faculty of Agriculture, Nagoya University, Japan.*

The rice-root nematodes, *Hirschmanniella oryzae* and *Hirschmanniella imamuri*, are widespread in Japan, occurring apparently in all the areas where rice is grown under irrigation. Sometimes hundreds of nematodes are found in 1 g of fresh rice root. Soil fumigation tests were conducted in order to determine the effect of nematocides on the population of the nematodes and the growth and yields of rice. The chemicals tested were ethylene dibromide (EDB 30% in kerosene), dichloropropene-dichloropropane (D-D mixture), and in later experiments 1,2-dibromo-3-chloropropane (DBCP). Ethylene dibromide and D-D were applied with hand injectors at the rate of 3 ml/injection in replicated and randomized plots 4 × 20 m at 15 cm depth and 30/30 cm intervals, 6 weeks before planting.

Soil fumigation reduced the number of *Hirschmanniella* spp. isolated from rice roots as illustrated by the fact that 7 weeks after treatment check plants yielded 352 nematodes/10 g of rice roots while plants grown in D-D-treated soil yielded 107 nematodes/10 g of roots and those grown in EDB-treated soil yielded 74 nematodes/10 g of roots.

Though kills of the nematodes were not necessarily satisfactory, the soil treatments resulted in remarkable changes in plant growth. Among the chemicals tested, EDB gave the most conspicuous changes in growth and appearance of rice plants. The growth of the rice treated with EDB was characterized by darker foliage color, shorter plants, thicker stands, and a remarkable increase in blade inclination angles (Fig. 1).

Number of tillers and panicles was increased, while length of blades and panicles was shortened. In EDB-treated plants, the culms were shorter with an increased number of nodes, and shortening of internodes in the upper part of the culm, and elongation of the internodes at the base of the culm was observed. This irregularity in the length of internodes often resulted in lodging of plants.

Chlorophyll contents in the leaves of rice plants were determined colorimetrically after Koski's method (Koski, 1950). Larger amounts of protochlorophyll and chlorophyll a and b were observed in the leaves of rice plants grown in soil treated with EDB than those grown in D-D-treated or untreated soil.

There were 9 times as many green rice leafhoppers on the foliage of rice plants grown in EDB-treated soil as there were on plants grown in untreated soil and 4 times as many leafhoppers on plants grown in D-D-treated soil as there were on plants grown in untreated soil. Fewer leaf spots caused by *Helminthosporium* occurred on rice plants grown in treated soil, but injury from the rice stem borer, *Chilo suppressalis*, was greater on plants grown in treated soil.

Though both chemicals used in the test evidently reduced the number of the nematodes, they did not always increase the yields of rice (Table 1).

TABLE 1. Yield of treated and untreated rice plants

Treatment		Number of grains/hill	Weight of grains/hill
EDB		652	16.6g
D-D		814	21.9
Check		1,060	26.7
L.S.D.	5%	72	2.7
	1%	120	4.4

Growth response of plants to fumigation is a complex phenomenon. Fumigation appears to influence yield in ways other than by reducing the number of plant-parasitic nematodes in the soil. Fumigants can kill other pathogens, affect the nutrient and physical status of the soil, and have direct effects on plant growth (Newhall, 1955; Atkins and Fielding, 1956; Martin and Pratt, 1958; Goffart and Heiling, 1958; Hollis, et al., 1959; Peachey and Winslow, 1962; Enik, Kort, and Luesink, 1964). Generally paddy fields have heavy soil with high water content, so that diffusion of fumigants is poor and fumigation often results in phytotoxicity. Although the above-mentioned experiments did not increase the yields of rice by soil fumigation, in other cases very high increases in yields, comparable to those in experiments by Atkins and Fielding (1956) and Hollis et al. (1959), were secured by fumigation. In most cases, high increases in the rice yields have been secured by D-D treatments, and less increase by DBCP treatments. EDB treatments generally reduced yields. However, different varieties of rice respond differently to the fumigants, and some varieties have consistently secured high yields with EDB treatments.

With nematocidal treatments it is generally impossible to estimate which part of the increase in yield

Fig. 1. Epinastic leaf-blade inclination of rice plant induced by EDB treatment. Right, rice plant grown for about 1 month in EDB-treated soil; left, check plant grown for about 1 month in untreated soil.

is due to the killing of nematodes and which part is caused by other effects of the treatments. Thereupon, tests were made to determine the effect of nematocides on the growth and yields of rice under conditions of nematode-free soil and different soil-water contents. From the results, beneficial side-effects of the fumigations were clearly observed in some cases, especially in D-D and DBCP treatments, and as was expected, great differences were recognized on the growth between the dosages and/or the soil-water conditions.

EFFECTS ON NITRIFICATION.—Non-spore-forming bacteria which oxidize ammonia to nitrites and nitrates appear to be relatively susceptible to soil fumigants, and nitrification is thereby inhibited for several weeks, whereas spore-forming ammonifiers which release ammonia from organic nitrogenous complexes quickly return to soil following treatment, and ammonification goes on almost uninterrupted for weeks. During this time ammonium accumulates from the decomposing organic fraction of the soil, and any ammonia added in fertilizer will remain as ammonium. Ammonium is not removed readily from the soil as is nitrate during rain or irrigation, thus a prolonged retention is found in fumigated soil (Warcup, 1957; Martin and Pratt, 1958; Good and Carter, 1965). Soil fumigation produced an initial increase in the ammonium nitrogen and a decrease in the nitrate nitrogen in the soil, but total nitrogen was increased, and this status was maintained for a few months at least in the paddy field. Rice plants can readily utilize ammonium-nitrogen, and may even prefer this form of nitrogen.

HORMONAL ACTION.—Fumigants, especially EDB, inhibit stem elongation, promote tillering, and increase the angle of blade inclination, which closely resemble the action of auxins and/or ethylene (Crocker, Zimmerman, and Hitchcock, 1932; Michener, 1938; Burg, 1962; Lyon, 1963). To test the hormonal action of the fumigants on plant growth, *Avena* (Nitsch and Nitsch, 1956) and lamina-joint (Maeda, 1965) tests were conducted, but no direct effect on the plant growth was found at concentrations from 10^{-4} to 10 ppm of EDB or D-D. According to Burg and Burg (1966) and Morgan and Gausman (1966), ethylene acts upon the auxin in the plant, thereby affecting growth. EDB or its degraded products may have a similar effect as ethylene and alters the metabolism and transport of auxins. However, contrary to some instances in dicotyledonous plants treated with ethylene, much more auxin was found in acid-ether fraction extracted from EDB-treated rice plants than that of untreated plants.

CONCLUSION.—Though not all the fumigants tested gave a satisfactory kill of the nematodes in the paddy field, they all produced differences in plant growth and appearance, such as foliage color, height, tillering, and angle of blade inclination; EDB produced the most conspicuous changes, such as darker green color, thicker stands of shorter plants, and an increase in blade inclination-angles. Numbers of tillers and panicles were increased, while lengths of blades, panicles, and stems were shortened. These abnormal growths of rice induced abundance of leafhoppers and other insect pests, but reduced *Helminthosporium* leaf spotting. Soil fumigation affected the nitrifying bacteria, so that the nitrification was retarded, but ammonium-nitrogen was increased. No direct hormonal growth-promoting effect of the fumigants was found in the *Avena* and lamina-joint tests, but much more auxin was found in EDB-treated rice than in the control. Some of these phenomena favorably affected plant growth while others had unfavorable effects.

ACKNOWLEDGMENTS.—The authors thank Dr. T. A. Toussoun, Institute for Fungus Research, for his useful remarks and suggestions in the preparation of the manuscript. We are also much indebted to Dr. S. D. Garrett, Botany School, University of Cambridge, for his encouraging support.

LITERATURE CITED

ATKINS, J. G., and M. J. FIELDING. 1956. A preliminary report on the response of rice to soil fumigation for the control of stylet nematodes, *Tylenchorynchus martini*. Plant Disease Reptr. 40:488–489.

BURG, S. P. 1962. The physiology of ethylene formation. Ann. Rev. Plant Physiol. 13:265–302.

BURG, S. P., and E. A. BURG. 1966. The interaction between auxin and its role in plant growth. Proc. Natl. Acad. Sci. Wash. 55:262–269.

CROCKER, W., P. W. ZIMMERMAN, and A. E. HITCHCOCK. 1932. Ethylene-induced epinasty of leaves and the relation of gravity to it. Cont. Boyce Thompson Inst. 4:177–218.

ENNIK, G. C., J. KORT, and B. LUESINK. 1964. The influence of soil disinfection with DD, certain components of DD and some other compounds with nematocidal activity on the growth of white clover. Neth. J. Plant Path. 70:117–135.

GOFFART, H., and A. HEILING. 1958. Nebenwirkungen bei der Nematodenbekämpfung mit Shell D-D und verwandten Mitteln. Nematologica 3:213–228.

GOOD, J. M., and R. L. CARTER. 1965. Nitrification lag following soil fumigation. Phytopathology 55:1147–1150.

HOLLIS, J. P., L. S. WHITLOCK, J. S. ATKINS, and M. J. FIELDING. 1959. Relations between nematodes, fumigation and fertilization in rice culture. Plant Disease Reptr. 43:33–40.

KOSKI, V. M. 1950. Chlorophyll formation in seedings of *Zea mays* L. Arch. Biochem. 29:339–343.

LYON, C. J. 1963. Auxin factor in branch epinasty. Plant Physiol. 38:145–152.

MAEDA, E. 1965. Rate of lamina inclination in excised rice leaves. Physiol. Plantarum 18:813–823.

MARTIN, J. P., and P. F. PRATT. 1958. Fumigants, fungicides, and the soil. J. Agr. Food Chem. 6:345–348.

MICHENER, H. D. 1938. The action of ethylene on plant growth. Am. J. Botany 25:711–720.

MORGAN, P. W., and H. W. GAUSMAN. 1966. Effect of ethylene on auxin transport. Plant Physiol. 41:45–52.

NEWHALL, A. G. 1955. Disinfection of soil by heat, flooding and fumigation. Botan. Rev. 21:189–250.

NITSCH, J. P., and C. NITSCH. 1956. Studies on the growth of coleoptile and first internode sections. A new, sensitive, straight-growth test for auxins. Plant Physiol. 31:94–111.

PEACHEY, J. E., and R. D. WINSLOW. 1962. Effects of soil treatments on populations of soil nematodes and on carrot crops grown for two years after treatment. Nematologica 8:75–79.

WARCUP, J. H. 1957. Chemical and biological aspects of soil sterilization. Soil Fertilizers 20:1–5.

Physical Soil Factors and Soil Fumigant Action

CLEVE A. I. GORING—*The Dow Chemical Company, Agricultural Products Research, Walnut Creek, California.*

Fumigants have been employed to control soil pests, and investigators have studied the effects of physical soil factors on their action for a century. Rules for using carbon disulfide to control *Phylloxera* were formulated at least as early as 1877 (Newhall, 1955). Many aspects of this subject have been described since then (Hagan, 1941; Hannesson, Raynor, and Crafts, 1945; Taylor, 1951; Newhall, 1955; Burchfield, 1960; Goring, 1962; Peachey and Chapman, 1966). Our present knowledge is very briefly reviewed in this paper.

NATURE OF SOIL.—Soil is composed of air, water, and solid phases, the latter including both mineral and organic fractions.

Moist soil in good tilth is a honeycomb type of structure consisting of tiny, highly branched, interconnecting air spaces bordered by a layer of water covering the solid phase. As the soil dries, the water layer becomes thinner and eventually exposes the solid phase. As the soil wets, bridging occurs across the thickened water layer. Spaces surrounded or dead-ended by water increase in number and decrease in size. The volume of interconnecting air space decreases and its tortuosity increases.

The mineral portion of the solid phase includes sand, silt, clays such as kaolinite, illite, and montmorillonite, and hydrated iron and aluminum oxides. The organic fraction includes nonhumic substances such as carbohydrates, proteins, fats, waxes, and resins associated with humic acids. The latter are amorphous, acidic, polymeric substances of aromatic structure. The clay minerals are complexed with the organic fraction.

The water phase contains inorganic cations and anions, many kinds of organic substances, and the living population of the soil except the larger organisms. Even the latter may be covered with a layer of water.

PHYSICAL PROPERTIES OF FUMIGANTS.—The physical properties of soil fumigants are shown in Table 1 (Goring, 1962; Peachey and Chapman, 1966). Boiling points vary from 3.6°C for methyl bromide to 199°C for 1,2-dibromo-3-chloropropane. Corresponding vapor pressures are 1,380 mm and 0.58 mm of Hg and water solubilities 1.6% and 0.123% at 20°C.

INTERACTIONS OF FUMIGANTS WITH SOILS.—*Physicochemical relationships.*—Fumigants distribute between the air, water, and solid phases, and also decompose in soil.

Typical sorption is illustrated by results for ethylene dibromide (Hanson and Nex, 1953; Wade, 1954). The fumigant is sorbed strongly by very dry soils. As moisture content increases, sorption decreases and is minimal near the wilting point. Sorption by the soil then increases with increasing moisture above the wilting point because more fumigant is dissolved in the soil water (Hanson and Nex, 1953; Wade, 1954; Call, 1957d).

Sorption by dry soils, clays, and organic matter is partly due to Van der Waal's forces. The isotherms fall within the BET-BDDT classification (Stark, 1948; Wade, 1954; Call, 1957d; Jurinak and Volman, 1957). Weak hydrogen bonding of fumigants like chloropicrin and methyl isothiocyanate to the OH groups of clays or the OH, COOH, or NH groups of organic matter probably also occurs. Furthermore, all of the fumigants are probably sorbed, or dissolved, or both, in lipophilic constituents of organic matter.

Fumigants are completely displaced from clays but not from organic matter by water, and sorption increases with increasing organic matter (Wade, 1954; Call, 1957a).

Water-to-air distribution ratios for the various fumigants are shown in Table 1 and are constant for a specified temperature but varying fumigant concentrations. They range from 1.8 for carbon disulfide to 163.8 for 1,2-dibromo-3-chloropropane (Call, 1957d; Goring, 1962; Peachey and Chapman, 1966) and are inversely related to rates of diffusion (Youngson and Goring, 1962; Peachey et al., 1965).

There is little information on organic matter-to-water distribution ratios. Estimates of 72 and 11 are given in Table 1 for 1,2-dibromo-3-chloropropane and ethylene dibromide. Organic matter-to-water ratios are also inversely related to rates of diffusion.

Both water-to-air and organic matter-to-water distribution ratios decrease with increasing temperature

TABLE 1. Some physical properties of soil fumigants

Fumigant	Boiling point at 760 mm Hg Pressure °C	Vapor pressure at 20°C mm Hg	Solubility in water at 20°C %	Distribution Ratios at 20°C Water-to-Air[*]	Organic Matter-to-Water[†]
Methyl bromide	3.6	1,380.0	1.60	4.1	–
Carbon disulfide	46.3	298.0	0.217	1.8	–
Cis 1, 3-dichloropropene	106.0	25.0	0.270	17.7	–
Trans 1, 3-dichloropropene	111.0	18.5	0.280	24.6	–
Chloropicrin	111.9	20.0	0.195	10.8	–
Methyl isothiocyanate	119.0	21.0	0.76	88.0	–
1, 2-dibromoethane	131.7	7.69	0.337	42.7	11[‡]
1, 2-dibromo-3-chloropropane	199.0	0.58	0.123	163.8	Circa 72[§]

[*] Ratio of weights of chemical in equal volumes of water and air.
[†] Ratio of chemical sorbed per unit weight of soil organic matter to chemical dissolved per unit weight of water.
[‡] Average for 4 soils. Calculated from table 4, Call (1957d) and from organic matter contents given in table 1, Call (1957a).
[§] Calculated from table 4, Youngson, Goring, Noveroske (1967) on the basis of an estimated 90% sorption of Fumazone from solution by the muck soil.

and so do coefficients for sorption by soils (Wade, 1954; Call, 1957a; Goring, 1967).

Fumigants decompose in soils (Hanson and Nex, 1953; Castro and Belser, 1966) because of nucleophilic displacement reactions with active groups of organic matter (Burchfield, 1960; Moje, 1960; Goring, 1967), because of hydrolysis in water (Castro and Belser, 1966), and because of decomposition by soil microorganisms (Castro and Belser, 1968). Rates of decomposition can probably be fitted approximately to first-order kinetics (Goring, 1967).

Toxicity to pests.—Toxicity of fumigants has been attributed to nucleophilic displacement reactions with active sites in the pest (Moje, 1960), but other mechanisms are possible. The dose required for control of the pest is the cumulative product of concentration × time and, with some exceptions, is relatively constant for a given pest, fumigant, and environmental situation (Schmidt, 1947; Call, 1957c; Goring, 1962; Call and Hague, 1962; Hague and Sood, 1963; Johnson and Lear, 1966). This is reasonable as long as uptake of toxicant is proportional to concentration (Marks, Thomason, and Castro, 1968; Goring, 1967).

Dose of fumigant required for control will vary greatly with type of pest, fumigant, and environmental situation. Generally, but not always, it will increase with decreasing temperature. It also will increase with decreasing moisture, but only under relatively dry soil conditions (Stark, 1948; Lear, 1951; Brande, Kips, and D'Herde, 1956; Harrison, 1957; Sasser and Uzzell, 1960; Youngson, Baker, and Goring, 1962; Youngson and Goring, 1962; Lear and Johnson, 1963).

MOVEMENT OF FUMIGANTS AND PEST CONTROL.—
Basic principles.—Fumigants are transported through soil by diffusion, or mass transfer, or both. Diffusion is characterized by random molecular movement, while mass transfer involves movement in response to an applied force such as irrigating soil. Mass transfer downward in response to gravity occurs with soil fumigants in liquid but apparently not in vapor form (Currie, 1961; Goring, 1962; Youngson, Baker, and Goring, 1962; Turner, 1965). Initial volatilization and

distribution of fumigant between air, water, and solid phases occurs over a relatively short period of time. However, for a fumigant with a vapor pressure as low as 1,2-dibromo-3-chloropropane, the period can be a significant fraction of the time required for diffusion.

The diffusion coefficients of fumigants through air vary to some extent (Goring, 1962), but because movement is much more rapid through air than water, the distribution ratios between air, water, and soil solids are of much greater significance to the diffusion process. Equally important is the tortuosity of the air space and rate of decomposition of the fumigant. Pest control at each point in the soil is a function of the cumulative product of concentration of fumigant × time.

The diffusion process has been described quantitatively by Call (1957b) and Hemwall (1962). Using Fick's law in conjunction with equations for sorption and decomposition, Hemwall was able to derive a differential equation for the diffusion process. He then modified it to its finite-difference form so it could be solved numerically. Using assumed values for water-to-air and soil-to-water ratios, and for rates of decomposition, Hemwall was able to calculate dosage patterns for several assumed soil conditions and methods of application. Many of his predicted nematode-control patterns were in good agreement with patterns actually obtained by Youngson and Goring (1962) for soil conditions and methods of application similar to those he assumed.

Application to uncovered and covered soil.—Pest-control patterns for fumigants injected into uncovered soil or applied under a soil cover have been reviewed by Goring (1962). Point, line, or planar applications give spherical, cylindrical, or rectangular parallelepiped patterns of control distorted downward relative to the injection locations. If the points or lines of application are sufficiently close together, or if the dosages are sufficiently high, the initial spherical or cylindrical patterns overlap and the final pattern of control is a rectangular parallelepiped.

A larger pattern of pest control is almost invariably

obtained below than above the injection location in uncovered soil (Allen and Raski, 1950; Siegel, Erickson, and Turk, 1951; Thorne, 1951; Ichikawa, Gilpatrick, and McBeth, 1955; Youngson and Goring, 1962; Okada and Mori, 1964). This is due to slower diffusion, which causes higher concentrations of fumigant for longer periods of time below than above the injection location (Hannesson, Raynor, and Crafts, 1945; Call, 1957c; Goring, 1962; Youngson and Goring, 1962). Even when soil is covered and pest control is obtained at the surface, the pattern is usually larger below than above the injection location (Youngson, Baker, and Goring, 1962).

Increased injection depth generally increases the size of the pest-control pattern (Wilhelm and Ferguson, 1953; Gilpatrick et al., 1956; Baines et al., 1959; Youngson, Baker, and Goring, 1962; Youngson and Goring, 1962; Okada and Mori, 1964). However, with sufficiently deep injection, control may not be obtained at the surface (Hannesson, Raynor, and Crafts, 1945; Gilpatrick et al., 1956; Goring, 1962; Youngson, Baker, and Goring, 1962). Nevertheless, in uncovered soils, yield response may be significantly improved by deep injection (Lembright, Healy, and Norris, 1968).

Uniformly decreased air space, increased moisture above or decreased moisture below the wilting point, decreased temperature, finer texture, and increased organic matter all tend to decrease rate of diffusion. Provided the air space is continuous and the rate of decomposition is not rapid, the decreased rate of diffusion should increase the cumulative concentration × time products at every point in the soil and, therefore, the size of the pest-control pattern (Call, 1957c; Call and Hague, 1962; Goring, 1962; Schmidt, 1947).

Decreased air space only in the top 6 inches will improve pest control in this soil zone as well as in the zone below 6 inches because of the decreased rate of loss of fumigant from the soil surface (Hannesson, Raynor, and Crafts, 1945; Call, 1957c; Goring, 1962); but, decreased air space below 6 inches restricts diffusion in this zone, increases rate of loss from the soil surface and may decrease overall pest control and yield (Turner, 1958).

Increased moisture above the wilting point is similar in its effect to decreased air space. Thus, high moisture in the top 6 inches and low moisture below 6 inches is far more desirable than the reverse situation (Hannesson, Raynor, and Crafts, 1945; Turner, 1958; Goring, 1962).

When the soil is covered, it is desirable to have relatively low moisture and high air space for the more slowly diffusing fumigants such as chloropicrin (Youngson, Baker, and Goring, 1962). But for methyl bromide, diffusion is so rapid when moisture is too low and air space too high that poor pest control may be obtained (Youngson, Baker, and Goring, 1962).

Decreasing temperature usually decreases pest control even though the fumigant persists longer in the soil (Goring, 1962). This is because of its decreased

basic toxicity. One example of an exception is nematode control by 1,3-dichloropropene. The size of the control pattern given by this fumigant does not decrease with decreasing temperature because its basic toxicity does not decrease (Brande, Kips, and D'Herde, 1956; Youngson and Goring, 1962; Lear and Johnson, 1963).

Smaller patterns of pest control are obtained in soils having a finer texture and higher content of organic matter, except in rare circumstances for the more rapidly diffusing fumigants such as methyl bromide and carbon disulfide (Stark, 1948; Allen and Raski, 1950; Siegel, Erickson, and Turk, 1951; Thorne, 1951; Baines et al., 1959; Youngson, Baker, and Goring, 1962). Reduced control is caused by too slow diffusion in conjunction with greater sorption and more rapid decomposition of fumigant as it diffuses (Goring, 1962), but penetration into the soil crumbs may also be restricted (Peachey and Chapman, 1966). In uncovered muck soils, maximum pest control is obtained when diffusion is enhanced by relatively low moisture and high temperature (Lewis and Mai, 1963; Bird and Smith, 1965). Similarly, pest control and yields are improved by fumigating uncovered heavy clay soils when moisture is relatively low and the soil is porous (Warren, 1958).

Application in irrigation water.—The movement of chemicals through soil in water has been reviewed by Kirkham (1964) and Goring (1967). Among soil fumigants, detailed studies were conducted for 1,2-dibromo-3-chloropropane by Youngson, Goring, and Noveroske (1967), and Johnson and Lear (1968). Qualitatively similar results would probably be obtained for other soil fumigants.

Other than the capacity of the soil to hold water, the most important factor influencing penetration and pest control by 1,2-dibromo-3-chloropropane is sorption by organic matter. Increased organic matter reduces penetration and pest control. Sorption by organic matter also causes the pattern of fumigant penetration and pest control to be smaller than the pattern of water movement, irrespective of whether the water infiltrates downward or subs sideways into a plant bed. Pest control at the centers of wide plant beds is not obtained by applying fumigant in water to the furrows.

Depth of pest control is increased with increasing amounts of water, whether the fumigant is applied in the water or whether it is injected or applied to the soil surface and then leached into the soil with water. In the latter case, considerable amounts of water are needed to leach it out of the top 6 inches of soil because of sorption by organic matter.

Depth of pest control is also increased by the capacity of water added to soil to move downward through the centers of the pores and not displace water initially present. This capacity is increased with increased rates of infiltration and, providing the soil is not saturated, with increased soil moisture prior to treatment.

RESEARCH NEEDED.—I will not attempt to list comprehensively the research needed in this field of investigation. However, more active research programs in certain areas would provide data needed for improving prediction of pest-control patterns. Specifically, data are needed on organic matter-to-water ratios. Also needed are data on rates of decomposition of fumigants and concentration × time products for more of the important pests. This information, taken together with water-to-air ratios, can be used in the equation derived by Hemwall (1962) to predict pest-control patterns for any method of injection and soil-environmental situation. Appropriate equations can also be developed for predicting pest-control patterns for fumigants applied in irrigation water.

Field data for uncovered soils are needed on the effect of deeper injection and wider spacings on pest control and yield. Hannesson, Raynor, and Crafts (1945) pointed out over 20 years ago that as injection depth is increased, injection spacing may also be increased without sacrificing overlap of control patterns. Subsoiling equipment now in use can place fumigants much deeper in soil than the conventional 6–8 inches, and such placement has given improved yields (Lembright, Healy, and Norris, 1968). More extensive investigation of the practice is warranted in order to establish the full scope of its value.

LITERATURE CITED

ALLEN, M. W., and D. J. RASKI. 1950. The effect of soil type on the dispersion of fumigants. Phytopathology 40:1043–1053.

BAINES, R. C., F. J. FOOTE, L. H. STOLZY, R. H. SMALE, and M. J. GARBER. 1959. Factors influencing control of the citrus nematode in the field with D-D. Hilgardia 29:359–381.

BIRD, G. W., and H. A. SMITH. 1965. Factors affecting control of onion bloat by fumigants containing 1,3-dichloropropene in organic soils in Southern New York. Plant Disease Reptr. 49:33.

BRANDE, J. VAN DEN, R. H. KIPS, and J. D'HERDE. 1956. Survey of the results of four years experiments on the chemical control of the potato root eelworm. Nematologica 1:81–87.

BURCHFIELD, H. P. 1960. Performance of fungicides on plants and in soil—physical, chemical, and biological considerations, Vol. 3, p. 477–520. *In* J. G. Horsfall and A. E. Dimond [ed.], Plant pathology, an advanced treatise. Academic Press, New York.

CALL, F. 1957a. Soil fumigation. IV. Sorption of ethylene dibromide on soils at field capacity. J. Sci. Food Agr. 8:137–142.

CALL, F. 1957b. Soil fumigation. V. Diffusion of ethylene dibromide through soils. J. Sci. Food Agr. 8:143–150.

CALL, F. 1957c. Soil fumigation. VI. The distribution of ethylene dibromide round an injection point. J. Sci. Food Agr. 8:591–596.

CALL, F. 1957d. The mechanism of sorption of ethylene dibromide on moist soils. J. Sci. Food Agr. 8:630–639.

CALL, F., and N. G. M. HAGUE. 1962. The relationship between the concentration of ethylene dibromide and nematocidal effects in soil fumigation. Nematologica 7:186–192.

CASTRO, C. E., and N. O. BELSER. 1966. Hydrolysis of *cis*- and *trans*-1,3-dichloropropene in wet soil. J. Agr. Food Chem. 14:69–70.

CASTRO, C. E., and N. O. BELSER. 1968. Biodehalogenation. Reductive dehalogenation of the biocides ethylene dibromide, 1,2-dibromo-3-chloropropane, and 2,3-dibromobutane in soil. J. Env. Sci. Technol. 2:779–783.

CURRIE, J. A. 1961. Gaseous diffusion in the aeration of aggregated soils. Soil Sci. 92:40–45.

GILPATRICK, J. D., S. T. ICHIKAWA, M. TURNER, and C. W. MCBETH. 1956. The effect of placement depth on the activity of Nemagon. Phytopathology 46:529–531.

GORING, C. A. I. 1962. Theory and principles of soil fumigation. Adv. Pest Control Res. 5:47–84.

GORING, C. A. I. 1967. Physical aspects of soil in relation to the action of soil fungicides. Ann. Rev. Phytopathol. 5:285–318.

HAGAN, R. M. 1941. Movement of carbon disulfide vapor in soils. Hilgardia 14:83–118.

HAGUE, N. D., and V. SOOD. 1963. Soil sterilization with methyl bromide to control soil nematodes. Plant Pathol. 12:88–90.

HANNESSON, H. A., R. N. RAYNOR, and A. S. CRAFTS. 1945. Herbicidal use of carbon disulfide. Calif. Univ. Agr. Exp. Sta. Bull. 693, 57 p.

HANSON, W. J., and R. W. NEX. 1953. Diffusion of ethylene dibromide in soils. Soil Sci. 76:209–214.

HARRISON, M. B. 1957. Fumigation of encysted golden nematode eggs and larvae under controlled environmental conditions. Phytopathology 47:610–613.

HEMWALL, J. B. 1962. Theoretical consideration of soil fumigation. Phytopathology 52:1108–1115.

ICHIKAWA, S. T., J. D. GILPATRICK, and C. W. MCBETH. 1955. Soil diffusion pattern of 1,2-dibromo-3-chloropropane. Phytopathology 45:576–578.

JOHNSON, D. E., and B. LEAR. 1966. The influence of nematocide exposure time on nematode control. Plant Disease Reptr. 50:915–916.

JOHNSON, D. E., and B. LEAR. 1968. Evaluating the movement of 1,2-dibromo-3-chloropropane through soil. Soil Sci. 105:31–35.

JURINAK, J. J., and D. H. VOLMAN. 1957. Application of the Brunauer, Emmett, and Teller equation to ethylene dibromide adsorption by soils. Soil Sci. 83:487–496.

KIRKHAM, D. 1964. Some physical processes causing movement of ions and other matter through soil. Overdruk uit de Mededelingen en de Landbouwhogeschool en de Opzoekingsstations van de Staat te Gent. Deel XXIX, N 1.

LEAR, B. 1951. Use of methyl bromide and other volatile chemicals for soil fumigation. Cornell Univ. Agr. Exp. Sta. Mem. 303, 48 p.

LEAR, B., and D. E. JOHNSON. 1963. Influence of temperature on the nematocidal activity of soil fumigants containing ethylene dibromide and dichloropropenes. Soil Sci. 95:322–325.

LEMBRIGHT, H. W., M. J. HEALY, and M. G. NORRIS. 1968. Improved soil fumigant utilization through deep placement. Down to Earth 24 (No. 1):10–12.

LEWIS, G. D., and W. F. MAI. 1963. Temperature and moisture in relation to the toxicity of 1,3-dichloropropene to *Ditylenchus dipsaci* (Kühn) Filipjev, in organic soil. Plant Disease Reptr. 47:1097–1101.

MARKS, C. F., I. J. THOMASON, and C. E. CASTRO. 1968. Dynamics of the permeation of nematodes by water, nematocides and other substances. Experimental Parasitology 22:321–337.

MOJE, W. 1960. The chemistry and nematocidal activity of organic halides. Adv. Pest Control Res. 3:181–217.

NEWHALL, A. G. 1955. Disinfestation of soil by heat, flooding and fumigation. Botan. Rev. 21:189–250.

OKADA, T., and T. MORI. 1964. The diffusion of soil fumigants. 2. Effect of injection depth on the diffusion pattern of a D-D mixture and its killing range for soybean nematodes. Hokkaido Nat. Agr. Exp. Sta. Res. Bull. 83:24–31.

PEACHEY, J. E., and M. R. CHAPMAN. 1966. Chemical control of plant nematodes. Tech. Commun. 36, Commonwealth Bur. Helminthol., St. Albans, 119 p. Commonwealth Agr. Bur. Farnham Royal, Bucks, England.

PEACHEY, J. E., D. N. GREET, D. J. HOOPER, and M. R. CHAPMAN. 1965. The effects of two soil types on diffusion of soil sterilants. Plant Pathol. 14:36–38.

SASSER, J. N., and G. UZZELL, JR. 1960. Methyl bromide fumigation of *Heterodera glycines* in North Carolina. Plant Disease Reptr. 44:728–732.

SCHMIDT, C. T. 1947. Dispersion of fumigants through soil. J. Econ. Entomol. 40:829–837.

SIEGEL, J. J., A. E. ERICKSON, and L. M. TURK. 1951. Diffusion characteristics of 1,3-dichloropropene and 1,2-dibromoethane in soils. Soil Sci. 72:333–340.

STARK, F. L., JR. 1948. Investigations of chloropicrin as a soil fumigant. Cornell Univ. Agr. Exp. Sta. Mem. 278, 61 p.

TAYLOR, A. L. 1951. Chemical treatment of the soil for nematode control. Adv. Agron. 3:243–264.

THORNE, G. 1951. Diffusion patterns of soil fumigants. Proc. Helminthol. Soc. Wash. 18:18–24.

TURNER, G. O. 1958. Techniques for increasing the efficacy of 1,3-dichloropropene soil fumigants in the control of the sugar beet and root-knot nematodes. J. Amer. Soc. Sugar Beet Technol. 10:80–86.

TURNER, N. J. 1965. A simple laboratory technique for determining the diffusion rates and patterns of lethal concentrations of nematocides in soil. Contrib. Boyce Thompson Inst. 23:19–20.

WADE, P. 1954. The sorption of ethylene dibromide by soils. J. Sci. Food Agr. 5:184–192.

WARREN, L. E. 1958. Do nematodes reduce your sugar beet crop? Spreckels Sugar Beet Bull. 22:14–15.

WILHELM, S., and J. FERGUSON. 1953. Soil fumigation against *Verticillium albo-atrum*. Phytopathology 43:593–596.

YOUNGSON, C. R., R. G. BAKER, and C. A. I. GORING. 1962. Diffusion and pest control by methyl bromide and chloropicrin applied to covered soil. J. Agr. Food Chem. 10:21–25.

YOUNGSON, C. R., and C. A. I. GORING. 1962. Diffusion and nematode control by 1,2-dibromoethane, 1,3-dichloropropene, and 1,2-dibromo-3-chloropropane in soil. Soil Sci. 93:306–316.

YOUNGSON, C. R., C. A. I. GORING, and R. L. NOVEROSKE. 1967. Laboratory and greenhouse studies on the application of Fumazone in water to soil for control of nematodes. Down to Earth 23 (No. 1):27–32.

Selective Killing of Soil Microorganisms by Aerated Steam

KENNETH F. BAKER—*Department of Plant Pathology, University of California, Berkeley.*

The destruction of microorganisms has been the dominant idea in soil treatments since 1880–1890, and recommendations have tended toward overkill rather than minimal treatments. Broad-spectrum, high-potency chemicals (e.g., formaldehyde, chloropicrin, methyl bromide) at high dosages have generally been used for field treatment, and steam at 212°–250° F/30 min for soil in containers (benches, beds, pots, flats).

Disturbing and unexpected results have indicated that there is more to soil treatment than merely killing organisms.

1. Treatment at too high a dosage or temperature often created a biological vacuum in which contaminant pathogens caused great crop loss.

2. Too small an application of chloropicrin was followed by a greatly accentuated attack of *Verticillium albo-atrum*, presumably by decreasing antagonistic activity.

3. Application of methyl bromide is effective against most pathogens but not against *V. albo-atrum*.

4. Application of PCNB inhibited *Rhizoctonia solani*, but the crop was lost to *Pythium* spp. whose presence was unsuspected.

5. Surface application of Vapam, followed by water to a 6-inch depth, protected the crop through part of the cycle, only to lose it just before harvest to *Pythium* and *Phytophthora* spp. that spread up the taproots from untreated soil below.

6. Application of CS_2 against *Armillaria mellea* in peach orchards and against Fusarium wilt of asparagus in California, provided better 5-year control than did more potent fungicides. The ultimate problem in any soil treatment is, in fact, subsequent contamination by pathogens.

There is now a marked trend toward minimal soil treatments with toxicants, and toward use of selective fungicides (e.g., CS_2, D-D, PCNB, Dexon, Nemagon). This specificity has, however, created the possibility, apparently so far observed largely in the laboratory (Georgopoulos and Zaracovitis, 1967; Kuiper, 1965), that pathogens can become resistant to the chemical. Organisms exposed to lethal agents, such as heat, which affect large metabolic targets, are less likely to develop resistance to them.

We have progressively "cooled it" in the last 27 years from 250°/6–8 hr in autoclaves to 212°/30 min in flowing steam, to 180° F/30 min with a moving soil mass (Baker and Roistacher, 1957), and finally to 140° F/30 min with aerated steam (Baker and Olsen, 1960).

COMPARISON OF CHEMICAL AND STEAM TREATMENT OF SOIL.—Severity of fumigant treatment of soil may be controlled by varying the amount injected. The volatile chemical diffuses out in expanding spheres (Fig. 1) from the point of injection, but since it is sorbed by the soil, its concentration decreases progressively outward from the point of injection (Goring, 1967). Treatment, therefore, is characteristically nonuniform through the soil mass, with overkill at the point of injection and undertreatment at the margins.

By comparison, the temperature in soil treated with steam is uniform throughout. Steam at the point of injection condenses until soil reaches the injection temperature, then passes by and condenses on the next cool soil (Fig. 2). B.t.u. are thus released only at the point to be heated. The only treatment variable is the time required for steam to permeate the soil; with proper equipment and adequate steam flow this can be very short.

The biological effect of treatment with steam is, therefore, more uniform than with chemicals, and the former is better suited to controlled manipulation of soil microflora than is the latter. In some cases (e.g., CS_2 against *A. mellea*) chemicals have, however, given excellent results.

APPROACHES TO BIOLOGICAL CONTROL OF SOIL-BORNE PATHOGENS.—Biological control of root pathogens must commonly occur, otherwise agriculture would not be possible. The occurrence of root disease shows that only imperfect protection is afforded. Some soils quickly suppress the invasion of some pathogens —e.g., most California soils with respect to *Fusarium oxysporum* f. sp. *pisi* (W. C. Snyder, unpublished data), soils at the Waite Agricultural Research Institute, South Australia, for a wheat-root-rotting *R. solani* (Baker et al., 1967). Soils of other areas may develop biological suppression several years after a pathogen is introduced—e.g., *Cercosporella herpotrichoides* and *Ophiobolus graminis* on wheat at Rothamsted, and *Streptomyces scabies* on potato in California (Glynne, 1965).

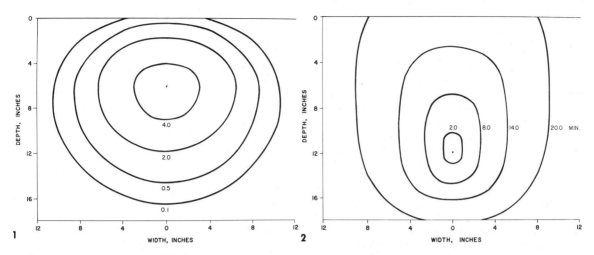

Figs. 1–2. Comparison of diffusion of a volatile chemical and steam through soil. Fig. 1. Diffusion of a volatile chemical, showing the isoconcentrations of chemical sorbed after one day. The concentrations decrease progressively outward from the point of injection, giving a nonuniform dosage. (Data from Hemwall, 1959.) Fig. 2. Diffusion of steam outward from the point of injection, showing 212° F isotherms at various periods of time. The temperature attained is uniform throughout the soil mass.

There have been two philosophies of manipulation of the microflora by soil treatment in biological control of root pathogens:

1. *Drastic soil treatment* to create a biological vacuum, followed by inoculation ("controlled colonization") with one or more selected antagonists (Fergu-

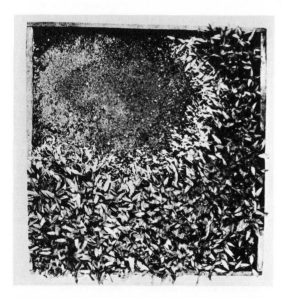

Fig. 3. Retardant effect of *Myrothecium verrucaria* on damping-off of pepper seddlings by *Rhizoctonia solani* in soil steamed at 212° F/30 min. Upper right, both fungi introduced into corner at time of seeding. Damping-off was slight. Lower left, same isolate of *R. solani* introduced into corner at same time as above. Damping-off is unchecked. (From Ferguson, 1958.)

son, 1958). These retardant organism(s) luxuriate in such treated soil and greatly decrease the chance of successful invasion by a pathogen (Fig. 3). This was a natural accommodation to the philosophy of overkill. The difficulties of the procedure are:

a) Antagonists have a degree of specificity for both different soils and pathogens (Olsen and Baker, 1968), and must be changed for each situation.

b) There must be mutual compatibility among the antagonists if several are used. This may be difficult to attain.

c) A single retardant microorganism is highly unstable, and may fluctuate between producing too little antibiotic to be effective, and producing so much as to be phytotoxic. *Myrothecium verrucaria* thus both inhibits *R. solani* and stunts seedlings until the cellulose is consumed, when both effects diminish (Ferguson, 1958). This is consistent with the philosophy that the greater the complexity of a biological community, the greater is its stability (Elton, 1958).

d) Microorganisms grown in large cultures must continually be reselected for antagonistic effect because of their well-known mutability under these conditions.

This method is not now used, and if it has future utility, it will probably not be in field soil but in situations employing a uniform soil-mix (now common in glasshouse culture), permitting use of a standardized group of antagonists. This would be particularly useful with inert soil-mixes naturally deficient in antagonistic microorganisms (Baker, 1967).

2. *Soil treatment at the minimal level necessary to destroy pathogens.* The oldest and most conspicuously successful biological control of soil-borne plant pathogens is afforded by the mushroom industry.

Careful control of temperature and aeration during the composting and pasteurization processes provides commercial control of fungus, bacterial, and nematode pathogens of the mushroom crop, as well as reduces "weed fungi" (Lambert and Ayers, 1952). Overheating of the compost during the operation may produce disastrous disease outbreaks.

Armillaria mellea in buried roots is destroyed by *Trichoderma viride* following low-dosage treatment with the weak fungicide, CS_2, but this may require 26 days. Without the chemical treatment, *Armillaria* in field soil is not weakened and killed by *Trichoderma*. Heating soil to 91.4° F/7 days with aerated steam also weakened *Armillaria* and enabled *Trichoderma* to destroy it. Since roots of citrus and peach tolerated these temperatures, there is an interesting possibility that heat sublethal to *Armillaria* may be used to cure infected trees (Darley, 1958; Darley and Wilbur, 1954; Moubasher, 1963).

A third example of effective thermal manipulation of soil microflora is afforded by the relationship between *Bacillus subtilis* and *R. solani* (Baker et al., 1967; Olsen, Flentje, and Baker, 1967; Olsen and Baker, 1968). Because of its thermal resistance, *B. subtilis* survived in soil treated with aerated steam, and inhibited *R. solani* introduced into the soil as a contaminant. This bacterium formed colonies along the mycelium of *R. solani*, utilizing exudates from the cells (but not the cell walls) and causing lysis; it increased in numbers at the expense of the fungus. *B. subtilis* exhibited a high level of specificity to different isolates of *R. solani* and to different soil types. The *Bacillus* thus increased in soil when a susceptible *R. solani* was present, reducing the quantity of the fungus and thereby reducing the amount of damping-off of seedlings; with other isolates of *R. solani*, this effect either was lacking or was slight. While this specificity is disadvantageous to commercial utilization of the method, it is to be expected in such a relationship.

Soils differ markedly in the specific antagonists present, as well as in the relative numbers active at a given time (Baker et al., 1967).

BIOLOGICAL ASPECTS OF DISEASE CONTROL BY AERATED STEAM.—It has been known for over 30 years that steaming of soil at 140° F/30 min will kill plant-pathogenic fungi, bacteria, and nematodes, and will inactivate nearly all viruses. Until the advent of aerated steam, however, this could only be achieved by injecting 212° F steam into a moving soil mass, a severe limitation to commercial application.

The biological principle involved in aerated-steam treatment is that pathogens are generally less resistant to certain unfavorable conditions, such as heat, than are many saprophytes (Baker and Olsen, 1960; Baker, 1962). Such soil treatment thus leaves a group of resident microorganisms adapted to the particular soil, to the environment, and to each other. They luxuriate because of reduced competition, and their relative numbers are further increased by the resulting breaking of spore dormancy (Warcup and Baker, 1963).

Actinomycetes were abundant in soil which had been treated at temperatures up to 160° F, bacteria up to 180° F (Baker, 1962). The total number of bacteria, fungi, and actinomycetes surviving treatment was inversely related to temperature in the range 120° to 212° F/30 min. Ten days later, however, they were directly related to treatment temperature, probably because of reduced competition (Olsen and Baker, 1968). As the treatment temperature is increased above 140° F, fewer and fewer kinds of microorganisms survive, and there is less chance of effective antagonists being present.

The disease-producing capacity of a fungus in a given soil can be determined by the inoculated-flat test (Ferguson, 1958). The rate of advance of the fungus when inoculated in one corner, through a flat of densely planted susceptible seedlings (as indicated by damping-off and by culturing) provides an index of active specific antagonists in the soil. Yolo clay loam treated at 212° F/30 min and thus inoculated with *R. solani* when planted with pepper seed, had 253.4 cm² and 17.8 cm of linear advance by the pathogen. That treated at 160° F had 64.5 cm² and 8.8 cm, that at 140° F had 3.2 cm² and 2.0 cm, and the untreated soil had only 0.3 cm linear advance of *R. solani*, but many localized spots (totaling 102 cm²) from resident *Pythium* sp. This soil thus had active antagonists to this strain of *R. solani* but not to *Pythium* sp., and antagonist activity was relatively unimpaired by treatment at 140° F, although drastically reduced at 160° F and largely eliminated at 212° F (Olsen and Baker, 1968). A test of this nature should be made with an untreated soil to determine the level of antagonism for the given pathogen before studies are begun.

The old steam treatments at 212° F/30 min had a very large margin (72° F) of safety, and even relatively careless treatment was usually successful. If portions of a bench were more compact or wetter than the rest, or the soil had large clods, a minimum of 140° F would almost certainly be reached in these spots in any case. It is obvious that the margin of safety decreases with the treatment temperature. This situation is not unique to soil treatments, since almost all aspects of modern agriculture operate on closer tolerances than in the past. Furthermore, modern practices facilitate narrower margins; modern soil-mixes essentially eliminate the possibility of clods, for example. Nevertheless, this safety margin causes some growers to prefer temperatures above 140° F.

Excessive heating of soil produces phytotoxic residues, particularly of water-soluble manganese, known to produce serious injury to tomatoes in England and to carnations in Pennsylvania and California. Certain fungicides may leave a toxic residue (e.g., methyl bromide is injurious to carnation and snapdragon), and most fumigants applied at high dosage release sorbed ions from the soil particles in a manner similar to heat treatment. One of the benefits of soil treatment with aerated steam is that at or below 160° F there is little or no phytotoxicity problem (Fig. 4;

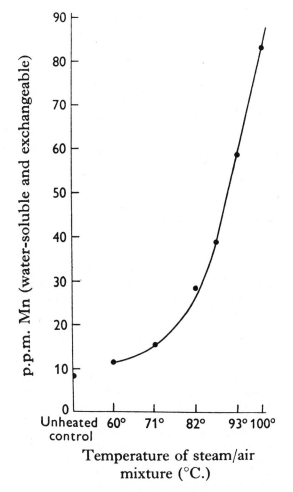

Fig. 4. Effect of treating English glasshouse tomato soil with aerated steam at different temperatures for 30 min on the amount of phytotoxic soluble and exchangeable manganese three days later. (From Dawson et al., 1965.)

Dawson et al., 1965, 1967). Grower experience generally shows that, following the change from 212° to 140°–160° F soil steaming, there is a marked improvement in plant growth. Many of the modern inert soil mixes (Baker and Matkin, 1967) present no phytotoxicity problem when overtreated. The soil must be moistened to a good planting tilth for three days preceding treatment in order to increase thermal sensitivity of spores and seeds; weed seeds are then killed by 140° F, which is 5°–10° F higher than seed treatments of equal duration. Although 212° F steaming will kill most weed seeds even in dry soil, such treatment has never been recommended because it may not kill fungus and bacterial spores. Experience shows that weed seeds present no problem in soil properly steamed at 140°–150° F.

A number of significant biological events seem to occur at about 140° F. Industrial use has long been made of this fact; milk, wine, and beer are thus pasteurized at 140°–143° F/30 min to free them of human pathogens or spoilage organisms. When soil is steamed at 140° F/30 min, the following situations develop:

1. Plant-pathogenic fungi, bacteria, and nematodes are destroyed, and nearly all viruses inactivated.

2. The fungus flora is drastically reduced in numbers and types, and the survivors are largely in genera recognized as potential antibiotic producers. The bacteria and actinomycetes are less affected by such treatment than are fungi.

3. Germination of ascospores and bacterial spores increases from breaking dormancy. This increases the relative proportion of these antagonists in the active soil population.

4. Fungistasis is largely destroyed, but is said to return in about three days and to reach normal levels in about eight days (Dobbs and Hinson, 1953).

5. Phytotoxin production in soil, which occurs commonly after treatment at 212° F/30 min, is practically eliminated.

6. Most weed seeds are killed if the soil is kept moist for three days prior to treatment.

7. The "weed molds" (e.g., *Peziza ostrachoderma*, *Trichoderma viride*, *Pyronema confluens*), prevalent on soil treated at 212°/30 min, are largely suppressed on soil treated at 140° F.

The fuel economy effected by raising the temperature only half as high at 140° as at 212° F is a commercial advantage in permitting more soil to be steamed with a small boiler, and the saving in fuel will largely offset the expense of supplying the necessary air. Additional advantages are decreased manual discomfort in handling equipment at the lower temperature, and the more rapid cooling (or even rapid evaporative cooling if air alone is used following treatment).

AERATED STEAM AS AN AID IN STUDYING SOIL MICROFLORA.—The complexity of the soil microflora makes it difficult to evaluate the role of any given organism(s) in studying interactions. Selective media (Nash and Snyder, 1962), bait techniques (Durbin, 1961), selective fungicides (Georgopoulos and Zaracovitis, 1967; Kreutzer, 1963), direct isolation (Warcup, 1959), selective heat treatment (Olsen and Baker, 1968), selective drying, and various combinations of these methods (Baker, 1961) have been used or suggested as means of determining the organisms present in soil and perhaps involved in the interaction.

It is a curious fact that most of the studies on biological control of root pathogens have been conducted with soils in which the pathogen occurred and produced disease. There have been few attempts to analyze instances where it is difficult or impossible to establish a given pathogen, where considerable time or high inoculum density is necessary for success, or where resistence of a crop or variety is destroyed by some treatment.

Eaton and Rigler (1946) found that corn was immune to *Phymatotrichum omnivorum* when grown in field soil, although the fungus grew along the roots.

When grown in the same soil after sterilization, corn was killed by the pathogen. Resistance was thought to be due to rhizosphere saprophytes, but was not further defined. Tveit and Moore (1954) similarly found that Brazilian oat varieties resistant to *Helminthosporium victoriae* lost this resistance if the seed was treated with hot water. *Chaetomium cochlioides* and *C. globosum*, which occurred naturally on the Brazilian seed, were responsible for its resistance when planted, and imparted resistance when inoculated on seed of susceptible varieties or in soil in which susceptible varieties were planted. These same species also protected oat seedlings against *Fusarium nivale* (Tveit and Wood, 1955). There is abundant evidence that pathogen-infected seed planted in field soil produces fewer diseased seedlings than when planted in steamed soil—e.g., Henry and Campbell (1938) with *Polyspora lini* and *Colletotrichum lini* on flax; Machacek and Wallace (1942) with *H. sativum* on wheat; Schippers and Schermer (1966) with *Verticillium albo-atrum* on *Senecio vulgaris*. The same situation obtains with gladiolus cormels (Ferguson, 1958). The biological control developed after a crop is grown in a soil for many years (Glynne, 1965) is another example.

Analysis of the numerous known examples of effective natural biological control of root pathogens offers a logical and promising, but little used, approach to understanding the phenomenon. It would be ironic to find that we should have been studying areas where a given disease does *not* occur rather than where it does, that is, where biological control may be working. Selective soil thermotherapy offers a means for doing this. The central idea of this concept (Baker et al., 1967; Olsen and Baker, 1968) is that microorganisms differ in their resistance to certain unfavorable conditions, such as heat. By careful manipulation of temperature and time of exposure, it is possible to successfully destroy portions of the microflora, and thus narrow the biological spectrum to be searched for the active antagonist(s).

Among organisms shown to be effective antagonists against plant pathogens, bacteria (particularly *Bacillus* spp.) and actinomycetes are both effective and favorable material for study. They are omnipresent in soils, and generally more uniformly distributed throughout a soil mass than are fungi. *Bacillus subtilis* has a fairly high thermal tolerance (Olsen and Baker, 1968) which expedites isolation and makes it possible, through selective thermal treatment, to increase its relative population in soil. *Bacillus* spp. may be produced in large quantity in broth culture, and by transfer to sterile distilled water, the cells may be converted to resting spores. These microorganisms provide excellent research material for studies in biological control, even if they should later prove not to be the most important agents involved.

Soils that exhibit no antagonism to a given pathogen in the untreated condition have not shown inhibition following treatment. Sterile or inert media (e.g., perlite, vermiculite, or sphagnum peat), or soil mined

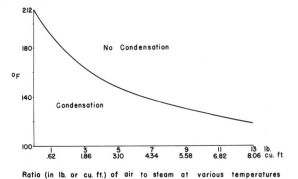

Fig. 5. The ratio of air, by volume and weight, required per comparable unit of steam to produce aerated steam of various temperatures.

at considerable depth (e.g., in deposits of windblown fine sand), lack antagonists. Such soils should not be expected to develop antagonism after mild chemical or heat treatments, which only select antagonists but do not create them.

It is clear that at least three levels of soil treatment should be used in experiments to determine pathogenicity of a microorganism to a host plant and its seed transmission:

1. Use of untreated soil from the area involved will contribute information on the importance of the pathogen in the given soil and effectiveness of seed transmission.

2. Use of the same soil sterilized (autoclaved) will indicate the potential maximum pathogenicity of the microorganism and will reveal seed carriage of it.

3. Use of the same soil treated at 140° F/30 min, in comparison with 1 and 2, will contribute information on the presence and importance of resident antagonists and provide a clue as to what they are. Demonstration of seed transmission of a pathogen consists of more than planting seeds in sterile agar or in autoclaved soil.

METHODS FOR TREATING SOIL WITH AERATED STEAM.—Mixing air with steam dilutes and lowers its temperature to any definite level (Fig. 5) (Baker, 1962, 1967). The simplest way to produce aerated steam is to join together a line of flowing steam and one of flowing air; a needle valve in each line provides precise temperature control.

The air is provided by a centrifugal, straight radial-blade blower producing a static pressure of up to 6 in (depending on the depth and porosity of the soil to be treated) and an output of 5–10 cfm for each cu ft soil to be treated, depending on the efficiency of the method used. Venturis were tried in England in 1954 and in Australia in 1961–1963 to provide the required air. The method was dropped, but seems now to be returning in England (Dawson et al., 1967).

The soil should reach the desired temperature in 30 min or less, and be held at that level for 30 min; the flow may be reduced after the temperature is attained (Baker, 1967). The thermometers *must be accurate in the range used.*

Aerated steam penetrates through soil in the same way and at the same rate as 212° F steam, and may, with slight modification, be used with the same kinds of equipment (Baker and Roistacher, 1957). "Blowouts" of soil may be eliminated by applying steam at the upper surface rather than from the bottom.

Aerated steam is now commercially used in the United States, Australia, New Zealand, and England; some of the installations have been in use for eight years. Aerated-steam treatment of soil is thus today a practice of demonstrated feasibility, economic desirability, and promising, although still largely unexplored, biological potentiality.

LITERATURE CITED

BAKER, K. F. 1961. Control of root-rot diseases, pp. 486–490. *In* Recent advances in botany. IX Internatl. Botan. Congress. Univ. Toronto Press, Toronto. 1766 p.

BAKER, K. F. 1962. Principles of heat treatment of soil and planting material. J. Australian Inst. Agr. Sci. 28: 118–126.

BAKER, K. F. 1967. Control of soil-borne plant pathogens with aerated steam. Proc. Wash. State Univ. Greenhouse Growers Inst. 1967:3–18.

BAKER, K. F., N. T. FLENTJE, C. M. OLSEN, and H. M. STRETTON. 1967. Effect of antagonists on growth and survival of *Rhizoctonia solani* in soil. Phytopathology 57:591–597.

BAKER, K. F., and O. A. MATKIN. 1967. Concepts involved in soil mixes for container crops. Proc. Wash. State Univ. Greenhouse Growers Inst. 1967:118–130.

BAKER, K. F., and C. M. OLSEN. 1960. Aerated steam for soil treatment. Phytopathology 50:82.

BAKER, K. F., and C. N. ROISTACHER. 1957. Heat treatment of soil. Principles of heat treatment of soil. Equipment for heat treatment of soil. Calif. Agr. Exp. Sta. Manual 23:123–196, 290–293, 302–303.

DARLEY, E. F. 1958. *Armillaria* and *Dematophora*. Proc. Pac. Coast Res. Conf. on Control of Soil Fungi 5:6–7.

DARLEY, E. F., and W. D. WILBUR. 1954. Some relationships of carbon disulfide and *Trichoderma viride* in the control of *Armillaria mellea*. Phytopathology 44:485.

DAWSON, J. R., R. A. H. JOHNSON, P. ADAMS, and F. T. LAST. 1965. Influence of steam/air mixtures, when used for heating soil, on biological and chemical properties that affect seedling growth. Ann. Appl. Biol. 56: 243–251.

DAWSON, J. R., A. A. T. KILBY, M. H. EBBEN, and F. T. LAST. 1967. The use of steam/air mixtures for partially sterilizing soils infested with cucumber root rot pathogens. Ann. Appl. Biol. 60:215–222.

DOBBS, C. G., and W. H. HINSON. 1953. A widespread fungistasis in soils. Nature 172:197–199.

DURBIN, R. D. 1961. Techniques for the observation and isolation of soil microorganisms. Botan. Rev. 27: 522–560.

EATON, F. M., and N. E. RIGLER. 1946. Influence of carbohydrate levels and root-surface microfloras on Phymatotrichum root rot in cotton and maize plants. J. Agr. Res. 72:137–161.

ELTON, C. S. 1958. The ecology of invasions by animals and plants. Methuen and Co. Ltd., London. 181 p.

FERGUSON, J. 1958. Reducing plant disease with fungicidal soil treatment, pathogen-free stock, and controlled microbial colonization. Ph.D. thesis, Univ. of California, Berkeley. 169 p.

GEORGOPOULOS, S. G., and C. ZARACOVITIS. 1967. Tolerance of fungi to organic fungicides. Ann. Rev. Phytopathol. 5:109–130.

GLYNNE, M. D. 1965. Crop sequence in relation to soil-borne pathogens, p. 423–435. *In* K. F. Baker and W. C. Snyder [ed.], Ecology of soil-borne plant pathogens. Univ. of Calif. Press, Berkeley and Los Angeles.

GORING, C. A. I. 1967. Physical aspects of soil in relation to the action of soil fungicides. Ann. Rev. Phytopathol. 5:285–318.

HEMWALL, J. B. 1959. A mathematical theory of soil fumigation. Soil Sci. 88:184–190.

HENRY, A. W., and J. A. CAMPBELL. 1938. Inactivation of seed-borne plant pathogens in the soil. Can. J. Res., (C) 16:331–338.

KREUTZER, W. A. 1963. Selective toxicity of chemicals to soil microorganisms. Ann. Rev. Phytopathol. 1:101–126.

KUIPER, J. 1965. Failure of hexachlorobenzene to control common bunt of wheat. Nature (London) 206: 1219–1220.

LAMBERT, E. B., and T. T. AYERS. 1952. An improved system of mushroom culture for better control of diseases. Plant Disease Reptr. 36:261–268.

MACHACEK, J. E., and H. A. H. WALLACE. 1942. Nonsterile soil as a medium for tests of seed germination and seed-borne disease in cereals. Can. J. Res., (C) 20: 539–557.

MOUBASHER, A. H. 1963. Selective effects of fumigation with carbon disulphide on the soil fungus flora. Trans. Brit. Mycol. Soc. 46:338–344.

NASH, S. H., and W. C. SNYDER. 1962. Quantitative estimations by plate counts of propagules of the bean root rot Fusarium in field soils. Phytopathology 52:567–572.

OLSEN, C. M., and K. F. BAKER. 1968. Selective heat treatment of soil, and its effect on the inhibition of *Rhizoctonia solani* by *Bacillus subtilis*. Phytopathology 58:79–87.

OLSEN, C. M., N. T. FLENTJE, and K. F. BAKER. 1967. Comparative survival of monobasidial cultures of *Thanatephorus cucumeris* in soil. Phytopathology 57:598–601.

SCHIPPERS, B., and A. K. F. SCHERMER. 1966. Effect of antifungal properties of soil on dissemination of the pathogen and seedling infection originating from Verticillium-infected achenes of *Senecio*. Phytopathology 56: 549–552.

TVEIT, M., and M. B. MOORE. 1954. Isolates of *Chaetomium* that protect oats from *Helminthosporium victoriae*. Phytopathology 44:686–689.

TVEIT, M., and R. K. S. WOOD. 1955. The control of Fusarium blight in oat seedlings with antagonistic species of *Chaetomium*. Ann. Appl. Biol. 43:538–552.

WARCUP, J. H. 1959. Distribution and detection of root-disease fungi, p. 317–326. *In* C. S. Holton et al., Plant pathology—problems and progress. 1908–1958. Univ. of Wisconsin Press, Madison. 588 p.

WARCUP, J. H., and K. F. BAKER. 1963. Occurrence of dormant ascospores in soil. Nature (London) 197:1317–1318.

Index

Actinomycetes, antagonists, 86
Aerated steam
—aid to soil microflora study, 237–238
—disease control by soil treatment: biological aspects, 236–237; temperature required, 237
—phytotoxicity lack, 236
Aeration
—pathogen survival, 19
—phytoalexin formation, 26
—plant exudation, 26
—root growth, 26
—seed exudation, 101
—soil: *Armillariella* growth, 79; fungal activity, 77–80, 89–90; *Fusarium* suppression, 85
Agaricus, 45
Agrobacterium tumefaciens, soil population studies, 22
Alanine, 99; chlamydospore germination, 112
Alfalfa (*Medicago*)
—meal, disease decrease, 4
—rootlet exchange, 208
Allium spp.
—*Sclerotium* sclerotia, germination stimulation, 130–131
—*see also* Onion
Allyl alcohol, effect on mycorrhizal fungi, 218
Alkyl sulphides, sclerotia germination stimulation, 132
Alkythiolsulphinates, 131
4-Aminobutyric acid, zoospore attraction, 109
Ammonium sulfate, conifer growth increase, 223–224
Anaerobiosis, *Verticillium* control, 32
Anastomosis, 63
—compatibility, 64
—environment effects, 64
—mosaic condition, 66
—phycomycetes, 72
—*Thanatephorus*: groups, 47–48; influencing factors, 48; process, 64
—vegetative hyphae, 65
—*Verticillium*, 57, 65
Antagonists
—actinomycetes, 238
—*Bacillus*, 238
—*Fomes* control, 192
—fumigation effect, 218–219
—*Poria* control, 188
—soil treatment for biological control, 234–239
Aphanomyces
—antheridial type, 51
—existence in soil, 3, 50
—host: zoospore attraction, 4
Aphanomyces cochlioides
—beet root exudate, 107
—zoospore chemotaxis, 108
Aphanomyces euteiches, zoospore encystment, 110
Apple (*Malus*), rootlet exchange, 208
Arginine, 99

—chlamydospore germination, 112
—zoospore accumulation, 104
Armillaria, 72
Armillaria mellea, 45, 184
—basidiospore role in stump infection: evidence for, 141; frequency, 146; inoculum dosage, 143; laboratory studies, 142–145; microbial interactions, 144–145; moisture content, 143; nutrient status, 144; stump experiments, 145–146; temperature, 143–144; variation with different tree species, 142–143
—biological control, 150
—compatibility factors, 64
—competitive ability: sawdust substrates, 151
—control by *Trichoderma*, 236
—disease centers, 182
—distribution, 179
—ecology: water films, 79
—epiphytic growth, 182
—ethanol effect, 124; glucose metabolism, 123; growth stages, 122; rhizomorph formation, 122–123
—fruiting body development, 141
—infection behavior in field: natural infection, 150; artificial infection, 150
—movement in soil, 3
—nutrients, pathogenesis, 4
—oil palm disease, 194; control, 195
—pathogenesis, 181–183
—rhizomorph formation, 180
—rhizomorph growth: food base, 148–149; soil type, 147–148
—ring barking, 150; effects, 151–152
—root sawdust breakdown, 150–151
—spore: dissemination, 180; formation, 180–181; numbers, 141
—sporophore formation, 180
—temperature, root infection, 179–180
—translocation, 183
—wood colonization, moisture, 143
—zone plates, 183
Armillariella elegans, 180, growth; soil water-areation interactions, 79
Armoracia. See Horseradish
Arthrobacter globiformis
—plant growth stimulation, 218
—growth in presence of fumigants, 218
Ascochyta imperfecta
—heterokaryosis, 65
—parasexuality, 65
—segregation in diploids, 56
Asparagine
—chlamydospore: germination, 95, 112–113; lysis, 95–96
—cottonseed exudate, 99
—disease severity, 96; *Rhizoctonia*, 100, 117
—zoospore chemotaxis, 109
Aspartic acid, 99